The Microbiology of Meat and Poultry

The Microbiology of Meat and Poultry

Edited by

ANDREW DAVIES
Department of Food Microbiology
Leatherhead Food Research Association
UK

and

RON BOARD
Former Professor
School of Biological Sciences
University of Bath
UK

BLACKIE ACADEMIC & PROFESSIONAL
An Imprint of Chapman & Hall
London · Weinheim · New York · Tokyo · Melbourne · Madras

Published by Blackie Academic & Professional, an imprint of Thomson Science, 2–6 Boundary Row, London SE1 8HN, UK

Thomson Science, 2–6 Boundary Row, London SE1 8HN, UK

Thomson Science, 115 Fifth Avenue, New York, NY 10003, USA

Thomson Science, Suite 750, 400 Market Street, Philadelphia, PA 19106, USA

Thomson Science, Pappelallee 3, 69469 Weinheim, Germany

First edition 1998

© 1998 Thomson Science

Typeset in 10/12pt Times by Type Study, Scarborough

Thomson Science is a division of International Thomson Publishing   I T P

Printed in Great Britain by St Edmundsbury Press Ltd, Bury St Edmunds, Suffolk

ISBN 0 7514 0398 9

A catalogue record for this book is available from the British Library

Library of Congress Catalog Card Number: 97-76803

∞ Printed on acid-free text paper, manufactured in accordance with ANSI/NISO Z39.48–1992 (Permanence of Paper)

Contents

3 Yeasts and moulds associated with meat and meat products 85
V.M. DILLON

4 Microbiological contamination of meat during slaughter and butchering of cattle, sheep and pigs 118
C.O. GILL

8 The microbiology of stored poultry 266
N.A. COX, S.M. RUSSELL and J.S. BAILEY

9 Chemical changes in stored meat 288
G.-J.E. NYCHAS, E.H. DROSINOS and R.G. BOARD

List of contributors

J.S. Bailey USDA, ARS, Russell Research Center, PO Box 5677, Athens, Georgia 30604-567, USA

R.G. Board South Lodge, Northleigh, Bradford on Avon, Wiltshire BA15 2RG, UK

N.M. Bolder DLO Institute for Animal Science and Health, Research Head Office, Edelhertweg 15, Postbus 65, 8200 AB Lelystad, The Netherlands

N.A. Cox Department of Poultry Science, The University of Georgia, Athens, Georgia 30602-2772, USA

A.R. Davies Department of Food Microbiology, Leatherhead Food Research Association, Randalls Road, Leatherhead, Surrey KT22 7RY, UK

V.M. Dillon Department of Botany, School of Plant Sciences, University of Reading, Whiteknights, PO Box 221, Reading, Berkshire RG6 5SA, UK

E.H. Drosinos Agricultural University of Athens, Department of Agricultural Industries, Laboratory of Microbiology of Foods, 75 Iera Odos Street, Votanikos 118 55, Greece

M.L. García-López Department of Food Hygiene and Food Technology, University of León, E-24071-León, Spain

C.O. Gill Agriculture and Agri-food Canada Research Centre, Bag Service 5000, Lacombe, Alberta T4L 1W1, Canada

W.H. Holzapfel Institute of Hygiene and Toxicology, Engessiestra 20, Karlsruhe D-76131, Germany

K.K.S. Nair Central Food Technological Research Institute, Mysore 570 013, India

D. Narasimha Rao Department of Meat, Fish and Poultry Technology, Central Food Technological Research Institute, Mysore 570 013, India

G.-J.E. Nychas Agricultural University of Athens, Department of Agricultural Industries, Laboratory of Microbiology of Foods, 75 Iera Odos Street, Votanikos 118 55, Greece

A. Otero Department of Food Hygiene and Food Technology, University of León, E-24071-León, Spain

M. Prieto Department of Food Hygiene and Food Technology, University of León, E-24071-León, Spain

P.Z. Sakhare Central Food Technological Research Institute, Mysore 570 013, India

L.H. Stanbridge Whitbread Technical Centre, Park Street, Luton LU1 3ET, UK

S.M. Russell Department of Poultry Science, The University of Georgia, Athens, Georgia 30602-2772, USA

Preface

The microbiology of meat began to attract attention almost immediately following the founding of bacteriology as a science in its own right. With the passing years there has been an ebb and flow of interest which can be related to contemporary practices in commerce. Thus there was much activity in meat microbiology in the 1930s when for the first time large amounts of meat were being shipped long distance – from Australia to the UK, for example. There was a pronounced renewal of interest in the 1950s when the growth of supermarkets called for joints of meat that had some of the characteristics of an item of grocery – a wrapped product with a predictable shelf-life under controlled conditions. In attempts to achieve this goal, food technologists created packaging systems that had two attributes, extension of shelf-life, a topic discussed in this book, and selection of associations of little known or, in some cases, previously unknown species of bacteria. In practice the present book provides critical reviews of the major groups of yeasts and bacteria that can grow on and spoil meat. Additionally the book focuses attention on the many subtle interactions that occur between micro-organisms and the wide range of chemicals found in meat – probes for determining the freshness of meat will surely be a product of this line of research. This book, probably for the first time, draws attention to the fact that the concerns/demands of supermarkets in developed countries have distracted attention from the practices that have evolved over hundreds of years which have enabled people in hot climates to have meat included in their daily diet. This topic is included in the book as well as chapters that contrast the microbiological problems generated by the processing of vast numbers of poultry per day with those of a beef, pig and sheep processing line. A reader may well pose the question – How can it be that a book on meat microbiology does not have a chapter(s) dealing with food-borne diseases? In the opinions of the editors, such a question is easily answered. Meat spoilage and the related microbiology is largely concerned with factors occurring after slaughter. Recent experience tells us that diseases such as BSE and food poisoning due to *E. coli* 0157 need to be considered against a very broad canvas. Indeed the topic of meat-borne diseases would have been trivialized had it been considered superficially in a couple of chapters in this book.

The editors wish to thank the authors for all their hard work in producing chapters of quality and Rose Gilliver of the publishers for endless patience.

A.R. Davies
R.G. Board
April 1997

1 The physiological attributes of Gram-negative bacteria associated with spoilage of meat and meat products

M.L. GARCÍA-LÓPEZ, M. PRIETO AND A. OTERO

1.1 Introduction

The slaughtering and butchering of food animals provide bacteria with an opportunity to colonize meat surfaces. A wide range of micro-organisms coming from different sources are introduced to surfaces which contain abundant nutrients and which have a high water availability. Only a few of the contaminants will be able to initiate growth, and only some of these will eventually spoil the meat by means of their biochemical attributes. Man has searched (until recently in an empirical manner), for ways to keep spoilage organisms away from meat, to reduce their growth rate, or to select those with low spoilage potential. Predominance of different groups of micro-organisms on meat depends on the characteristics of the meat, the environment in which meat is stored as well as the processing that meat may undergo.

Gram-negative bacteria constitute the greatest spoilage potential for meat and meat products. When fresh meat is chill-stored aerobically, members of the genera *Pseudomonas*, *Acinetobacter*, *Psychrobacter* and *Moraxella* display the fastest growth rates and hence the greatest spoilage potential. Species of *Shewanella* and Enterobacteriaceae need conditions more favourable than those of the above genera in order to develop and produce spoilage metabolites. Depending upon conditions, the shelf-life of fresh meat is in the range of days before signs of spoilage (off-odours and slime) are evident. An extension of shelf-life is achieved by hindering the growth of Gram-negative organisms relative to that of Gram-positive ones (Micrococcaceae and lactic acid bacteria). To achieve this, environmental and or product conditions (atmosphere, a_w, salt and nitrite concentrations, temperature, etc.) that favour growth of Gram-positive bacteria in meat are selected.

The number of micro-organisms on fresh meat surfaces change during chill storage following a typical microbial growth pattern. Counts of bacteria in meat are in the range 10^2–10^5 cfu/cm^2, but only around 10% are able to initiate growth (Nychas *et al.*, 1988). The initial lag phase is attributed to microbial adaptation to changing conditions (chill temperatures and surface desiccation). Ensuing logarithmic growth takes place after cells have accommodated to the new environmental setting and adapted their metabolism. When numbers exceed 10^7 cells per cm^2, the first spoilage signs are detected, as off-odours. Another typical spoilage sign, bacterial slime, is noticeable with cell density around 10^8 cells per cm^2 (Gill, 1982). Shortly after, growth declines and the stationary phase is reached. Dominance of a single microbial group is due to its higher growth rates under specific conditions, and this higher rate is either because of metabolism advantages (substrate transformation and affinity) or tolerance to factors (psychrotrophism, pH, a_w). Thus, it seems that there are no interactions (or they are indirect) between microorganisms until one of the genera present reaches its maximum cell density (Gill and Molin, 1991).

At the beginning of chill storage, the role played by the composition of *post-mortem* meat is selective more than limiting. High water content favours microbial growth and, although there exist plenty of growth-sustainable nutrients, psychrotrophs prefer to use low-molecular-weight compounds rather than complex proteins and lipids (Gill, 1982).

During development of *post-mortem* rigor, muscular fibres degrade glycogen (first aerobically and then anaerobically) in order to obtain the necessary ATP to maintain cellular structures and osmotic balance. As a consequence, proportions of many of low-molecular-mass substances change during conversion of muscle into meat (Table 1.1). When oxygen is depleted, anaerobic routes are used and lactic acid becomes the end-product of glycolysis, and its accumulation in turn causes the pH to fall. In normal circumstances, pH reaches the value of 5.8–5.5, equivalent to 0.9–1% of lactic acid in muscle.

Since nutrient composition does not stop bacteria from growing, other factors have to be used either to inhibit microbial activity (Eh, a_w, temperature, atmosphere) or to promote growth of non-spoilage bacteria.

Table 1.1 Chemical composition of typical adult mammalian muscle post *rigor-mortis*

Component	Post-rigor (%)
Water	75
Protein	19
Lipid	3
Creatine phosphate	–
Creatine	0.7
ATP	–
IMP	0.3
Glycogen	0.1
Glucose	0.1
Glucose-6-P	0.2
Lactic acid	0.9
Amino acids	0.4
Carnosine, anserine	0.3

Surface desiccation takes place during refrigeration of carcasses. The evaporation of water from the surface causes the a_w to drop during the first four hours (or more in connective tissue overlying fat) below limits (0.95) for meat spoilers such as *Pseudomonas* and *Moraxella* (Grau, 1979). Diffusion from the inner parts counteracts this effect, which can only be used to delay microbial growth in the first 24 hours after dressing. Temperature is the main factor used to decrease growth of spoilage bacteria on meat, by increasing both lag phase and generation time. Psychrotrophs are especially favoured by the normal practice of refrigerating meat.

1.2 Gram-negative spoilage bacteria in meat and meat products

1.2.1 Fresh meat

It is generally recognized that Gram-negative, motile and non-motile aerobic rods and coccobacilli belonging to the genera *Pseudomonas*, *Moraxella*, *Psychrobacter* (formerly *Moraxella*-like) and *Acinetobacter* are the major components of the spoilage flora of raw meat stored aerobically under refrigeration (Molin and Ternström, 1982, 1986; Shaw and Latty, 1982, 1984, 1988; Prieto *et al.*, 1992a, 1992b; Drosinos and Board, 1995a). Certain species of psychrotrophic Enterobacteriaceae commonly occur on chilled meat. These organisms, which are able to grow aerobically on adipose tissue and on muscle tissue of high pH (> 6), appear to be more prevalent on pork and lamb (Grau, 1981; Dainty and Mackey, 1992). Their growth is favoured by temperatures of $\geqslant 4\,°C$ (Blickstad and Molin, 1983). Isolation of *Flavobacterium*, *Alcaligenes*, *Vibrio*, *Aeromonas* and *Alteromonas* is reported less frequently (Patterson and Gibbs, 1977; Nottingham, 1982; Blickstad and Molin, 1983).

The initial flora of poultry skin is partially eliminated during scalding. A significant proportion of the subsequent contaminants are Gram-negative bacteria (Daud *et al.*, 1979). Although the incidence of psychrotrophs in the initial flora (10^3–10^4 per cm^2) is variable, the finished carcasses are generally contaminated with large numbers of species capable of survival and even growth in chilled water. For this reason, it appears that the psychrotrophic flora on poultry is more likely to be less variable than that on other meats (Gill, 1986). At the time of spoilage, the predominant organisms on eviscerated poultry are pseudomonads and to a lesser extent *Acinetobacter* and probably *Psychrobacter*. *Shewanella putrefaciens* may also be present. This bacterium, which is a potent spoilage organism, is considered an important part of the spoilage association even though it may not be numerically dominant (McMeekin, 1977). Other Gram-negative bacteria (*Flavobacterium* and Enterobacteriaceae) have been recovered on many occasions from spoiled chicken and turkeys (McMeekin, 1975; Daud *et al.*,

1979; Lahellec and Colin, 1979). A relationship has been observed between the occurrence of the above genera and different sites of the poultry carcass. *Pseudomonas* is dominant on the skin of the breast (pH 5.7–5.9) and leg (pH 6.4–6.7), but the remaining genera are restricted to that of the breast apparently because of its lower pH value (Barnes and Impey, 1968; McMeekin, 1975, 1977).

The spoilage of aerobically stored meat, as with other proteinaceous foods, is preceded by a phase of variable duration during which bacteria make use of carbohydrates particularly glucose and glucose-6-P as a carbon and energy source (Gill and Newton, 1978).

Glucose is present in *post-mortem* beef meat at concentrations in the range 0.1–0.5% (Gill, 1986); it is readily used by microbial cells growing on the surface of meat. At the outset, a diffusion gradient develops from within the muscle. This maintains an adequate glucose concentration at the meat surface and hence ensures bacteria continue to metabolize carbohydrates thereby delaying the utilization of other compounds (Gill, 1986). Only when the demand from large numbers of bacteria (typically more than 10^7 cells/cm^2) cannot be met, do they attack amino acids, and cause a rise in the concentration of ammonia and pH. Gram-negative bacteria appear to be particularly fitted to use low-molecular-weight compounds at refrigeration temperatures and in meat of normal pH (5.5–5.8). They have higher growth rates than would-be competitors and thus outnumber them on meat surfaces.

Residual glucose values in meat *post-mortem* can be low as a result of stress, starvation or fright prior to slaughter of animals. Such circumstances deplete the glycogen concentration in live animals. Due to its organoleptic characteristics, meat from stressed animals is referred to as DFD (Dark, Firm, Dry) meat. As *post-mortem* glycolysis is curtailed by low substrate concentration, lactic acid is not produced in the normal amounts. As glucose levels are lower than normal, the resulting meat spoils rapidly because glucose scarcity prompts bacteria to the early use of amino acids. The addition of glucose to DFD meat does delay the onset of spoilage because bacteria can increase their maximum cell density without attacking amino acids (Lambropoulou *et al.*, 1996).

In the first phase of growth on meat, the metabolism of glucose by pseudomonads and other Gram-negative bacteria does not give rise to offensive off-odours. Growth is supported by carbohydrates, and their catabolism releases a complex mixture of substances containing short-chain fatty acids, ketones and alcohols, that exhibit a variety of fruity and sweety odours (Dainty, 1996).

The second phase begins when glucose is depleted, and the micro-organisms begin to use amino acids for energy. This occurs when flora reach numbers of 10^7 bacteria/cm^2. Volatiles responsible for the spoilage odours in this phase are well characterized (McMeekin, 1982). Amino acids (cysteine, cystine and methionine) are precursors of hydrogen sulphide, methylsulphide and dimethylsulphide. These compounds are generated by *Pseudomonas* spp. and Enterobacteriaceae, and produce odours described as putrid and sulphury. Deamination of these amino acids gives rise to pyruvate, ammonia, H_2S, CH_3SH and $(CH_3)_2S$. Dimethylsulphide may also be metabolized from methylation of methylsulphide. It is interesting to note that *Pseudomonas* spp. degrade amino acids by deamination, while Enterobacteriaceae also possess the ability to decarboxylate amino acids (McMeekin, 1982).

Methylamine, dimethylamine and trimethylamine have also been commonly detected their formation is associated with the growth of *Ps. fluorescens* as well as non-fluorescent pseudomonads. Among the pseudomonads, *Ps. fragi* is considered to be mainly responsible for the production of ethyl esters having a sweet, fruity odour. Other sulphur-containing compounds generated in more advanced stages of spoilage by pseudomonads have also been detected. Amino acids are the source of these products and their production occurs when the numbers of bacteria are greater than 10^7 cfu/cm^2.

Two main end products of amino acid decarboxylation have been identified: cadaverine, from lysine, and putrescine from ornithine or arginine. High correlations between putrescine production and pseudomonads counts, and between cadaverine and Enterobacteriaceae counts have been obtained (Dainty and Mackey, 1992). Nevertheless, their use as spoilage indicators is not feasible as the detection occurs when bacterial numbers exceed $10^7/cm^2$. Other significant but less studied amines are spermidine, spermine, histamine and tyramine.

Experiments done in pure cultures with highly proteolytic strains of *Proteus* and *Pseudomonas* (*Ps. fragi*, *Ps. fluorescens*) demonstrated their high protease activity against myofibrils and sarcoplasmic proteins (Dainty *et al.*, 1983). This is in agreement with results from taxonomic studies (Molin and Ternström, 1982; Shaw and Latty, 1982), which reveal production of extracellular enzymes by *Ps. fragi* clusters 1 and 2. Bacteria may use proteins if they are the sole source of carbon and energy, but not when easy-to-assimilate compounds are available. Although some species of different genera (*Pseudomonas*, *Proteus*, *Aeromonas*) are easily shown to be proteolytic or lipolytic on laboratory media, there is clear evidence that these phenomena do not contribute to spoilage as other factors are detected first. It seems that release of exoproteases from bacteria only occurs in the stationary phase, when cells have attained their maximum density and amino acids have been depleted. Analysis of myofibrillar and sarcoplasmic proteins exhibited no change until numbers exceeded 10^{10} cells/cm^2 and long after off-odours and slime had been detected (McMeekin, 1982).

Even though many psychrotrophic bacteria produce lipases, the role of bacteria in the lipolytic and oxidative changes of meat is generally overlooked. Some experiments have shown however, the ability of *Ps. fragi* to increase the level of free fatty acids in meat as well as in a culture medium. Pseudomonads and psychrobacters use the same compounds on fatty surfaces as they do on lean meat, and produce the same metabolites. Nevertheless, the rate of diffusion of low-weight-molecular compounds is slower and water activity is lower, thereby reducing their growth rate on fatty surfaces *vis-à-vis* that on lean meat. Spoilage is thus first noticed on lean parts of the carcass.

1.2.2 Vacuum stored meat

The storage life and keeping quality may be extended by modifying the gaseous atmosphere surrounding the meat. Vacuum and modified atmosphere packaging (MAP) are the two methods commonly used in wholesale marketing to modify the gas atmosphere. Both of these procedures and conventional overwrapped (aerobic) trays are also used in retail marketing (Hood and Mead, 1993). Vacuum packaging is the preferred method for the storage and distribution of large pieces of chilled primals or wholesale cuts. Within the vacuum packs, the residual oxygen is rapidly consumed (below 1%) by tissue and microbial respiration, and CO_2 increases to about 20%. Completely anaerobic conditions are rarely achieved, since all films in commercial use have a certain oxygen permeability. Thus, during storage, aerobic Gram-negative bacteria are replaced by the slow growing Gram-positive bacteria (Dainty *et al.*, 1983; Egan and Roberts, 1987; Dainty and Mackey, 1992). Lactic acid bacteria are the most frequently isolated bacteria from this kind of product since they are tolerant to CO_2 and low temperatures. These bacteria metabolize glucose – as they do in aerobic meat – to produce lactic, isobutanoic, isopentanoic and acetic acids. This gives meat a sour (cheesy, acid) taste and smell. The accumulation of these acids occurs mainly during the stationary phase and at some point in time the meat is rejected. Other spoilage phenomena, such as proteolysis or lipolysis are very limited or nonexistent because the Gram positive bacteria have very limited proteolytic activity.

For a variety of reasons (high initial contamination levels, film permeability, storage temperature, etc.), Gram-negative bacteria (Enterobacteriaceae and even *Pseudomonas*) may on occasions form large populations on vacuum packed beef cuts of normal pH (Dainty *et al.*, 1983; Gill and Penney, 1988). On vacuum packed pork, substantial numbers of enterobacteria may be present throughout storage at -1.5 and 3 °C (Gill and Harrison, 1989). These bacteria and *Shewanella putrefaciens* have been reported to grow readily on fat and skin tissues of vacuum packed pork, irrespective of the pH of the muscle tissue. Growth of Enterobacteriaceae on vacuum packed lamb has also been observed (Gill and Penney, 1985). When pH is ≥ 6, growth of *Shewanella putrefaciens*, *Alcaligenes*, *Aeromonas* spp. or some species of Enterobacteriaceae may cause spoilage.

1.2.3 Modified atmosphere packed meat

Carbon dioxide is used in the MAP of meats because it inhibits microbial growth. Generally, it is used in combination with N_2 and/or O_2. The percentages used vary from 10% to 40% in the case of CO_2 and 90% to 60% for oxygen. In general, the higher the CO_2 concentration, the better in terms of inhibition of spoilage organisms. A long shelf-life may be attained in 100% CO_2. However, a product may undergo chemical changes that are detrimental to meat quality (Gill and Molin, 1991). Prevalence of slow growing lactic acid bacteria is responsible for the extended storage life. However, depending on several factors (pH, storage temperature, initial numbers, packaging materials, etc.), Enterobacteriaceae and *Aeromonas* spp. may grow and cause spoilage (Gill and Penney, 1985, 1988; Gill and Harrison, 1989; McMullen and Stiles, 1993). According to several authors (Asensio *et al.*, 1988; Ordóñez *et al.*, 1991; Dainty and Mackey, 1992), Enterobacteriaceae and *Pseudomonas* are more prevalent on MAP than on vacuum packed meat, especially on pork, their growth being favoured by storage at *ca.* 5 °C, and by prior conditioning in air.

The growth of some genera of lactic acid bacteria (e.g. *Leuconostoc*) may be favoured if oxygen is available. Even so spoilage organisms, such as pseudomonads, enterobacteria and *Brochothrix* can also compete effectively, and their numbers are higher than those in vacuum packed meat. Contamination during slaughter and meat conditioning has a large influence, since if conditions are not stringent, spoilage can be caused by several groups of bacteria. Lactic acid bacteria and *Brochothrix* are also dominant on poultry stored in vacuum packs, CO_2, and nitrogen, sometimes accompanied by cold-tolerant coliforms, *Sh. putrefaciens* and *Pseudomonas* (Hood and Mead, 1993; Kakouri and Nychas, 1994).

1.2.4 Meat products

Comminuted meats (fresh minced meat, certain types of fresh sausages and burger-type products) tend to have a short shelf-life because of the quality of the raw ingredients (usually with a high load of micro-organisms), as well as the effect of comminution. The spoilage flora of minced meat stored in air is dominated by *Pseudomonas* and to a lesser extent by Enterobacteriaceae (von Holy and Holzapfel, 1988; Lambropoulous *et al.*, 1996). Different treatments (addition of preservatives, vacuum and modified atmosphere packaging) select other organisms such as lactic acid bacteria, *Brochothrix* and yeasts. Even so both groups of Gram-negative bacteria are normally present (von Holy and Holzapfel, 1988; Nychas and Arkoudelos, 1990;

Drosinos and Board, 1995b; Lambropoulou *et al.*, 1996). The microflora of fresh sausages resembles that of minced meat though type of meat, preservatives and storage temperature influence the selection of the dominant types. On some occasions (low sulphite levels, and high storage temperatures), Enterobacteriaceae may be involved in spoilage of burger-type products.

One of the major uses of mechanically recovered meat (MRM) is the extension of ground meat products. Significant growth of *Pseudomonas*, '*Achromobacter*' and *Flavobacterium* may occur in MRM (Swingler, 1982).

In most cases, spoilage by Gram-negative bacteria of certain cooked but uncured meats (whole and restructured joints and poultry, ready meals) results from post-processing contamination. Products stored unpacked or packed in air permeable films tend to develop a spoilage flora dominated by *Pseudomonas* (<5 °C) or environmental Enterobacteriaceae (higher temperatures). Occasionally *Janthinobacterium lividum* has been associated with formation of slime on roast beef. In vacuum and modified atmosphere packs stored at high temperatures (10 °C), Enterobacteriaceae may become a significant proportion of the spoilage microflora (Penney *et al.*, 1993).

Factors regulating microbial growth undergo gradual change during the drying and ripening of meat products. Temperature, a_w, pH, and concentration of additives such as salt, nitrite and nitrate, modify the sensory properties of meat and select particular microorganisms. Micrococcaceae, lactic acid bacteria (*Carnobacterium, Leuconostoc*), *Vibrio*, Enterobacteriaceae, and non-motile Gram-negative bacteria (*Psychrobacter*) are the major groups present on cured raw meat products.

Bacterial spoilage of cured meats has been reviewed by Gardner (1983) and by Borch *et al.* (1996). Raw cured meats include ham and bacon. The occurrence of *Vibrio* spp. as spoilage organisms in bacon is widely recognized. Gardner (1981) demonstrated that there are three groups of halophilic vibrios on bacon: *Vib. costicola*, '*Vib. costicola* subsp. *liquefaciens*' and an unidentified group (probably *Vib. costicola*) which is the most frequently found in the slime on spoiled Wiltshire bacon. Other Gram-negative bacteria (*Acinetobacter, Aeromonas, Alcaligenes, Janthinobacterium* and members of Enterobacteriaceae) have been isolated from the surface of bacon, but their role in the spoilage has been difficult to prove though Gardner (1982) includes *Acinetobacter* among the surface spoilage flora of refrigerated bacon. Internal taints such as 'pocket' taint may be caused by *Vibrio, Alcaligenes* and, on occasions, *Proteus inconstans* (now *Providencia alcalifaciens* and *Providencia stuartii*) (Gardner, 1983). Both *Providencia* and halophilic species of *Vibrio* have been isolated from bone taints of Wiltshire bacon. *Providencia*, particularly *Prov. rettgeri*, appears to be the major cause of internal taints in unpumped hams (Gardner, 1983). Spoilage (souring) of packed raw cured meats is mainly due to lactic acid bacteria. H_2S production can result from the growth of *Vibrio* and members of Enterobacteriaceae on vacuum packed bacon held at high temperatures. *Providencia* ('*Proteus inconstans*') is responsible for 'cabbage odour' due to the production of methane thiol from methionine. This type of spoilage is associated with bacon of high pH, and low salt content stored at high temperature (Gardner, 1982, 1983). The spoilage microflora of bacon in MAP appears to be similar to that of vacuum packed. *Proteus* and *Providencia* are also involved in the spoilage of sweet-cured bacon (Varnam and Sutherland, 1995).

Spoilage of dry-cured hams by Gram-negative bacteria has also been reported. Blanco *et al.* (1994) isolated *Burkholderia* (*Pseudomonas*) *cepacia* from a spoiled sample ('potato defect' taint) of Parma ham. The strain produced the odour compound responsible for the 'potato defect'. Cantoni *et al.* (1994) and Papa (1994) found that the putrefactive type spoilage of raw hams was due to Enterobacteriaceae (mainly to *Prot. vulgaris*).

There are many types of cooked cured meats (ham, luncheon meats, various sausages, etc.). Most vegetative bacteria are inactivated by heat treatment, but post-process contamination occurs during slicing, portioning or skinning. Although a wide variety of Gram-negative bacteria are often recovered from these products, they are unable to compete with lactic acid bacteria in vacuum or MA packages under refrigerated storage (Borch *et al.*, 1996). Other processed meats such as fermented sausages, dried meats, canned uncured meats, shelf-stable canned cured meats, etc. are not normally spoiled by Gram-negative bacteria.

Since frozen meat and meat products ($< -12\ ^{\circ}\mathrm{C}$) do not allow the growth of micro-organisms, bacterial spoilage is related to the number and type of micro-organisms before freezing as well as to the thawing conditions. It is often stated that thawed-frozen meat is more perishable than fresh meat, especially because of the drip exuded from thawed meat. However, Lowry and Gill (1985) concluded that 'where handling before freezing, storage and thawing have been satisfactory, thawed meat is as microbiologically sound as those that have never been frozen and as such will spoil in exactly the same manner for any given storage conditions'.

1.2.5 Offals

Although a great variety of bacteria are isolated from offals or 'variety' meats (edible offal or glandular meat) after refrigerated storage in air, pseudomonads commonly become dominant with Enterobacteriaceae being favoured by storage at 3°C rather than at 0°C. Occasionally, *Acinetobacter* and *Aeromonas* are predominant. *Flavobacterium*, *Moraxella* and *Alcaligenes* have also been found (Hanna *et al.*, 1982a, b). Spoilage of thawed offals is likely to follow the spoilage pattern of that of the fresh ones. Thus temperature abuse favours the development of Enterobacteriaceae (Lowry and Gill, 1985). Lactic acid bacteria and streptococci predominate on spoiled kidneys and livers (beef and pork) in vacuum packs; *Flavobacterium*, *Alteromonas*, *Moraxella*, *Acinetobacter* and *Sh. putrefaciens* may also grow to significant numbers. Enterobacteriaceae appear to form a substantial fraction of the flora during the storage of vacuum packed pork livers and sweetbreads (Gill and Jeremiah, 1991).

1.2.6 Irradiated products

Irradiation of poultry, pork and sausages with doses in the range up to 10 kGy is permitted (WHO, 1994) in a number of countries as a means of extending their refrigerated shelf-life and for inactivation of non-spore-forming pathogenic bacteria and foodborne parasites (protozoa and helminths). Most Gram-negative bacteria (i.e. *Pseudomonas* and Enterobacteriaceae) are easily destroyed by low irradiation doses, but *Psychrobacter immobilis*, *Moraxella* and *Acinetobacter* are more resistant. The spoilage flora of irradiated meat and poultry depends on several factors (initial flora, intrinsic and extrinsic factors, radiation dose and atmosphere in the pack). *Moraxellae* (most of the strains probably *Psychrobacter*) are among the primary flora of irradiated poultry stored in air at low temperatures (ICMSF, 1980). When the atmosphere is anaerobic (vacuum and MA pack), the relatively radiation-resistant lactobacilli become dominant in poultry, pork and beef. *Aeromonas* spp. were isolated from temperature-abused vacuum packed irradiated pork by Lebepe *et al.* (1990). Nonmotile organisms (probably *Psychrobacter immobilis*) were dominant on stored irradiated poultry and responsible for the spoilage of irradiated Vienna sausages (Shaw and Latty, 1988).

1.3 Taxonomy and physiology of Gram-negative bacteria associated with spoilage of meat and meat products

1.3.1 Gram-negative aerobic motile rods

Genus Pseudomonas. The genus *Pseudomonas* has been subdivided into five groups on the basis of nucleic acids similarity studies (Palleroni, 1993). The first rRNA similarity group (Group I) includes both fluorescent (*Ps. fluorescens* biovars I to IV, *Ps. putida* biovar A and *Ps. lundensis*) and nonfluorescent species (*Ps. fragi* biovars 1 and 2) which are responsible for low temperature aerobic spoilage of meats. In practice, these usually account for more than 50% and sometimes up to 90% of the spoilage flora (Molin and Ternström, 1982, 1986; Shaw and Latty, 1982, 1984, 1988; Prieto *et al.*, 1992a). *Pseudomonas fragi* is the most common pseudomonad on spoiled meat with an incidence ranging between 56.7 and 79%. *Pseudomonas lundensis* may be considered as the second component of the *Pseudomonas* association of spoiled meat. It is a new species proposed by Molin *et al.* (1986) to accommodate strains attached to one of the major clusters defined by numerical analysis in the studies of Molin and Ternström (1982) and Shaw and Latty (1982). Biovars I and III of *Ps. fluorescens* are also frequently reported on meat, their incidence being lower than that of *Ps. lundensis* (5–13%). *Pseudomonas putida*, which can occur on red meats (less than 5% of the pseudomonad population) does not seem to be of great importance in meat spoilage though the species, *Ps. stutzeri* (nonfluorescent member of Group I), *Burkholderia (Pseudomonas) cepacia* (member of Group II) and *Ps. fluorescens* have been associated with spoilage of loins packed in MA and in permeable films (Ahmad and Marchello, 1989a, b). *Burkholderia cepacia* has also been associated with the spoilage of dry cured ham.

The ability of *Pseudomonas* spp. to grow in refrigerated meat is due partly to the metabolism of glucose to 2-oxo-gluconate or gluconate via the Entner–Doudoroff metabolic pathway (Farber and Idziak, 1982; Nychas *et al.*, 1988). These compounds are not readily assimilable by other microorganisms, and pseudomonads can build up an extracellular energy reserve for use when glucose is depleted. 2-oxo-gluconate or gluconate have been proposed recently as good spoilage indicators (Dainty, 1996). Drosinos and Board (1994) have found metabolic differences among *Pseudomonas* spp. from meat. They proposed that dominance of *Ps. fragi* over *Ps. lundensis* and *Ps. fluorescens* is due to its ability to metabolize creatine and creatinine under aerobic conditions. In broth, pseudomonads use amino acids and lactic acid as the next choice of substrate. Apparently, lactic acid was used in broth culture only after glucose had been depleted (Drosinos and Board, 1994). There are contradictory reports on this issue (Molin, 1985). In relation to their spoilage metabolites, pseudomonads produce dimethylsulphide, but not H_2S. This feature distinguishes them from the Enterobacteriaceae. The inability of *Ps. fragi* to metabolize creatine and creatinine under modified atmospheres in meat broth in the laboratory (Drosinos and Board, 1994) could partly explain the failure of these organisms to become dominant in MAP.

Genus Shewanella.   The taxonomic status of *Shewanella putrefaciens* is not clear. This species has been assigned to the genera *Pseudomonas* (*Ps. putrefaciens*) and *Alteromonas* (*Alt. putrefaciens*). MacDonell and Colwell (1985) classified this organism in a new genus – *Shewanella* – which was established around this species. It also includes *Sh. hanedai* (*Alteromonas hanedai*) and *Sh. benthica*. Two new species (*Sh. alga* and *Sh. colwelliana–Alt. colwelliana*) have been proposed also. The validity of the new genus is supported by the analysis of the 5S rRNA sequence and by data on polar lipids, fatty acids and isoprenoid quinones. However, the inclusion of *Shewanella* in the family Vibrionaceae is still controversial.

Shewanella putrefaciens is the major cause of greening in high pH (> 6) meats. It uses amino acids such as cysteine and serine even when glucose is available. In broth, *Shewanella* is only able to use amino acids after glucose depletion. Metabolism of sulphur-containing amino acids (cysteine, cystine) leads to the accumulation of H_2S, which interacts with myoglobin to give sulphmyoglobin with detrimental effects on meat colour (Newton and Rigg, 1979). Such production is possible only when conditions favour its growth, although its dominance of spoilage is not a prerequisite. Control of this spoilage organism is easily achieved by means of a combination of low pH and low temperatures. Most strains are unable to initiate growth even at pH values of 6.0 when temperature is 2 °C.

Genus Alteromonas.   The role of *Alteromonas* spp. (other than '*Alt. putrefaciens*') in the deterioration of meat is uncertain even though several authors (Grau, 1986; Kraft, 1986) include these bacteria among the spoilage flora of meat stored aerobically or in vacuum packs.

The genus *Alteromonas*, which initially consisted of four species of marine bacteria, was later amended to include 10 additional species. Van Landschoot and De Ley (1983) showed that the genus is highly heterogeneous and suggested that two species (*Alt. communis* and *Alt. vaga*) should be separated from the other alteromonads and assigned to a new genus, *Marinomonas*. Gauthier *et al.* (1995) proposed that a new genus, *Pseudoalteromonas*, should be created to accommodate 11 species that were previously assigned to *Alteromonas* and restrict the genus *Alteromonas* to a single species (*Alt. macleodii*).

Genera Alcaligenes *and* 'Achromobacter'.   The genus *Achromobacter* was not accepted in the most recent edition of Bergey's *Manual of Systematic Bacteriology* (Baumann and Schubert, 1984) and the taxonomic status of the genus *Alcaligenes* is questionable. De Ley *et al.* (1986) proposed the family Alcaligenaceae with three species of the genus *Alcaligenes* (*Alc. faecalis*, *Alc. xyloxidans* subsp. *xyloxidans* and subsp. *denitrificans* and *Alc. piechaudii*) and four species of the genus *Bordetella*. On the basis of phylogenetic data, another three species (*Alc. eutrophus*, *Alc. paradoxus* and *Alc. latus*) will be excluded from the genus in the near future. Strains associated with abattoirs or meat spoilage are reported as *Alcaligenes* spp. or '*Achromobacter*' spp.

Genus Janthinobacterium.   The colonies formed by this genus are commonly gelatinous or rubbery. It is closely related to *Alcaligenes* and some nonfluorescent pseudomonads. The type species, *J. lividum*, which produces a water insoluble purple pigment (violacein) is occasionally involved in the spoilage of certain meat products.

1.3.2   Gram-negative nonmotile aerobic rods

Genus Flavobacterium.   The taxonomy of the flavobacteria was reviewed in the Prokaryotes (Holmes, 1992). More recently, Bernardet *et al.* (1996) provided an amended description of the genus *Flavobacterium* and proposed new combinations for seven described species. In addition, a new species, *F. hydatis* was proposed for *Cytophaga aquatilis*. The emended genus *Flavobacterium* contains bacteria that are motile by gliding, produce yellow colonies and are widely distributed in soil and freshwater habitats. Because of the taxonomic and nomenclatural changes that have been proposed, it is difficult to establish the identity of the *Flavobacterium* spp. associated with meat spoilage.

Genera Moraxella, Acinetobacter *and* Psychrobacter.   Motile and nonmotile, nonpigmented, nonfermentative Gram-negative saprophytic bacteria were classified in the first edition of Bergey's *Manual of Systematic Bacteriology* as *Achromobacter* (Bergey *et al.*, 1923). Thornley (1967) grouped the nonmotile ones in the genus *Acinetobacter*. Later, this genus was modified; the nonmotile, oxidase-negative strains were assigned to *Acinetobacter* spp., and the nonmotile, oxidase positive bacteria to *Moraxella* because of their resemblance to *Moraxella*-like spp. More recently, Juni and Heym (1986) described a new species, *Psychrobacter immobilis*, which embraces some of the oxidase-positive strains which are unrelated to the true moraxellas, as shown by DNA transformation assay. Since 1996, three new *Psychrobacter* spp. (*Psy. frigidicola, Psy. urativorans*, and *Psy. glacincola* have been described. Rossau *et al.* (1991) proposed the creation of a new family, Moraxellaceae, on the basis of DNA–rRNA hybridization studies, to accommodate the above cited micro-organisms which were previously included in the family Neisseriaceae. This new family is divided into two main groups, *Acinetobacter* and another supercluster with four subgroups: the authentic *Moraxella* spp., *M. osloensis, M. atlantae* and a heterogeneous group containing *M. phenylpyruvica, Psychrobacter immobilis*, and allied organisms. Catlin (1991) proposed the new family Branhamaceae for *Moraxella* and *Branhamella*.

The genus *Acinetobacter* is biochemically and genetically heterogeneous. DNA hybridization studies have identified 18 phenotypically distinct hybridization groups (genospecies) and species names have been proposed for seven of these groups. *Acinetobacter johnsonii* was the species found by Shaw and Latty (1988) on poultry and aerobically stored red meats. *Acinetobacter lwofii* was the predominant species isolated from spoiled meat by Gennari *et al.* (1992).

Several authors (Shaw and Latty, 1988; Gennari *et al.*, 1989, 1992; Prieto *et al.*, 1992b) concluded that most of the strains isolated from proteinaceous foods and formerly identified with *Moraxella* and *Moraxella*-like micro-organisms were *Psychrobacter immobilis*. Prieto *et al.* (1992b) isolated this species and *M. phenylpyruvica* throughout the storage life of lamb carcasses. It should be noted that the latter species appears to be closely related to *Psy. immobilis*. In fact, Bowman *et al.* (1996) have proposed that *M. phenylpyruvica* should be transferred to the genus *Psychrobacter* as *Psy. phenylpyruvicus*.

It is often assumed that strains of the family Moraxellaceae form a significant portion of any spoilage flora on aerobically stored meat. It has also been reported however, that their importance is overstated as they often occur only as a minor part of the microflora and have a low spoiling potential (Eribo and Jay, 1985; Gennari *et al.*, 1989; Prieto *et al.*, 1992b). Nevertheless, it has been suggested that acinetobacters and *Psy. immobilis* could play a lipolytic role when they form large populations or in irradiated foods.

This group of bacteria were characterized as poor competitors with a limited enzymatic arsenal (Nychas *et al.*, 1988). They cannot metabolize hexoses but use amino acids and organic acids as carbon and energy sources. The substrates used by *Psychrobacter* are not known (Dainty and Mackey, 1992). *Acinetobacter* use amino acids first, and then lactate. They often occur on meats together with *Pseudomonas*, mainly on surfaces of fat, or on meats with intermediate pH. Their incidence declines as storage progresses, when conditions become stringent. Even though they use amino acids, their metabolic end products are not offensive. In pure cultures, off-odours described as fishy are produced. Their commercial importance could come from their capacity to restrict (under conditions of maximum cell density) oxygen availability to pseudomonads and *Sh. putrefaciens*, enhancing their spoilage potential as they start to attack amino acids and produce H_2S.

In taxonomic studies, strains identified with *Psychrobacter immobilis* were capable of producing acid from a large number of carbohydrates. As with other Gram-negative bacteria, the group is unable to use many compounds as a carbon source. Atypical nonmotile variants of *Pseudomonas fragi* are not uncommon on spoiled meat (Shaw and Latty, 1988; Prieto *et al.*, 1992b). These variants of *Ps. fragi* are closely related to their motile relatives, as they share many metabolic properties. Likewise they do not disappear from stored meat as spoilage progresses. In poultry meat, strains of *Moraxella/Acinetobacter* were isolated more frequently from leg (higher pH) than breast (McMeekin, 1975, 1977). They also appear to prefer fat surfaces (Shaw and Latty, 1988).

1.3.3   Facultatively anaerobic Gram-negative rods

Family Enterobacteriaceae.   A total of 29 genera (14 'traditional' and 15 'additional') are included in this family (Brenner 1992). Of these, *Citrobacter, Enterobacter, Hafnia, Klebsiella, Kluyvera, Proteus, Providencia* and *Serratia* are associated with meat spoilage (Table 1.2). *Escherichia coli* and

Table 1.2 Genera and species of Enterobacteriaceae found in significant numbers in spoiled meat and meat products by several authors

Genera	Species	Meat products
Citrobacter	*Citrobacter* spp. *freundii* *koseri* (*C. diversus*)	Vacuum and MA packed beef lamb and poultry, air stored meat and meat products
Enterobacter	*Enterobacter* spp. *aerogenes* *cloacae* complex *agglomerans/Erwinia herbicola* complex	Lamb, pork, beef, high pH red meat (in air, MA and vacuum), ground meat (air and vacuum), poultry, offals (in PVC and vacuum), fresh sausages, raw cured meats (packed)
Hafnia	*Hafnia* spp. *alvei*	Pork, high pH red meat, ground meat, poultry, premarinated beef and raw cured meats (in vacuum). Red meats and poultry (in air)
Klebsiella	*Klebsiella* spp. *pneumoniae* subsp. *pneumoniae* *pneumoniae* subsp. *ozaenae*	Beef (in vacuum), poultry and red meats (in air)
Kluyvera	*Kluyvera* spp.	Beef (in air)
Proteus	*Proteus* spp. *vulgaris* *mirabilis*	Bacon (vacuum packed), raw hams, cured meats
Providencia	*Providencia* spp. *alcalifaciens* *stuartii* *rettgeri*	Internal taints (bacon and ham), raw cured hams and bacon (in vacuum and MA)
Serratia	*Serratia* spp. *liquefaciens* *marcescens*	Lamb, pork, beef, poultry and high pH meat (in vacuum), high pH meat, poultry, fat, ground pork and fresh sausages (in air)

Yersinia have also been reported. Table 1.2 shows the main genera and species of Enterobacteriaceae involved in the spoilage of meat and meat products.

Three species of *Citrobacter* are currently recognized: *C. freundii*, *C. koseri* (also called *C. diversus* and *Levinia malonatica*) and *C. amalonaticus*. The classification and nomenclature of the genus *Enterobacter* have undergone major changes in the last decade but the situation is still confused. There are 14 named species of *Enterobacter* and probably additional species will be added from the *Ent. cloacae* and the *Erwinia herbicola–Ent. agglomerans* complexes. The latter comprises 12 or more DNA hybridation groups some of which have been assigned to new genera (i.e. *Pantoea–Pan. agglomerans*). Transfer of *Ent. aerogenes* to the genus *Klebsiella* has also been proposed.

Escherichia, which currently contains five species, is a typical member of the family. A strong relationship both genetic and phenotypic exists between this genus (mainly *E. coli*) and *Shigella*.

The genus *Hafnia* contains only one species (*Haf. alvei*) though two separate genospecies are recognized. This bacterium has been described under several names including *Enterobacter alvei*. The genus *Klebsiella* contains five species, with *K. pneumoniae* including three subspecies. Two of these, *K. pneumoniae* subsp. *pneumoniae* and *K. pneumoniae* subsp. *ozaenae*, are often isolated from spoiled meat and meat products. It is difficult to separate this genus from *Enterobacter*. Strains of *Klebsiella* are nonmotile, but nonmotile strains of *Enterobacter* also occur. *Kluyvera* is not commonly associated with spoilage of meat. However, Kleeberger *et al.* (1980) reported that *Kluyvera* (not identified to species) was dominant on beef stored at 7 °C and 15 °C. The species currently recognized are *Kluy. cryocrescens* and *Kluy. ascorbata*.

The taxonomy of *Proteus* and *Providencia* has undergone marked changes in the last years. At present, four species are recognized in the genus *Proteus*, two of these being associated with meat spoilage (*Prot. vulgaris* biogroups and *Prot. mirabilis*). The swarming phenomenon is characteristic of *Proteus*. Three of the five species of *Providencia* (*Prov. alcalifaciens*, *Prov. stuartii* and *Prov. rettgeri*) are involved in the spoilage of meat products. *Providencia alcalifaciens* and *Prov. stuartii* were previously included as two strains (A+B) in one species of *Proteus* (*Proteus inconstans*) and *Prov. rettgeri* was also formerly listed as a *Proteus* sp. Ten species are listed within the genus *Serratia*. *Serratia liquefaciens* is frequently and *Serr. marcescens* only occasionally involved in meat spoilage. Production of extracellular enzymes, salt tolerance and relatively low minimal growth temperature are characteristics of these species. In vacuum packed meat, *Yersinia* spp. and *Yer. enterocolitica* may form a significant part of the microflora. The latter species is pathogenic to humans, the infection usually being waterborne and foodborne.

Although most attention is generally paid to the pathogenic properties of particular genera of Enterobacteriaceae, some members of the family constitute an important spoilage group when conditions favour their growth over that of pseudomonads. This group includes *Serr. liquefaciens*, *Haf. alvei* and *Ent. agglomerans*. Similarly to pseudomonads, they use glucose although some (e.g. *Enterobacter*) appear to have secondary preferences for metabolic intermediates such as glucose-6-P. Amino acids are degraded after carbohydrates with the release of amines, organic sulphides and H_2S. A characteristic of this group is their ability to produce H_2S, but not dimethylsulphide, a feature that distinguishes them from pseudomonads.

DFD meat is not suitable for vacuum packaging due to its high pH, and its tendency to spoil due to H_2S, which combines with myoglobin to form sulphmyoglobin (greening). Several species are known to grow on DFD under anaerobiosis: *Serratia liquefaciens*, *Haf. alvei* and *Yersinia* spp. require pH 6 for growth. These organisms spoil meat due to production of H_2S and greening, the extent of which is accentuated in meat rich in myoglobin. Inhibition of these bacteria is achieved by combined use of pH, temperature and anaerobiosis, but they can grow if these conditions change.

Even though *Enterobacter* has higher affinity for glucose than lactic acid bacteria, pH conditions and metabolic characteristics (ability of lactic acid bacteria to get energy from the use of arginine) preclude *Enterobacter* from becoming dominant (McMeekin, 1982). *Enterobacter* use serine concurrently with glucose. Once glucose has been depleted, glucose 6-P, lysine, arginine and threonine are used. *Serratia liquefaciens* produces acetic acid. *Aeromonas* cause spoilage when traces of oxygen are present (0.1%). *Providencia* (*Prov. rettgeri*) in cases of temperature abuse can stand salt concentration up to 8%. Together with *Salinivibrio* (*Vibrio*) *costicola*, are known to spoil hams (bone taint). It is able to metabolize methionine to methanethiol (cabbage odour).

Family Vibrionaceae. The most recent edition of Bergey's Manual (Bauman and Schubert, 1984) included four genera in the family *Vibrionaceae*: *Vibrio*, *Aeromonas*, *Plesiomonas* and *Photobacterium*. In recent years, this family has undergone marked revision and a number of other genera have been classified in the family. *Aeromonas* has been placed in a new family, Aeromonadaceae, and *Plesiomonas shigelloides* included within the genus *Proteus* as *Prot. shigelloides* (MacDonell and Colwell, 1985). On the basis of DNA–DNA hybridization studies, Janda (1991) recognized 13 species (also called hybridization groups) of *Aeromonas*, some of which have been subdivided into subspecies. Additional hybridization groups have also been proposed.

Both *Aeromonas* and *Vibrio* have been involved in spoilage of meat and meat products. Under aerobic conditions, motile aeromonads, mostly *Aer. hydrophila* and *Aer. caviae*, are unable to compete with faster growing organisms such as pseudomonads. When low levels of oxygen are available, however, they may become significant contaminants on meat of high pH.

Changes in the nomenclature and classification of *Vibrio* make it difficult to know which *Vibrio* spp. are implicated in the spoilage of cured meats. It would appear that the three groups of *Vibrio* strains studied by Gardner (1981) can now be assigned to *Vib. costicola*. Recently, Mellado *et al.* (1996) proposed that this species be transferred to a new genus, *Salinivibrio*, as *Sal. costicola*.

Spoilage of products is carried out by bacteria belonging to the genus *Vibrio*, which are favoured by their high salt concentrations. They spoil Wiltshire hams and bacon causing 'rib taint'. Their metabolic contribution to spoilage and sensory faults of products seem to be very low, even though they seem to have a strong spoilage potential. Vibrios are psychrotrophs which can reduce nitrate and nitrite, and some ('*Vib. costicola* subsp. *liquefaciens*') degrade casein and gelatin. Some strains also produce H_2S, and others can form slime due to dextran production.

1.4 Origin of Gram-negative bacteria in meat and meat products

The microbiology of red meat and poultry is determined by the conditions under which the animals are reared, slaughtered and processed. The most critical stages for meat contamination are the slaughter procedures but a considerable amount of contamination is also possible during subsequent operations. With cattle and sheep, the major source of the psychrotrophic spoilage bacteria appears to be the hides and fleece of animals contaminated by soil and water. *Pseudomonas, Acinetobacter* and *Moraxella* were the most common psychrotrophs found by Newton *et al.* (1977, 1978) on hides and fleece, as well as on meat. Both habitats and vegetation are important reservoirs of the majority of Gram-negative bacteria associated with meat spoilage (Table 1.3).

In poultry, the psychrotrophic flora is carried principally on the feathers, but is also found on the skin. Most of these bacteria are destroyed during scalding. Levels of psychrotrophs and Enterobacteriaceae can increase

Table 1.3 Habitats of Gram-negative bacteria associated with meat spoilage

Aerobic rods	Habitat	Facultatively anaerobic rods	Habitat	Origin
Acinetobacter	Ubiquitous, soil, water and sewage, human skin	*Citrobacter*	Soil, water and sewage man and animals	NF
Alcaligenes	Ubiquitous, soil and water	*Enterobacter*	Soil, water, sewage and plants	NF
Alteromonas	Marine environments	*Hafnia*	Soil, water and sewage mammals and birds	NF/F
Flavobacterium	Widely distributed in nature, especially in water	*Klebsiella*	Soil, vegetation and water, wild and domestic animals, humans	NF/F
Janthinobacterium	Soil and water	*Kluyvera*	Soil, water and sewage	
Moraxella	Mucosal surfaces	*Proteus*	Intestine of humans and animals, manure, soil and polluted waters	NF/F
Pseudomonas	Ubiquitous, fresh and sea water, soil, plants, etc.	*Providencia*	Soiled bedding (faeces and urine) water and environment	NF
Psychrobacter	Aquatic habitats, fish and poultry	*Serratia*	Plants, water and soil, small mammals	NF
Shewanella	Aquatic and marine habitats	*Aeromonas*	Aquatic environments, widely distributed in the environment	
		Vibrio	Aquatic and marine habitats	
		Salinivibrio	Hypersaline environments	

F, faecal origin; NF, not of faecal origin; NF/F, both (Mossel *et al.*, 1995).

during defeathering. In addition, the water used in washing and chilling carries a psychrotrophic population which may recontaminate the poultry carcasses. Further recontamination occurs at subsequent stages of processing (Bremner and Johnston, 1996). Lahellec and Colin (1979) showed that pseudomonads form a very small proportion of the psychrotrophs on the outside of live chicken and turkeys and that *Acinetobacter* and to a lesser extent *Flavobacterium* were dominant. Contamination with pseudomonads occurred during processing from water, hands and materials, and they became dominant among the psychrotrophic flora at the end of chilling.

The scalding treatments applied to pigs largely destroys the Gram-negative organisms, but the carcasses are then recontaminated from the processing equipment. Gill and Bryant (1992) demonstrated that spoilage bacteria (*Pseudomonas*, *Acinetobacter* and *Moraxella*) grew to high numbers in the accumulated detritus of the dehairing equipment and contaminated the circulating waters. Furthermore, they observed that the composition of the flora was largely unaltered after the singeing operations. *Aeromonas* spp. (*Aer. hydrophila* and *Aer. caviae*) also grew well in this niche, the organisms being then spread throughout the dressing and breaking lines where they grew further (Gill and Jones, 1995). These authors isolated both species from most of the samples obtained from the equipment in pig slaughtering plants.

Patterson and Gibbs (1978) reported that Gram-negative bacteria (non Enterobacteriaceae) were widely distributed in abattoirs (lairage, slaughter hall, chill room and boning room), *Pseudomonas* being present at most sites. Members of the family Enterobacteriaceae involved in meat spoilage were isolated from all sites except carcass wash water and air samples in the lairages and boning room. Both groups of bacteria are successful colonizers of wet environments in the structural and work surfaces within the abattoir (Newton *et al.*, 1978). Nortjé *et al.* (1990) investigated the particular contribution of each link in the production chain to the microbial profile of the final products (carcasses and minced meat). In the abattoir, Enterobacteriaceae and *Pseudomonas* were the dominant psychrotrophs, Enterobacteriaceae and micrococci at the wholesaler, and micrococci and pseudomonads at the retailers. They concluded that Enterobacteriaceae are common psychrotrophs in the meat production chain possibly originating from the abattoir and wholesale environments. Using the contamination index, Gustavsson and Borch (1993) studied the contamination of beef carcasses by psychrotrophic *Pseudomonas* and Enterobacteriaceae during slaughter, chilling and cutting. Rapid chilling was identified as a critical processing step. Dehiding and chilling in cold storage rooms were also implicated as critical processing steps, with respect to aerosol contamination and surface cross contamination. *Pseudomonas fluorescens* was dominant in atmosphere samples, as well as those obtained from meat and the processing environment. Other Gram-negative spoilage bacteria

(*Ps. fragi*, *Ps. lundensis*, *Acinetobacter* and *Psychrobacter*) were detected on meat and/or in atmospheric and environmental samples. Carcass wash water was identified by Sierra (1991) as the origin of the fluorescent pseudomonads found on freshly dressed lamb carcasses. The association of these bacteria with free water on surfaces has also been reported by Drosinos and Board (1995a). A wide spectrum of Gram-negative bacteria (*Pseudomonas*, *Acinetobacter*, *Serratia*, *Enterobacter*, *Proteus* and *Vibrio*) were recovered by von Holy *et al.* (1992) from environmental samples in a meat processing plant which manufactured vacuum-packed, Vienna sausages. They concluded that the psychrotrophic nature and simple nutritional requirements of the genera enabled them to persist and/or multiply in/on water, condensate, soil, equipment surfaces, brine solutions and moist floors. Although *Pseudomonas* do not grow in brines, they survive in this environment (Gardner, 1982). The probable source of *Vibrio* in cured meats is curing brines. Gardner (1981) discussed the origins of these bacteria and identified the following as possible sources of contamination: salt used in brine manufacture, fish meals included in the diet of pigs and even agonal bacteraemia.

1.5 Methods of isolation and identification

The preliminary differentiation of the main groups of Gram-negative meat-spoiling bacteria is based on a simplified key (Table 1.4).

1.5.1 Isolation and identification of species of Pseudomonas

Isolation.   *Pseudomonas* spp. grow well on non-selective media (i.e., plate count agar and blood agar) and on routine primary isolation media (i.e. MacConkey and Eosine Methylene Blue Agar) when incubated at a temperature suitable for their growth (Jeppesen, 1995).

When meat and meat products are analyzed, strains of *Pseudomonas* are commonly isolated from plates of Plate Count Agar (PCA), Tryptone Glucose Yeast Extract Agar, or Standard One Nutrient Agar, after incubation at 4–30 °C for 2–10 days (Molin and Ternström, 1982, 1986; Gennari and Dragotto, 1992; Prieto *et al.*, 1992a; von Holy *et al.*, 1992; Hamilton and Ahmad, 1994). PCA plus 1% (w/v) NaCl has also been used for their isolation (Shaw and Latty, 1982; 1984).

Specific media for the isolation of *Pseudomonas* spp. are available. The medium B of King *et al.* (1954) is frequently used for the isolation of fluorescent *Pseudomonas* spp. Pyoverdin production is enhanced and the characteristic colonies can be identified under UV light.

A selective medium, with cephaloridine, fucidin and cetrimide as selective agents, (CFC agar) was described and tested with poultry meat (Mead

Table 1.4 Simplified key for the differentiation of the main groups of Gram-negative bacteria which are commonly related with the spoilage of meat and meat products

Test	Pseudomonas	Shewanella	Alteromonas/ Pseudo-alteromonas	Alcaligenes	Flavobacterium	Moraxella	Acinetobacter	Psychrobacter	Entero-bacteriaceae	Vibrio/ Salinivibrio	Aeromonas
Cell shape	Straight or curved rod	Straight or curved rod	Straight or curved rod	Rod or cocci	Straight or curved rod	Short rod	Rod or cocci	Rod or cocci	Straight rod	Curved or straight rod	Short rod
Flagellar arrangement	Polar	Polar	Polar	Peritrichous	None	None	None	None	Peritrichous/ None	Polar	Polar
Oxidase test	+	+	+	+	+	+	–	+	–	+	+
Motility	+	+	+	+	–	–	–	–	+/–	+	+
Utilization of glucose in the Hugh & Leifson medium	Oxidative	Oxidative	Oxidative	Oxidative	Oxidative	Oxidative	Oxidative	Oxidative	Fermentative	Fermentative	Fermentative
G + C content of DNA (mol %)	55–64	44–47	37–50	52–68	30–42	40–47	38–47	41–47	38–60	38–51	57–63
Arginine dihydrolase	+	–	–	–	–	–	–	–	+/–	–/+	+/–
DNAse	–	+	+	–	+/–	–	–	–	+/–	+/–	+
Pigmentation of colonies	None	Pink	None	None	Yellow	None	None	None	None/Others	None	None
Sensitivity to the vibriostatic compound O 129					+/–					+	–
Na⁺ required for (or stimulates) growth	–	+	+	–	–	–	–	–	–	+	–
Enterobacterial common antigen									+	–	–

and Adams, 1977). It allows the enumeration of both pigmented and non-pigmented *Pseudomonas* spp. Mead and Adams (1977) found that the medium effectively suppressed Gram-positive bacteria and supported growth of *Pseudomonas* spp. (including *Ps. aeruginosa*), whilst inhibiting most other Gram-negative bacteria. *Shewanella putrefaciens* was only partly inhibited. CFC medium can be used for enumeration of *Pseudomonas* spp. in various types of food (Mead, 1985). The medium is inoculated by surface plating and incubated for 48 h at 25 °C. There are two types of colonies: pigmented and non-pigmented, 2–5 mm in diameter. Confirmatory tests are not usually required but in cases of doubt flooding the plates with oxidase reagent can be used to distinguish *Pseudomonas* spp. from other organisms which may be present (i.e. in some cases *Ser. liquefaciens* and yeasts).

Members of Enterobacteriaceae, present on meat packed under vacuum or in modified atmospheres grow on CFC medium thereby giving inflated counts of *Pseudomonas*. In order to differentiate between these two groups of organisms, Stanbridge and Board (1994) modified the CFC medium by adding arginine (1% w/v) and phenol red (0.002% w/v). The pseudomonads produce ammonia from arginine (a pink colour develops in and around the colonies), whereas Enterobacteriaceae generally do not use this substrate and produce yellow colonies. However, with high Enterobacteriaceae counts and with a spiral plating machine as inoculator, the acid produced by the Enterobacteriaceae caused yellowing of the medium and neutralized the alkaline drift of pseudomonads (Stanbridge and Board, 1994).

Purification of strains is commonly achieved by streaking onto Nutrient Agar. Strains are easily maintained for 1–3 months on Nutrient Agar slopes at 5 °C. Lyophilization or liquid nitrogen techniques have to be used for long term storage.

Preliminary characterization of isolates.   Each isolate is tested for oxidase reaction, Gram reaction and cell morphology, as well as for production of acid from glucose (modified Hugh and Leifson medium). Isolates which are oxidase-positive, Gram-negative rods and produce acid from glucose oxidatively are deemed to be *Pseudomonas* spp.

Identification of isolates.   The identification of *Pseudomonas* spp. is problematic. Simple dichotomous determinative keys are of limited use. Taking into account the description of species in the Bergey's *Manual of Systematic Bacteriology* (Palleroni, 1984) and the later description of *Ps. lundensis* (Molin *et al.*, 1986), different systems employing tables including multiple characters have been suggested. However, many of the strains of meat origin show phenotypic properties that do not fit the description of recognized species. From various studies of numerical taxonomy of *Pseudomonas* strains of meat origin (Molin and Ternström, 1982, 1986;

Shaw and Latty, 1982, 1984), several simplified schemes for the differentiation of clusters and species have been described.

The scheme of Molin and Ternström (1982) includes eight characters (fluorescent pigments, gelatinase, acid from cellobiose and maltose, utilization of saccharate, trehalose, meso-inositol and benzylamine). From the data of Molin and Ternström (1986) and Molin *et al.* (1986), a simplified scheme with 16 biochemical tests (acid from maltose, assimilation of D-arabinose, DL-carnitine, creatine, deoxycholate, D-glucuronate, 4-hydroxybenzonate, hydroxy-L-proline, inosine, meso-inositol, malonate, D-mannitol, mucate, D-quinate, D-saccharate and D-xylose) has been devised (Drosinos and Board, 1995a).

Shaw and Latty (1982) employed 12 characters (fluorescent pigments, and utilization as carbon sources for growth of: benzylamine, butylamine, creatine, malonate, hippurate, mannitol, mucate, saccharate, pimelate, rhamnose, and mesaconate). They described a computer-assisted probabilistic identification technique with 18 carbon source utilization tests, to be complemented with extra carbon source tests for strains not correctly identified.

With databases from the phenotypic description of species and taxa of *Pseudomonas*, software for the probabilistic identification of field isolates has been described (Prieto, 1994).

The different species and biovars of fluorescent *Pseudomonas* can be reasonably identified with a scheme of 15 tests (denitrification, levan production, phenazine pigment, and assimilation of L-arabinose, D-xylose, trehalose, mucate, erythritol, myo-inositol, mesaconate, ethanol, anthranilate, histamine, trigonelline and quinate) and a computer assisted probabilistic method (Gennari and Dragotto, 1992). Some additional tests (assimilation of nicotinate, mannitol, L-tryptophan and D-galactose) are necessary for the differentiation of biovars A and B of *Ps. putida*.

The oxidative capacity of 95 organic substrates included in the Biolog GN microplates (Biolog, Hayward, USA) is less discriminating between the different species of *Pseudomonas* than the conventional tests of carbon source utilization (Ternström *et al.*, 1993).

A semi-automated system of identification based on electrophoresis of intracellular proteins (AMBIS) is useful for the differentiation of reference strains belonging to biovars I and III of *Pseudomonas fluorescens* (Rowe and Finn, 1991), although some strains of the other three biovars of this species cluster at values of relatively high similarity.

After standardization of cultural and chemical techniques, the fatty acid profiles (mainly 2- and 3-hydroxy fatty acids) can be used for assigning strains of *Pseudomonas* to one of the six major groups described by Stead (1992).

Rapid detection of added reference strains of *Ps. fluorescens*, *Ps. fragi* and *Ps. aeruginosa* (on meat surfaces at levels higher than 10^4–10^5 cfu/cm^2) is possible with different enzyme-linked immunosorbent assays using polyclonal antibodies against live cells, or against protein F of the cell envelope of *Ps. fluorescens* AH-70 (González *et al.*, 1994, 1996). A similar assay, that employs antibodies against heat-killed cells of *Ps. fluorescens* (and adsorbed with *Ps. aeruginosa*) has been used for the specific detection of *Ps. fluorescens* at levels over 3×10^5 bacteria per ml of an homogenate of meat (Eriksson *et al.*, 1995).

Several rRNA targeted probes for the detection of *Pseudomonas* spp. (mainly those belonging to Group I) have been described (Braun-Howland *et al.*, 1993; Ludwig *et al.*, 1994). A variety of hybridization methods are available.

A polymerase chain reaction (PCR) for the detection of *Pseudomonas* and other Gram-negative and Gram-positive bacteria has been designed (Venkitanarayanan *et al.*, 1996). It employs two primers (23 bases and 20 bases, respectively) selected from the 23S rDNA sequence of *Ps. aeruginosa*. PCR products are detected by gel electrophoresis. It seems that levels of spoilage bacteria over 10^4–10^5 cfu/cm^2 on meat surface can be detected, without interference from 'meat' DNA (Venkitanarayanan *et al.*, 1996). Also, a PCR protocol, with genus specific primers, has been designed for the assessment of the initial contamination-level of *Pseudomonas* (van der Vossen and Hofstra, 1996).

Finally, in order to establish the phylogenetic relationships between new isolates and previously defined taxa, the degree of relatedness of their genomes may be used. Several methods are available: (i) DNA-rRNA hybridization (Palleroni *et al.*, 1973; De Vos *et al.*, 1989); (ii) DNA–DNA hybridization (Ursing, 1986); (iii) direct comparison of rRNA sequences (Ludwig *et al.*, 1994); (iv) comparison of macrorestriction patterns by pulsed-field gel electrophoresis (Grothues and Tümmler, 1991); and (v) gel electrophoresis of stable low molecular weight components of the rRNA pool (Höfle, 1992).

1.5.2 *Isolation and identification of* Shewanella, Alteromonas *and* Alcaligenes

The old hydrogen sulphide-producing pseudomonads are now assigned to the species *Sh. putrefaciens*. With strains of meat origin, other useful characters for their differentiation from other pseudomonads are pink pigmentation, inability to produce arginine dihydrolase, and ability to produce DNAse (Molin and Ternström, 1982). Phenotypically, strains of *Sh. putrefaciens* from fish are quite separate from the type strain, and differ from poultry isolates in their ability to reduce trimethylamineoxide and to assimilate succinate (Stenström and Molin, 1990).

Strains of Gram-negative, heterotrophic, aerobic bacteria with single polar flagellum, which differ from members of the genus *Pseudomonas* mainly in DNA G+C content (37–50 mol %, compared with 55–64 mol % for *Pseudomonas* spp.) are commonly assigned to the genus *Alteromonas*.

Gram-negative, oxidase-positive, strictly aerobic rods motile with peritrichous flagella or cocci which are unable to attack carbohydrates aerobically have been commonly assigned to the genus *Alcaligenes* (Busse and Auling, 1992). For a reliable identification of isolates with *Alcaligenes*, chemotaxonomic or phylogenetic methods have to be used. The presence of both ubiquinone with eight isoprenoid units in its side chain (Q-8) and hydroxy-putrescine as the characteristic polyamine are considered the minimum requirements for a reliable allocation to the genus *Alcaligenes* (Busse and Auling, 1992). Assignment to the species is done mainly after analysis of fatty acid profiles, although a simplified scheme with 36 cultural and biochemical attributes may be used (Busse and Auling, 1992). DNA–DNA hybridization studies allow a definitive allocation to a defined species (Sneath, 1989).

1.5.3   Isolation and identification of Flavobacterium

In general, non-fermentative, Gram-negative, nonmotile, rod-shaped bacteria which produce yellow-pigmented colonies are placed in the genus *Flavobacterium* (Holmes, 1992). These organisms are easily isolated without the need for enrichment, from nutrient agar-type media (after 3–4 days at 20–25 °C). One of such media for the isolation of organisms of this genus from food was proposed by McMeekin *et al.* (1971). Further identification is rarely done with strains of meat origin. The genus includes only strains with low G+C content of the DNA (30–42%). The last emended description (based on DNA–DNA hybridization, DNA–rRNA hybridization, Fatty Acid Methyl Ester profiles, and PAGE of whole-cell proteins) of the genus *Flavobacterium* includes Gram-negative rods that exhibit gliding motility (Bernardet *et al.*, 1996). A scheme based on 27 phenotypic characters for the differentiation of 10 species is included in this description (Bernardet *et al.*, 1996).

1.5.4   Isolation and identification of Acinetobacter, Moraxella *and* Psychrobacter

Members of the three genera are easily isolated on standard laboratory media, such as Trypticase Soy Agar or Brain Heart Infusion Agar. They are isolated also on Plate Count Agar used for total viable counts. Media made selective with crystal violet and bile salts (Medium M and Medium B, Gennari *et al.*, 1992) improved the recovery of strains when these constitute a minority of the microflora. On both media B and M, colonies of *Acinetobacter* and *Psychrobacter* are convex, opaque and light blue (colony colour increases progressively after prolonged incubation at 5 °C).

The oxidase test and four other biochemical tests (use of aminovalerate, 2-keto-D-gluconate, glycerol and fructose as carbon sources for growth; Shaw and Latty, 1988) and examination of cell morphology (Gennari *et al.*, 1992) are used for the differentiation of *Acinetobacter, Moraxella–Psychrobacter* and nonmotile *Pseudomonas*. The identification of *Acineto-bacter* and *Moraxella–Psychrobacter* strains of meat origin is reliably achieved with the API 20E system (Eribo *et al.*, 1985).

The identification of meat isolates with (geno)species of *Acinetobacter* on the basis of biochemical characters alone is difficult. Two matrices consist-ing of 22 and 10 diagnostic characters respectively, are available for the identification of the majority of genomic species (Kämpfer *et al.*, 1993). Even so definitive identification can only be done with DNA-based methods (DNA–DNA hybridization, ribotyping, or restriction fragment length poly-morphism analysis of PCR amplified DNA; Gerner-Smidt, 1992; Novak and Kur, 1995; Vaneechoutte *et al.*, 1995).

The DNA transformation assay with *Psychrobacter immobilis* hyx7 (Juni and Heym, 1980) is the best method for the differentiation of *Psychrobac-ter* and *Moraxella phenylpyruvica* strains from the true *Moraxella*. On the basis of biochemical tests (acid production from melibiose, L-arabinose, cel-lobiose, maltose; oxidative utilization of glucose; phenylalanine deaminase; urease production; ability to utilize carbon sources), different phenotypic groups are found among the *Psychrobacter* and allied organisms isolated from meat and meat products (Gennari *et al.*, 1992; Prieto *et al.*, 1992b). Dif-ferentiation of the two species of *Psychrobacter, immobilis* and *phenylpyru-vicus* associated with meat and meat products is based on various phenotypic, genotypic and 16S ribosomal DNA phylogenetic analysis (Bowman *et al.*, 1996).

Useful methods for the identification of species of the true moraxellae are: the DNA transformation assay of Juni (1978) for *Moraxella osloensis*, determination of the 16S ribosomal DNA sequence (Enright *et al.*, 1994) and DNA–rRNA hybridization analysis (Rossau *et al.*, 1991).

1.5.5   *Isolation and identification of Enterobacteriaceae*

Of the many selective media devised for the isolation of Enterobacteriaceae (Blood and Curtis, 1995), Violet Red Bile Glucose Agar is the most com-monly used. It includes bile salts and crystal violet as selective agents, as well as glucose and a pH indicator as a differential system. Inoculated and solidified medium in petri dishes is overlayed with 10 ml of the same medium before incubation (at $32 \pm 2\,°C$ for 24–48 h). For specific purposes, incubation at $4\,°C$ for 10 d, $37 \pm 1\,°C$ or 42–44 °C is used. The recovery of injured cells can be assisted by shaking dilutions of a food homogenate for 1 h at room temperature before plating. Stressed populations from several sources can be resuscitated by being spread on the surface of

Tryptone Soya Peptone Glucose Yeast Extract Agar. After 6 h at room temperature, this medium is overlayed with VRBGA and the plates are incubated as usual (Mossel *et al.*, 1995). Confirmatory tests for isolates from such colonies are: Gram stain (–ve), oxidase test (–ve), and fermentation of glucose in a mineral salts medium covered with sterile mineral oil and incubated at 37 °C for 24 h (ICMSF, 1978).

The majority of strains of Enterobacteriaceae isolated from meat and meat products are easily identified at the species level by means of several rapid systems incorporating multiple biochemical tests. Results of the API 20 E (24 h of incubation), and API Rapid E (4 h of incubation), which include 20 tests (13 common to both systems) can be interpreted with a computer assisted program (Cox and Bailey, 1986). The Minitek system, of 35 tests, needs to be complemented with supplementary ones (motility, and several additional biochemical reactions) (Stiles and Ng, 1981; Holmes and Humphry, 1988). The AutoMicrobic System (AMS) – an automated system – is helpful for identification purposes. With the Enterobacteriaceae-plus (EBC+) 'biochemical card' (29 biochemical tests and one positive-control broth) and the AMS, 98–99% of strains of meat origin can be identified for species (Bailey *et al.*, 1985). The Gram-negative identification (GNI) database, to be used for the interpretation of the GNI card (which has now replaced the EBC+), includes information for the identification of 99 species of Enterobacteriaceae and other Gram-negative bacteria.

Several numerical identification software packages are available for comparison of the phenotypic features of isolates with strains included in several databases. One of these is described by Miller and Alachi (1996).

1.5.6   Isolation and identification of Aeromonas

More than one plating medium is recommended for the isolation of *Aeromonas*. With samples of meat and meat products, the best results are obtained with: (i) Bile Salts–Irgasan–Brilliant Green Agar, (ii) Ampicillin Sheep Blood Agar with 30 mg/l of Ampicillin (ASBA 30), and/or (iii) Starch Ampicillin Agar (SAA). The media are incubated at 25–30 °C for up to 72 h. Apart from the selective agents (bile salts or ampicillin) these media include differential systems (xylose, blood, starch). Other selective media are also available (Pin *et al.*, 1994; Gobat and Jemmi, 1995; Jeppesen, 1995). For the recovery of injured cells, enrichment in alkaline peptone water (pH 8.7 ± 0.1) at 28 °C is recommended before plating. Presumptive identification of typical colonies (showing haemolytic activity, starch hydrolysis and/or nonfermenting xylose) is simple. Additional tests for the assignment of isolates to genus are: oxidase (it can be carried out directly on some media), fermentation of glucose and other carbohydrates, and resistance to the vibriostatic agent O/129.

The identification to genospecies level may be done with DNA–DNA hybridization studies, random amplified polymorphic DNA polymerase chain reaction (RAPD–PCR) (Oakey *et al.*, 1996), or comparison of profiles of cellular fatty acid methyl esters (FAME) with those of the established hybridization groups (Huys *et al.*, 1995). PCR primers and hybridization probes for the rapid detection of some species of *Aeromonas* have also been described (Dorsch *et al.*, 1994).

1.5.7 Isolation and identification of Vibrio

Moderately halophilic *Vibrio* are isolated from cured meats by plating on media with 4–6% NaCl. Enrichment for 48 h at 22 °C in brain heart infusion broth containing 6% NaCl and 10 ppm crystal violet is necessary for isolation from certain uncured meats. Presumptive tests are Gram reaction (+ve), catalase, oxidase (+ve) and utilization of glucose (Hugh and Leifson medium). Maintenance medium is complex and includes 10% (w/v) salts. It is possible to differentiate *Salinivibrio* (*Vibrio*) *costicola* from other species of *Vibrio* on the basis of the response to different concentrations of NaCl, arginine decarboxylase and G+C content (Mellado *et al.*, 1996).

References

Ahmad, H.A. and Marchello, J.A. (1989a) Effect of gas atmosphere packaging on psychrotrophic growth and succession on steak surfaces. *J. Food Sci.*, **54**, 274–6.

Ahmad, H.A. and Marchello, J.A. (1989b) Microbial growth and successions on steaks as influenced by packaging procedures. *J. Food Protect.*, **52**, 236–9.

Asensio, M.A., Ordóñez, J.A. and Sanz, B. (1988) Effect of carbon dioxide and oxygen enriched atmospheres on the shelf-life of refrigerated pork packed in plastic bags. *J. Food Protect.*, **51**, 356–60.

Bailey, J.S., Cox, N.A. and Fung, D.Y.C. (1985) Identification of Enterobacteriaceae in foods with the AutoMicrobic System. *J. Food Protect.*, **48**, 147–9.

Barnes, E.M. and Impey, C.S. (1968) Psychrophilic spoilage bacteria of poultry. *J. Appl. Bacteriol.*, **31**, 97–107.

Baumann, P. and Schubert, R.H.W. (1984) Family II. Vibrionaceae Veron 1965, 4245[AL], in *Bergey's Manual of Systematic Bacteriology*, Vol. I, (eds N.R. Krieg and J.G. Holt), Williams and Wilkins, Baltimore, pp. 516–17.

Bergey, D.H., Harrison, F.C., Breed, R.S., Hammer, B.W. and Huntoon, F.M. (1923). *Bergey's Manual of Determinative Bacteriology*, 1st edn, Williams and Wilkins, Baltimore.

Bernardet, J.F., Segers, P., Vancanneyt, M. *et al.* (1996) Cutting a gordian knot: emended classification and description of the genus *Flavobacterium*, emended description of the family Flavobacteriaceae, and proposal of *Flavobacterium hydatis* nom. nov. (Basonym, *Cytophaga aquatilis* Strohl and Tait 1978). *Int. J. System. Bacteriol.*, **46**, 128–48.

Blanco, D., Barbieri, G., Mambriani, P. *et al.* (1994) Study of the 'potato defect' of raw dry-cured ham. *Industria Conserve*, **69**, 230–6.

Blickstad, E. and Molin, G. (1983) Carbon dioxide as a controller of the spoilage flora of pork, with special reference to temperature and sodium chloride. *J. Food Protect.*, **46**, 756–63.

Blood, R.M. and Curtis, G.D.W. (1995) Media for 'total' Enterobacteriaceae, coliforms and *Escherichia coli*. *Int. J. Food Microbiol.*, **26**, 93–116.

Borch, E., Kant-Muermans, M.L. and Blixt, Y. (1996) Bacterial spoilage of meat and cured meat products. *Int. J. Food Microbiol.*, **33**, 103–20.

Bowman, J.P., Cavanagh, J., Austin, J.J. and Sanderson, K. (1996) Novel *Psychrobacter* species from Antarctic ornithogenic soils. *Int. J. System. Bacteriol.*, **46**, 841–8.

Braun-Howland, H., Vescio, P.A. and Nierzwicki-Bauer, B. (1993) Use of a simplified cell blot technique and 16S rRNA-directed probes for identification of common environmental isolates. *Appl. Environ. Microbiol.*, **59**, 3219–24.

Bremner, A. and Johnston, M. (1996) *Poultry Meat Hygiene and Inspection*, WB Saunders, London.

Brenner, D.J. (1992) Introduction to the family Enterobacteriaceae, in *The Prokaryotes. A Handbook on Habitats, Isolation and Identification of Bacteria*, 2 edn, (eds A. Balows, H.G. Trüper, M. Dworkin, W. Harder and K.H. Schleifer), Springer Verlag, New York, pp. 2673–95.

Busse, H.-J. and Auling, G. (1992) The genera *Alcaligenes* and '*Achromobacter*', in *The Prokaryotes,* 2nd edn, (eds A. Balows, H.G. Trüper, M. Dworkin, W. Harder and K.-H. Schleifer), Springer Verlag, New York, pp. 2544–55.

Cantoni, C., Bersani, C. and Roncaglia, P.L. (1994) Putrefactive-type spoilage of raw ham. *Ingegneria Alimentare le Conserve Animali*, **10**, 40–4.

Catlin, B.W. (1991) Branhamaceae fam. nov., a proposed family to accommodate the genera *Branhamella* and *Moraxella*. *Int. J. System. Bacteriol.*, **41**, 320–3.

Cox, N.A. and Bailey, J.S. (1986) Enterobacteriaceae identification from stock cultures and high moisture foods with a four-hour system (API Rapid E). *J. Food Protect.*, **49**, 605–7.

Dainty, R.H. (1996) Chemical/biochemical detection of spoilage. *Int. J. Food Microbiol.*, **33**, 19–33.

Dainty, R.H. and Mackey, B.M. (1992) The relationship between the phenotypic properties of bacteria from chill-stored meat and spoilage processes. *J. Appl. Bacteriol.*, **73**, Symp. Suppl. 21, 103S–114S.

Dainty, R.H., Shaw, B.G. and Roberts, T.A. (1983) Microbial and chemical changes in chill-stored red meats, in *Food Microbiology: Advances and Prospects*, (eds T.A. Roberts and F.A. Skinner), Academic Press, London, pp. 151–78.

Daud, H.B., McMeekin, T.A. and Thomas, C.J. (1979) Spoilage association of chicken skin. *Appl. Environ. Microbiol.*, **37**, 399–401.

De Ley, J., Segers, P., Kersters, K. *et al.* (1986) Intra- and intergeneric similarities of the *Bordetella* ribosomal ribonucleic acid cistrons. Proposal for a new family, Alcaligenaceae. *Int. J. System. Bacteriol.*, **35**, 405–11.

De Vos, P., Van Landschoot, A., Segers, P. *et al.* (1989) Genotypic relationships and taxonomic localization of unclassified *Pseudomonas* and *Pseudomonas*-like strains by deoxyribonucleic acid:ribosomal ribonucleic acid hybridizations. *Int. J. System. Bacteriol.*, **39**, 35–49.

Dorsch, M., Ashbolt, N.J., Cox, P.T. and Goodman, A.E. (1994) Rapid identification of *Aeromonas* species using 16S rDNA targeted oligonucleotide primers: a molecular approach based on screening of environmental isolates. *J. Appl. Bacteriol.*, **77**, 722–6.

Drosinos, E.H. and Board, R.G. (1994) Metabolic activities of pseudomonads in batch cultures in extract of minced lamb. *J. Appl. Bacteriol.*, **77**, 613–20.

Drosinos, E.H. and Board, R.G. (1995a) Microbial and physicochemical attributes of minced lamb: sources of contamination with pseudomonads. *Food Microbiol.*, **12**, 189–97.

Drosinos, E.H. and Board, R.G. (1995b) A survey of minced lamb packaged in modified atmospheres. *Fleischwirtschaft*, **75**, 327–30.

Egan, A.F. and Roberts, T.A. (1987) Microbiology of meat and meat products, in *Essays in Agricultural and Food Microbiology*, (eds J.R. Norris and G.L. Pettipher), Wiley, New York, pp. 167–97.

Enright, M.C., Carter, P.E., MacLean, I.A. and McKenzie, H. (1994) Phylogenetic relationships between some members of the genera *Neisseria*, *Acinetobacter*, *Moraxella*, and *Kingella* based on partial 16S ribosomal DNA sequence analysis. *Int. J. System. Bacteriol.*, **44**, 387–91.

Eribo, B.E. and Jay, J.M. (1985) Incidence of *Acinetobacter* spp. and other Gram-negative, oxidase-negative bacteria in fresh and spoiled ground beef. *Appl. Environ. Microbiol.*, **49**, 256–7.

Eribo, B.E., Lall, S.D. and Jay, J.M. (1985) Incidence of *Moraxella* and other Gram-negative oxidase-positive bacteria in fresh and spoiled ground beef. *Food Microbiol.*, **2**, 237–40.

Eriksson, P.V., Paola, G.N., Pasetti, M.F. and Manghi, M.A. (1995) Inhibition enzyme-linked immunosorbent assay for detection of *Pseudomonas fluorescens* on meat surfaces. *Appl. Environ. Microbiol.*, **61**, 397–8.

Farber, J.M. and Idziak, E.S. (1982) Detection of glucose oxidation products in chilled fresh beef undergoing spoilage. *Appl. Environ. Microbiol.*, **44**, 521–4.

Gardner, G.A. (1981) Identification and ecology of salt-requiring *Vibrio* associated with cured meats. *Meat Sci.*, **5**, 71–81.

Gardner, G.A. (1982) Microbiology of processing: bacon and ham, in *Meat Microbiology*, (ed M.H. Brown), Applied Science, London, pp. 129–78.

Gardner, G.A. (1983) Microbial spoilage of cured meats, in *Food Microbiology: Advances and Prospects*, (eds T.A. Roberts and F.A. Skinner), Academic Press, London, pp. 179–202.

Gauthier, G., Gauthier, M. and Christen, R. (1995) Phylogenetic analysis of the genera *Alteromonas*, *Shewanella* and *Moritella* using genes coding for small-subunit rRNA sequences and division of the genus *Alteromonas* into two genera, *Alteromonas* (emended) and *Pseudoalteromonas* gen. nov., and proposal of twelve new species combination. *Int. J. System. Bacteriol.*, **45**, 755–61.

Gennari, M., Alacqua, G., Ferri, F. and Serio, M. (1989) Characterization by conventional methods and genetic transformation of Neisseriaceae (genera *Psychrobacter* and *Acinetobacter*) isolated from fresh and spoiled sardines. *Food Microbiol.*, **6**, 199–210.

Gennari, M. and Dragotto, F. (1992) A study of the incidence of different fluorescent *Pseudomonas* species and biovars in the microflora of fresh and spoiled meat and fish, raw milk, cheese, soil and water. *J. Appl. Bacteriol.*, **72**, 281–8.

Gennari, M., Parini, M., Volpon, D. and Serio, M. (1992) Isolation and characterization by conventional methods and genetic transformation of *Psychrobacter* and *Acinetobacter* from fresh and spoiled meat, milk and cheese. *Int. J. Food Microbiol.*, **15**, 61–75.

Gerner-Smidt, P. (1992) Ribotyping of the *Acinetobacter calcoaceticus–Acinetobacter baumanii* complex. *J. Clinical Microbiol.*, **30**, 2680–5.

Gill, C.O. (1982) Microbial Interaction with Meats, in *Meat Microbiology*, (ed M.H. Brown), Applied Science, London, pp. 225–64.

Gill, C.O. (1986) The control of microbial spoilage in fresh meats, in *Meat and Poultry Microbiology*, (eds A.M. Pearson and T.R. Dutson), Macmillan, Basingstone, pp. 49–88.

Gill, C.O. and Bryant, J. (1992) The contamination of pork with spoilage bacteria during commercial dressing, chilling and cutting of pig carcasses. *Int. J. Food Microbiol.*, **16**, 51–62.

Gill, C.O. and Harrison, J.C.L. (1989) The storage life of chilled pork packaged under carbon dioxide. *Meat Sci.*, **26**, 313–24.

Gill, C.O. and Jeremiah. L.E. (1991) The storage life of non-muscle offals packaged under vacuum or carbon dioxide. *Food Microbiol.*, **8**, 339–53.

Gill, C.O. and Jones, T. (1995) The presence of *Aeromonas*, *Listeria* and *Yersinia* in carcass processing equipment at two pig slaughtering plants. *Food Microbiol.*, **12**, 135–41.

Gill, C.O. and Molin, G. (1991) Modified atmospheres and vacuum packaging, in *Food Preservatives*, (eds N.J. Russell and G.W. Gould), Blackie, Glasgow, pp. 172–99.

Gill, C.O. and Newton, K.G. (1978) The ecology of bacterial spoilage of fresh meat at chill temperatures. *Meat Sci.*, **2**, 207–17.

Gill, C.O. and Penney, N. (1985) Modification of in-pack conditions to extend the storage life of vacuum packaged lamb. *Meat Sci.*, **14**, 43–60.

Gill, C.O. and Penney, N. (1988) The effect of the initial gas volume to meat weight ratio on the storage life of chilled beef packaged under carbon dioxide. *Meat Sci.*, **22**, 53–63.

Gobat, P.F. and Jemmi, T. (1995) Comparison of seven selective media for the isolation of mesophilic *Aeromonas* species in fish and meat. *Int. J. Food Microbiol.*, **24**, 375–84.

González, I., Martín, R., García, T. *et al.* (1994) Detection of *Pseudomonas fluorescens* and related psychrotrophic bacteria in refrigerated meat by a sandwich ELISA. *J. Food Protect.*, **57**, 710–14.

González, I., Martín, R., García, T. *et al.* (1996) Polyclonal antibodies against protein F from the cell envelope of *Pseudomonas fluorescens* for the detection of psychrotrophic bacteria in refrigerated meat using an indirect ELISA. *Meat Sci.*, **42**, 305–13.

Grau, F.H. (1979) Fresh meats: Bacterial association. *Archiv für Lebensmittelhygiene*, **30**, 87–92.

Grau, F.H. (1981) Role of pH, lactate, and anaerobiosis in controlling the growth of some fermentative Gram-negative bacteria on beef. *Appl. Environ. Microbiol.*, **42**, 1043–50.

Grau, F.H. (1986) Microbial ecology of meat and poultry, in *Meat and Poultry Microbiology*, (eds A.M. Pearson and T.R. Dutson), Macmillan, Basingstone, pp. 49–88.

Grothues, D. and Tümmler, B. (1991) New approaches in genome analysis by pulse-field gel electrophoresis: Application to the analysis of *Pseudomonas* species. *Molec. Microbiol.*, **5**, 2763–76.

Gustavsson, P. and Borch, E. (1993) Contamination of beef carcasses by psychrotrophic *Pseudomonas* and Enterobacteriaceae at different stages along the processing line. *Int. J. Food Microbiol.*, **20**, 67–83.

Hamilton, M.A.E. and Ahmad, M.H. (1994) Isolation and characterization of *Pseudomonas* from processed chicken in Jamaica. *Letts. Appl. Microbiol.*, **18**, 21–3.

Hanna, M.O., Smith, G.C., Savell, J.W. *et al.* (1982a) Microbial flora of livers, kidneys and hearts from beef, pork and lamb: effects of refrigeration, freezing and thawing. *J. Food Protect.*, **45**, 63–73.

Hanna, M.O., Smith, G.C., Savell, J.W. *et al.* (1982b) Effects of packaging methods on the microbial flora of livers and kidneys from beef and pork. *J. Food Protect.*, **45**, 74–81.

Höffle, M.G. (1992) Rapid genotyping of pseudomonads by using low-molecular-weight RNA profiles, in *Pseudomonas. Molecular Biology and Biotechnology*, (eds E. Galli, S. Silver and B. Witholt), American Society for Microbiology, Washington, pp. 116–26.

Holmes, B. (1992) The genera *Flavobacterium*, *Sphingobacterium* and *Weeksella*, in *The Prokaryotes. A Handbook on Habitat, Isolation and Identification of Bacteria*, 2nd edn, (eds A. Balows, H.G. Trüper, M. Dworkin, W. Harder and K.-H. Schleifer), Springer Verlag, New York, pp. 3620–30.

Holmes, B. and Humphry, P.S. (1988) Identification of Enterobacteriaceae with the Minitek system. *J. Appl. Bacteriol.*, **64**, 151–61.

Hood, D.E. and Mead, G.C. (1993) Modified atmosphere storage of fresh meat and poultry, in *Principles and Applications of Modified Atmosphere Packaging of Food*, (ed R.T. Parry), Blackie Academic and Professional, London, pp. 269–98.

Huys, G., Kersters, I., Vancanneyt, M. *et al.* (1995) Diversity of *Aeromonas* sp. in Flemish drinking water production plants as determined by gas-liquid chromatographic analysis of cellular fatty acid methyl esters (FAMEs). *J. Appl. Bacteriol.*, **78**, 445–55.

ICMSF (1978) *Microorganisms in foods. 1. Their significance and methods of enumeration*, 2nd edn, University of Toronto Press, Toronto.

ICMSF (1980) *Microbial ecology of foods. Vol. 2. Food commodities*, Academic Press, New York.

Janda, J.M. (1991) Recent advances in the study of the taxonomy, pathogenicity, and infectious syndromes associated with the genus *Aeromonas*. *Clini. Microbiol. Revs.*, **4**, 397–410.

Jeppesen. C. (1995) Media for *Aeromonas* spp., *Plesiomonas shigelloides* and *Pseudomonas* spp. from food and environment. *Int. J. Food Microbiol.*, **26**, 25–42.

Juni, E. (1978) New methodology for identification of nonfermenters: transformation assay, in *Glucose Nonfermenting Gram-negative Bacteria in Clinical Microbiology*, (ed G.L. Gilardi), CRC Press, West Palm Beach, pp. 203–11.

Juni, E. and Heym, G.A. (1980) Transformation assay for identification of psychrotrophic achromobacters. *Appl. Environ. Microbiol.*, **40**, 1106–14.

Juni, E. and Heym, G.A. (1986) *Psychrobacter immobilis* gen. nov. sp. nov.: genospecies composed of gram-negative, aerobic, oxidase-positive coccobacilli. *Int. J. System. Bacteriol.*, **36**, 388–91.

Kakouri, A. and Nychas, G.J.E. (1994) Storage of poultry meat under modified atmospheres or vacuum packs: possible role of microbial metabolites as indicator of spoilage. *J. Appl. Bacteriol.*, **76**, 163–72.

Kämpfer, P., Tjernberg, I. and Ursing, J. (1993) Numerical classification and identification of *Acinetobacter* genomic species. *J. Appl. Bacteriol.*, **75**, 259–68.

King, E.O., Ward, M.K. and Raney, D.E. (1954) Two simple media for the demonstration of pyocyanin and fluorescein. *J. Lab. Clinical Med.*, **44**, 301–7.

Kleeberger, A., Schaefer, K. and Busse, M. (1980) The ecology of enterobacteria on slaughterhouse meat. *Fleischwirtschaft*, **60**, 1529–31.

Kraft, A.A. (1986) Psychrotrophic microorganisms, in *Meat and Poultry Microbiology*, (eds A.M. Pearson and T.R. Dutson), Macmillan, Basingstoke, pp. 191–208.

Lahellec, C. and Colin, P. (1979) Bacterial flora of poultry: changes due to variations in ecological conditions during processing and storage. *Arch. für Lebensmittelhygi*, **30**, 95–8.

Lambropoulou, K.A., Drosinos, E.H. and Nychas, G.J.E. (1996) The effect of glucose supplementation on the spoilage microflora and chemical composition of minced beef stored aerobically or under modified atmosphere at 4 °C. *Int. J. Food Microbiol.*, **30**, 281–91.

Lebepe, S., Molins, R.A., Charoen, S.P. *et al.* (1990) Changes in microflora and other characteristics of vacuum-packaged pork loins irradiated at 3.0 kGy. *J. Food Sci.*, **55**, 918–24.

Lowry, P.D. and Gill, C.O. (1985) Microbiology of frozen meat and meat products, in *Microbiology of Frozen Foods*, (ed R.K. Robinson), Applied Science, London, pp. 109–68.

Ludwig, W., Dorn, S., Springer, N. *et al.* (1994) PCR-based preparation of 23S rRNA-targeted group-specific polynucleotide probes. *Appl. Environ. Microbiol.*, **60**, 3236–44.

MacDonell, M.T. and Colwell, R.R. (1985) Phylogeny of the Vibrionaceae, and recommendation for two new genera, *Listonella* and *Shewanella*. *System. Appl. Microbiol.*, **6**, 171–82.

McMeekin, T.A. (1975) Spoilage association of chicken breast muscle. *Appl. Microbiol.*, **29**, 44–7.

McMeekin, T.A. (1977) Spoilage association of chicken leg muscle. *Appl. Environ. Microbiol.*, **33**, 1244–6.

McMeekin, T.A. (1982) Microbial Spoilage of Meats, in *Developments in Food Microbiology – 1*, (ed R. Davies), Applied Science, London, pp. 1–39.

McMeekin, T.A., Patterson, J.T. and Murray, J.G. (1971) An initial approach to the taxonomy of some Gram-negative yellow pigmented rods. *J. Appl. Bacteriol.*, **34**, 699–716.

McMullen, L.M. and Stiles, M.E. (1993) Microbial ecology of fresh pork stored under modified atmosphere at −1, 4.4 and 10 degrees. *Int. J. Food Microbiol.*, **18**, 1–14.

Mead, G.C. (1985) Enumeration of pseudomonads using cephaloridine-fucidin-cetrimide agar (CFC). *Int. J. Food Microbiol.*, **2**, 21–6.

Mead, G.C. and Adams, B.W. (1977) A selective medium for the rapid isolation of pseudomonads associated with poultry meat spoilage. *Br. Poultry Sci.*, **18**, 661–70.

Mellado, E., Moore, E.R.B., Nieto, J.J. and Ventosa, A. (1996) Analysis of 16S rRNA gene sequence of *Vibrio costicola* strains: description of *Salinivibrio costicola* gen. nov., comb. nov. *Int. J. System. Bacteriol.*, **46**, 817–21.

Miller, J.M. and Alachi, P. (1996) Evaluation of new computer-enhanced identification program for microorganisms: adaptation of BioBASE for identification of members of the family Enterobacteriaceae. *J. Clinical Microbiol.*, **34**, 179–81.

Molin, G. (1985) Mixed carbon source utilization of meat-spoiling *Pseudomonas fragi* 72 in relation to oxygen limitation and carbon dioxide inhibition. *Appl. Environ. Microbiol.*, **49**, 1442–7.

Molin, G. and Ternström, A. (1982) Numerical taxonomy of psychrotrophic pseudomonads. *J. Gen. Microbiol.*, **128**, 1249–64.

Molin, G. and Ternström, A. (1986) Phenotypically based taxonomy of psychrotrophic *Pseudomonas* isolated from spoiled meat, water, and soil. *Int. J. System. Bacteriol.*, **36**, 257–74.

Molin, G., Ternström, A. and Ursing, J. (1986) *Pseudomonas lundensis*, a new bacterial species isolated from meat. *Int. J. System. Bacteriol.*, **36**, 339–42.

Mossel, D.A.A., Corry, J.E.L., Struijk, C.B. and Baird, R.M. (1995) *Essentials of the Microbiology of Foods. A Textbook for Advanced Studies*, Wiley, Chicester.

Newton, K.G., Harrison, J.C.L. and Smith, K.M. (1977) Coliforms from hides and meat. *Appl. Environ. Microbiol.*, **33**, 199–200.

Newton, K.G., Harrison, J.C.L. and Wauters, A.M. (1978) Sources of psychrotrophic bacteria on meat at the abattoir. *J. Appl. Bacteriol.*, **45**, 75–82.

Newton, K.G. and Rigg, W.J. (1979) The effect of film permeability on the storage life and microbiology of vacuum-packed meat. *J. Appl. Bacteriol.*, **47**, 433–41.

Nortjé, G.L., Nel, L., Jordaan, E. *et al.* (1990) A quantitative survey of a meat production chain to determine the microbial profile of the final product. *J. Food Protect.*, **53**, 411–17.

Nottingham, P.M. (1982) Microbiology of carcass meat, in *Meat Microbiology*, (ed M.H. Brown), Applied Science, London, pp. 13–65.

Novak, A. and Kur, J. (1995) Genomic typing of acinetobacters by polymerase chain reaction amplification of the recA gene. *FEMS Microbiol. Letts.*, **130**, 327–32.

Nychas, G.J. and Arkoudelos, J.S. (1990) Microbiological and physicochemical changes in minced meats under carbon dioxide, nitrogen or air at 3 °C. *Int. J. Food Sci. Technol.*, **25**, 389–98.

Nychas, G.J., Dillon, V.M. and Board, R.G. (1988) Glucose, the key substrate in the microbiological changes occurring in meat and certain meat products. *Biotechnol. Appl. Biochem.*, **10**, 203–31.

Oakey, H.J., Ellis, J.T. and Gibson, L.F. (1996) Differentiation of *Aeromonas* genomospecies using random amplified polymorphic DNA polymerase chain reaction (RAPD-PCR). *J. Appl. Bacteriol.*, **80**, 402–10.

Ordóñez, J.A., De Pablo, B., Pérez de Castro, B. *et al.* (1991) Selected chemical and microbiological changes in refrigerated pork stored in carbon dioxide and oxygen enriched atmospheres. *J. Agric. Food Chem.*, **39**, 668–72.

Palleroni, N.J. (1984) Genus I. *Pseudomonas*, in *Bergey's Manual of Systematic Bacteriology*, Vol. 1, (eds N.R. Krieg and J.G. Holt), Williams and Wilkins, Baltimore, pp. 141–99.

Palleroni, N.J. (1993) *Pseudomonas* classification. A new case history in the taxonomy of Gram-negative bacteria. *Antonie van Leeuwenhoek*, **64**, 231–51.

Palleroni, N.J., Kunisawa, R., Contopoulou, R. and Doudoroff, M. (1973) Nucleic acid homologies in the genus *Pseudomonas. Int. J. System. Bacteriol.*, **23**, 333–9.

Papa, F. (1994) Cause and prevention of bacterial alteration in cured ham. *Industrie Alimentari*, **33**, 636–8.

Patterson, J.T. and Gibbs, P.A. (1977) Incidence and spoilage potential of isolates from vacuum-packaged meat of high pH value. *J. Appl. Bacteriol.*, **43**, 25–38.

Patterson, J.T. and Gibbs, P.A. (1978) Sources and properties of some organisms isolated in two abattoirs. *Meat Sci.*, **2**, 263–73.

Penney, N., Hagyard, C.J. and Bell, R.G. (1993) Extension of shelf-life of chilled sliced roast beef by carbon dioxide packaging. *Int. J. Food Sci. Technol.*, **28**, 181–91.

Pin, C., Marín, M.L., García, M.L. *et al.* (1994) Comparison of different media for the isolation and enumeration of *Aeromonas* spp. in foods. *Letts. Appl. Microbiol.*, **18**, 190–2.

Prieto, M. (1994) Integrated software for probabilistic identification of microorganisms. *Computer Applic. Biosciences*, **10**, 71–3.

Prieto, M., García-Armesto, M.R., García-López, M.L. *et al.* (1992a) Species of *Pseudomonas* obtained at 7 °C and 30 °C during aerobic storage of lamb carcasses. *J. Appl. Bacteriol.*, **73**, 317–23.

Prieto, M., García-Armesto, M.R., García-López, M.L. *et al.* (1992b) Numerical taxonomy of Gram-negative, nonmotile, nonfermentative bacteria isolated during chilled storage of lamb carcasses. *Appl. Environ. Microbiol.*, **58**, 2245–9.

Rossau, R., Van Landschoot, A., Gillis, M. and De Ley, J. (1991) Taxonomy of Moraxellaceae fam. nov., a new bacterial family to accommodate the genera *Moraxella, Acinetobacter*, and *Psychrobacter* and related organisms. *Int. J. System. Bacteriol.*, **41**, 310–19.

Rowe, M.T. and Finn, B. (1991) A study of *Pseudomonas fluorescens* biovars using the Automated Microbiology Identification System (AMBIS). *Letts. Appl. Microbiol.*, **13**, 238–42.

Shaw, B.G. and Latty, J.B. (1982) Numerical taxonomic study of *Pseudomonas* strains from spoiled meat. *J. Appl. Bacteriol.*, **52**, 219–28.

Shaw, B.G. and Latty, J.B. (1984) A study of the relative incidence of different *Pseudomonas* groups on meat using a computer-assisted identification technique employing only carbon source tests. *J. Appl. Bacteriol.*, **57**, 59–67.

Shaw, B.G. and Latty, J.B. (1988) A numerical taxonomic study of non-motile non-fermentative Gram-negative bacteria from foods. *J. Appl. Bacteriol.*, **65**, 7–21.

Sierra, M. (1991) Flora superficial de canales de ovino recién obtenidas: microorganismos patógenos, alterantes e indicadores. Ph.D. Thesis, Universidad de León, León.

Sneath, P.H.A. (1989) Analysis and interpretation of sequence data for bacterial systematics: the view of a numerical taxonomist. *System. Appl. Microbiol.*, **12**, 15–31.

Stanbridge, L.H. and Board, R.G. (1994) A modification of the *Pseudomonas* selective medium, CFC, that allows differentiation between meat pseudomonads and Enterobacteriaceae. *Letts. Appl. Microbiol.*, **18**, 327–8.

Stead, D.E. (1992) Grouping of plant-pathogenic and some other *Pseudomonas* spp. by using cellular fatty acid profiles. *Int. J. System. Bacteriol.*, **42**, 281–95.

Stenström, I.-M. and Molin, G. (1990) Classification of the spoilage flora of fish, with special reference to *Shewanella putrefaciens. J. Appl. Bacteriol.*, **68**, 601–18.

Stiles, M.E. and Ng, L.-K. (1981) Biochemical characteristics and identification of Enterobacteriaceae isolated from meats. *Appl. Environ. Microbiol.*, **741**, 639–45.

Swingler, G.R. (1982) Microbiology of meat industry by-products, in *Meat Microbiology*, (ed M.H. Brown), Applied Science, London, pp. 197–294.

Ternström, A., Lindberg, A.M. and Molin, G. (1993) Classification of the spoilage flora of raw and pasteurized bovine milk, with special reference to *Pseudomonas* and *Bacillus*. *J. Appl. Bacteriol.*, **75**, 25–34.

Thornley, M.J. (1967) A taxonomic study of *Acinetobacter* and related genera. *J. Gen. Microbiol.*, **49**, 211–57.

Ursing, J. (1986) Similarities of genome deoxyribonucleic acids of *Pseudomonas* strains isolated from meat. *Current Microbiol.*, **13**, 7–10.

Van der Vossen, J.M.B.M. and Hofstra, H. (1996) DNA base typing, identification and detection systems for food spoilage microorganisms: development and implementation. *Int. J. Food Microbiol.*, **33**, 35–49.

Vaneechoutte, M., Dijkshoorn, L., Tjernberg, I. *et al.* (1995) Identification of *Acinetobacter* genomic species by amplified ribosomal DNA restriction analysis. *J. Clinical Microbiol.*, **33**, 11–15.

Van Landschoot, A. and De Ley, J. (1983) Intra and intergeneric similarities of the rRNA cistrons of *Alteromonas*, *Marinomonas* (gen. nov.) and some other Gram-negative bacteria. *J. Gen. Microbiol.*, **129**, 3057–74.

Varnam, A.H. and Sutherland, J.P. (1995) *Meat and Meat Products. Technology, Chemistry and Microbiology*. Chapman & Hall, London.

Venkitanarayanan, K.S., Khan, M.I., Faustman, C. and Berry, B.W. (1996) Detection of meat spoilage bacteria by using the polymerase chain reaction. *J. Food Protect.*, **59**, 845–8.

Von Holy, A. and Holzapfel, W.H. (1988) The influence of extrinsic factors on the microbiological spoilage pattern of ground beef. *Int. J. Food Microbiol.*, **6**, 269–80.

Von Holy, A., Holzapfel, W.H. and Dykes, G.A. (1992) Bacterial populations associated with Vienna sausage packaging. *Food Microbiol.*, **9**, 45–53.

WHO (1994) *Safety and nutritional adequacy of irradiated foods*, WHO, Geneva.

2 The Gram-positive bacteria associated with meat and meat products

WILHELM H. HOLZAPFEL

2.1 Introduction

The primary contamination of the meat surface of healthy animals is decisively influenced by the abattoir environment and the condition of the animal. Varying levels of both Gram-positive and Gram-negative bacteria constitute the initial microbial population. Adaptation and resistance to conditions on and around the meat surface (e.g. refrigeration, antimicrobial factors, reduction of a_w and air flow, etc.) will determine which groups among the initial contaminants will eventually survive. In addition to the Gram-negative group, 'Achromobacter', Pseudomonas and some Enterobacteriaceae, the Gram-positive one is initially represented by micrococci, followed by lactic acid bacteria and Brochothrix thermosphacta. Other saprophytic Gram-positive bacteria (e.g. Kurthia and nontoxinogenic staphylococci) may constitute minor groups, whilst pathogenic and toxinogenic representatives may originate either from the gut of slaughtered animals, from diseased animals, or may be due to cross contamination from workers' hands and skin. Examples are Staphylococcus aureus, Listeria monocytogenes, group A streptococci, several Clostridium spp. (Cl. perfringens A and C, Cl. bifermentans, Cl. botulinum A, B, E and F, Cl. novyi, Cl. sordellii). The mere presence of these bacteria may constitute a health risk and should be taken into account in practical hygienic measures.

Starting from the carcass directly after slaughtering, each step in handling, chilling, drying, processing, packaging and storage will determine which of the initial contaminating groups of bacteria will eventually survive and dominate the microbial population. Under aerobic conditions and refrigeration temperatures down to 0 °C the psychrotrophic pseudomonads, due to their higher growth rate, may typically dominate the microbial population of fresh and unprocessed meats (Egan and Roberts, 1987). However, the relatively high tolerance of most meat-associated Gram-positive bacteria (exception, the micrococci) against limiting factors such as a reduced a_w, refrigeration temperatures and reduced pH, allows a higher survival rate and longer persistence as compared to most Gram-negative bacteria in the meat environment (Table 2.1).

Table 2.1 Growth/survival criteria of Gram-positive bacteria associated with meat, compared to *Escherichia coli* and pseudomonads in the meat environment (only approximate data are given, derived from Baird-Parker, 1990; Keddie and Shaw, 1986; Müller, 1996; Reuter, 1996; Seeliger and Jones, 1986; Sneath and Jones, 1986)

Bacterium	Temp. (°C)		pH		a_w (min.)
	min.	max.	min.	max.	
Gram-negative bacteria:					
E. coli	7.0	44	4.4	9.0	0.95
Pseudomonas spp.[a]	−5.0	32	5.3	8.5	0.98
Gram-positive bacteria:					
Bacillus spp. (mesophilic)	5	45	4.5	9.3	0.95[b]/0.90
Bacillus spp. (thermophilic)	20	65–70	5.3	9.0	
Brochothrix thermosphacta	0	30	4.6	9.0	0.94
Clostridium spp. (mesophilic)	20[b]/10	45	4.4	9.6	0.97[b]/0.94
Kurthia spp.	5	45	5.0	8.5**	ca. 0.95
Lactobacillus spp.	2	45	3.7	7.2*	0.92
Leuconostoc spp.	1	40	4.2	8.5	0.93
Listeria spp.	1	45	5.5	9.6	0.94
Pediococcus	8	50–53	4.2	8.5	0.90
Staphylococcus (aerobic)	6	48–50	4.0	9.8	0.83
Staphylococcus (anaerobic)	8**	45	4.5**	8.5**	0.91

[a] = psychrotrophic species; [b] = for sporulation; * = some strains of *Lactobacillus sake* are known to grow at pH 8.5 (Min, 1994); ** = estimated values.

Within the typically mixed microbial population of a meat ecosystem a number of Gram-positive bacteria have a strong competitive advantage with respect to vacuum packaging, emulsifying and curing. The lactic acid bacteria (LAB) in particular often make up the typical spoilage association of such products. In some products (e.g. fermented sausages) their metabolic activities may even be desirable.

2.2 Taxonomy and physiology

Comparative sequence analysis of the 16S ribosomal ribonucleic acid (rRNA) reveals reliable phylogenetic relationships of bacteria. On this basis at least 17 major lines of descent are indicated (Fig. 2.1), two of which comprise the Gram-positive bacteria. Most Gram-positive bacteria relevant to meat and meat products form part of the so-called 'Clostridium' branch, representatives of which are characterized by a DNA base composition of less than 54 mol% G+C (Schleifer and Ludwig, 1995). Among these the clostridia and generally also the lactic acid bacteria (LAB) lack a complete cytochrome system. With some exceptions only, the LAB do not produce catalase, although a pseudocatalase may be formed by some LAB in the presence of low sugar concentrations. The genera *Brochothrix*, *Kurthia*, *Listeria*, *Staphylococcus* and most *Bacillus* spp. are catalase-positive and show

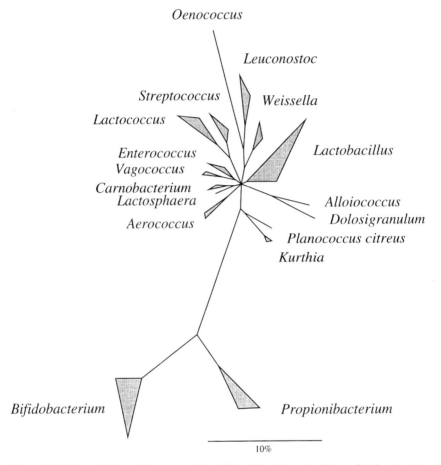

Figure 2.1. Major phylogenetic groups of the 'Clostridium' branch of bacteria, showing the lactic acid bacteria and related genera associated with meat systems, with low mol% G+C in the DNA (modified according to Schleifer and Ludwig, 1995).

either aerobic or microaerophilic growth in food substrates. Some key physiological characteristics of these genera are summarized in Table 2.2. In addition to adaptation to the substrate, features such as growth ability at restrictive conditions of pH, temperature and a_w will determine which strains will eventually grow and dominate a meat ecosystem (see Table 2.1 for summary of key factors for some Gram-positive bacteria).

2.2.1   The lactic acid bacteria

The most important rod-shaped LAB of meat systems are the genera *Carnobacterium*, *Lactobacillus* and *Weissella* (*W. paramesenteroides* shows a coccoid cell morphology); the coccus-shaped genera comprise *Enterococcus*,

Table 2.2 Key phenotypic features of some Gram-positive bacteria associated with meat and meat products.

Characteristics	Brochrothrix	Listeria	Kurthia	Staphylococcus	Carnobacterium	Lactobacillus (homo-fermentation)	Leuconostoc/ Weissella +hetero-fermentation Lactobacillus	Pediococcus	Bacillus	Clostridium
Strictly aerobic	–	–	+	–	–	–	–	–	d	–
Strictly anaerobic	–	–	–	–	–	–	–	–	–	+*
Facultative/ microaerophilic	+	+	–	+	+	+	+	+	+*	–*
Catalase	+	+	+	+	–	–	–	–	+*	–
Lactate as major fermentation product from glucose	+	+	no acid	+	+	+	–	+	d	–
Endospores	+	–	–	–	+	+	–	+	+	+
Diamino acid in peptidoglycan	mDAP	mDAP	Lys	Lys	mDAP	Lys, mDAP	Lys, Orn	Lys	mDAP	mDAP*
Mol% G+C of DNA	36	36–38	36–38	30–39	33–37	32–51	36–55	38–44	32–50[a]	24–54

* Exceptions are known.
[a] several 'Bacillus' spp. with mol% G+C > 55% are to be reclassified (Rainey et al., 1994; Fritze and Claus, 1995).

Leuconostoc, Pediococcus and *Tetragenococcus*, and, more rarely, *Aerococcus, Lactococcus* and *Vagococcus.* With the exception of *Streptococcus thermophilus* practically all *Streptococcus* spp. are considered as pathogens. This genus however has little relevance in terms of growth and hence spoilage in the meat ecosystem.

In addition to the fermentable carbohydrates glucose, glycogen, glucose-6-phosphate and small amounts of ribose, meat and meat products provide a number of vital growth factors such as available amino acids and vitamins that support the growth of several of the 'fastidious' LAB. Some lactobacilli, *Carnobacterium* spp., *Leuconostoc* spp. and *Weissella* spp. are especially well adapted to this ecosystem.

The genus *Lactobacillus* presently comprises at least 64 species which are subdivided into three groups (Kandler and Weiss, 1986; Hammes *et al.*, 1992; Hammes and Vogel, 1995; Holzapfel and Wood, 1995):

- the obligately homofermentative lactobacilli (formerly '*Thermobacterium*') which lack both glucose-6-phosphate dehydrogenase and 6-phosphogluconate-dehydrogenase and most of which grow at 45 °C but not at 15 °C;
- the facultatively heterofermentative lactobacilli ('*Streptobacterium*') which possess both dehydrogenases mentioned above, but use the Embden–Meyerhof–Parnas–pathway of glucose fermentation by preference (Kandler, 1983); and
- the obligately heterofermentative lactobacilli ('*Betabacterium*') lacking fructose 1,6-diphosphate aldolase (and therefore with a similar pathway of glucose fermentation as the genera *Leuconostoc* and *Weissella*) (see Table 2.2).

With the exception of *Lb. farciminis*, the so-called thermophilic lactobacilli are rarely associated with meat systems. On the other hand, several representatives of the heterofermentative and the facultatively heterofermentative lactobacilli may typically dominate the microbial population especially of vacuum packaged and processed meat products. First and foremost the facultatively heterofermentative species *Lb. sake* and *Lb. curvatus* are found in most meat systems and are probably the most frequently encountered lactobacilli and even LAB; their relative proportion in different meat systems, ranging from vacuum-packaged meat to fermented meat products, is illustrated by literature data summarized in Table 2.3 (and compared to data from plant systems). Formerly grouped as 'atypical' streptobacteria by Reuter (1970a, b), *Lb. sake* and *Lb. curvatus* (both producing DL-lactate from glucose) have been shown to be of major economic importance in meat products, either causing spoilage of (vacuum) packaged processed products, or acting as main (desirable) fermentative organisms in

Table 2.3 Relative proportion (in % of total LAB) of different lactobacilli species in meat systems as compared to plant ecosystems (fermented plant foods and phyloplane of plants). A zero (0) indicates a relative proportion < 0.2%.

Species	Fresh Meat (4 °C)[a]		Vacuum packed processed meats[a]	German market[a]	Fermented sausages					Plant systems[a,***]
	Vacuum packed	Irradiated (5 kGy)			Italian traditional[b,***]	Italian industrial[b,***]	Naples Style[c]	Spanish[d]	Greek[e]	
No. of strains	71	114	473	421	111	200	60	254	348	139
heteroferm. lactobacilli	0	0	0.9	5.2	0	0	1.7	0	9.8	–
Carnobacterium spp.	2.8	0	0	0	19.9	0	0	0	0	0
alimentarius	9.9	0	0	2.6	0	0	3.5	0	0	0
amylophilus	0	0	0	?	0	0	1.7	0	0	0
*bavaricus**	21.1	0	5.9	3.0	18.1	2.0	0	11.0	1.4(?)	5.8
*casei***	0	0	0.2	0	15.3	28.0	1.7	0	0.29	7.9
curvatus	22.5	3.5	16.9	22.6	0	0	25.0	26.0	25.8	13.0
coryniformis	0	0	0	0	0	6.0	0	0	0.29	0.7
farciminis	4.2	1.8	1.1	4.8	0	0	1.7	0	2.9	3.6
homohiochii	0	0	3.8	0	16.2	8.0	0	0	0	3.6
plantarum	0	0	0	6.0	0	0	0	8.0	9.8	52.5
sake	33.8	94.7	68.0	50.8	29.7	57.0	65.0	55.0	21.8	3.6
yamanashiensis	0	0	0	0	–	–	0	0	0	7.2
Leuconostoc spp.	2.8	0	3.2	3.3	–	–	0	0	24.7	–
Enterococcus spp.	0	0	0	1.7	–	–	0	0	0	–
Lactobacillus bavaricus + *Lb. curvatus* + *Lb. sake*	76.4	98.2	90.8	73.4	45.0	85.0	90.0	92.0	49	22.4

a = Holzapfel (1996); b = Comi *et al.* (1993); c = Pirone *et al.* (1990); d = Hugas *et al.* (1993); e = Samelis *et al.* (1994); * *Lb. bavaricus* is a 'subjective synonym' of *Lb. sake*; ** *Lb. paracasei* is to be rejected and substituted with *Lb. casei* (Dicks *et al.*, 1996); *** Only lactobacilli were considered.

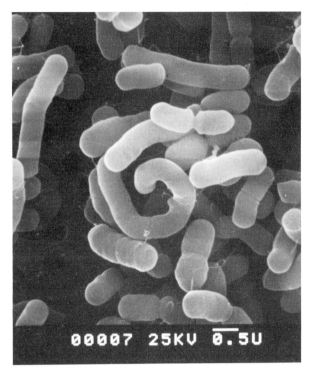

Figure 2.2. Electron micrograph showing typical curved morphology of a *Lactobacillus curvatus* strain isolated from spoiled processed meat. Bar = 0.5 μm.

dry sausages (Hammes *et al.*, 1992; Lücke, 1996a, b). As in shown in Figs. 2.2 and 2.3, *Lb. curvatus* may be distinguished from *Lb. sake* by its typical 'curved' morphology and arrangement in spiral chains on fresh isolation, as compared to the 'coccobacilli' typical of *Lb. sake*. In addition, freshly isolated strains of *Lb. curvatus* may show motility. Strains of *Lb. bavaricus* [a synonym for L(+)-lactic acid producing strains of either *Lb. sake* or *Lb. curvatus* (Hammes *et al.*, 1992; Holzapfel, 1996; Kagermeier-Callaway and Lauer, 1995)] may vary in their typical morphology; Fig. 2.4 shows a morphology typical of *Lb. curvatus* for a strain of *Lb. bavaricus* isolated from spoiled Vienna type sausages. Other *Lactobacillus* spp. (see also Table 2.3) such as *Lb. farciminis* and the facultatively heterofermentative *Lb. alimentarius* and (in contrast to earlier information in the literature) more rarely also *Lb. casei* (formerly *Lb. paracasei*, see Dicks *et al.*, 1996) and *Lb. plantarum* are also commonly associated with meat systems, although in lower frequency and numbers than *Lb. curvatus* and *Lb. sake*. *Lactobacillus bavaricus* was shown to be a L(+)-lactate producing synonym for *Lb. sake*. Strains of *Lb. bavaricus* may vary in their typical morphology (compare Fig. 2.4 as rather typical of *Lb. curvatus*).

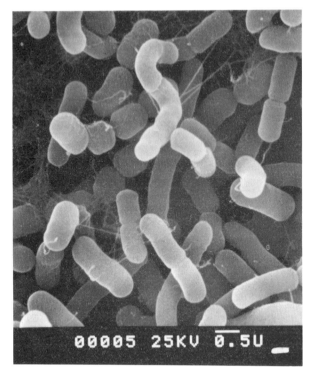

Figure 2.3 Electron micrograph showing typical coccobacillus morphology of a *Lactobacillus sake* strain isolated from spoiled processed sausages. Bar = 0.5 μm.

Persistence and competitive ability of these lactobacilli and several other species of the genera *Leuconostoc* (*Leuc. amelibiosum*, *Leuc. carnosum*, *Leuc. gelidum*), *Weissella* (*W. viridescens*, *W. halotolerans*) and *Carnobacterium* (*Cb. divergens*, *Cb. piscicola*) in processed meat systems are explained by their fermentative ability of the carbohydrates in meat under chill conditions and reduced redox potential as well as by their adaptation to the meat substrate. Whilst the leuconostocs appear to grow most rapidly on chilled fresh meat (Borch and Agerhem, 1992), *Lb. curvatus* and *Lb. sake*, on account of their higher tolerance of elevated salt concentrations and nitrite, typically dominate raw fermented sausage and pasteurized emulsified meat products (Holzapfel and Gerber, 1986; Holzapfel, 1996; Lücke, 1996b) (see also Table 2.3). Some of these features apply also to *P. pentosaceus* and *P. acidilactici*, the major representatives of the *Pediococcus* associated with fermented meat products. However, due to their inability to grow at temperatures <7 °C, they rarely play a role in the spoilage association of chilled meat products (Lücke, 1996a, b; Simpson and Taguchi, 1995). Their typical morphology, arrangement in pairs and

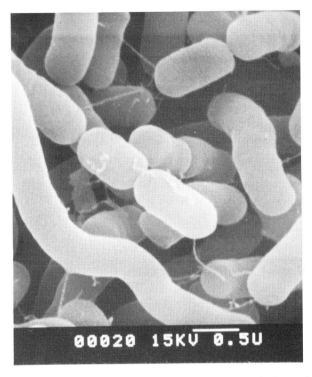

Figure 2.4 Electron micrograph showing curved shape (less typical morphology) of a 'Lactobacillus bavaricus' (syn. Lb. sake) strain isolated from spoiled Vienna sausages. Bar = 0.5 μm.

tetrads but not in chains allow their early differentiation from the other (catalase-negative) LAB upon isolation. By its high salt tolerance (growth in presence of 18% NaCl or $< a_w$ 0.90) *Pediococcus halophilus* is phenotypically different from other pediococci. Comparative 16S rRNA sequence analysis showed also that this species differs phylogenetically and its transfer to a new genus *Tetragenococcus* was proposed (Collins *et al.*, 1990). *Tetragenococcus halophilus* may be associated with cured meat products and concentrated brines used for the manufacture of bacon. The status of the genus *Aerococcus* is still not clear. Some phylogenetic relationship seems to exist between *Aerococcus viridans* and *Pediococcus urinaeequi* (Aguirre and Collins, 1992), and it was proposed that the latter be transferred to the genus *Aerococcus*. It differs from other pediococci and most LAB by growing in the absence of fermentable carbohydrates and also by the high pH value of 8.5–9.0 needed for optimum growth (Simpson and Taguchi, 1995).

On the basis of hybridization studies, Schillinger *et al.* (1989) showed that *Leuc. paramesenteroides* is more closely related to the heterofermentative lactobacilli, *Lb. confusus*, *Lb. halotolerans*, *Lb. kandleri*, *Lb. minor* and *Lb. viridescens* than to *Leuconostoc sensu stricto*. LAB, including *Leuconostoc*-like organisms, isolated from fermented Greek sausage were studied by Collins *et al.* (1993) who proposed that *Leuc. paramesenteroides* and related species should be reclassified in a new genus *Weissella*. They also described a new species, *Weissella hellenica*, for the isolates obtained from fermented sausage. These former *Lactobacillus* spp. all show an interpeptide bridge atypical of the peptidoglycan of the lactocbacilli. In Table 2.4 this phenomenon and the new nomenclature is explained.

Comparative 16S rRNA sequence analysis revealed a phylogenetically coherent group of the formerly 'atypical', acetate-sensitive lactobacilli, '*Lactobacillus carnis*' (Shaw and Harding, 1985) and '*Lactobacillus*' *divergens* (Holzapfel and Gerber, 1983) isolated from vacuum-packaged, refrigerated meat. These were classified in a new genus, *Carnobacterium* (Wallbanks *et al.*, 1990). Physiological features such as inability to grow at pH 4.5, but to grow well at pH 9.0 may be used to distinguish *Carnobacterium* from *Lactobacillus*; in addition *Carnobacterium* contains *meso*-DAP in the cell wall in contrast to most lactobacilli. Although small amounts of CO_2 are produced from glucose, fermentation is via the glycolytic pathway, with L(+)-lactic acid as major end product (De Bruyn *et al.*, 1987, 1988). '*Lactobacillus carnis*' was shown by Collins *et al.* (1987) to be homologous to '*Lactobacilus piscicola*', a salmonid fish pathogen (Hiu *et al.*, 1984), and transferred to *Cb. piscicola*; subsequently Collins *et al.* (1991) suggested that this species be renamed *Cb. maltaromicus*, justified by 100% 16S rRNA sequence similarity to '*Lactobacillus maltaromicus*' (Miller *et al.*, 1974). *Carnobacterium* shows a closer phylogenetic relationship to *Enterococcus* and *Vagococcus* (*vide infra*) than to the lactobacilli (see also Fig. 2.1).

For comprehensive information on the phenotypic characterization, phylogeny and general taxonomy of the LAB reference is made to Pot *et al.* (1994), Schleifer and Ludwig (1995), Stiles and Holzapfel (1997) (for

Table 2.4 Assignment of *Lactobacillus* and *Leuconostoc* species to the new genus *Weissella*.

Paramesenteroides group[a]	New designation as genus *Weissella* (Collins *et al.*, 1993)	Peptidoglycan: interpeptide bridge
Leuconostoc paramesenteroides	*Weissella paramesenteroides*	Lys-Ala$_2$ (Lys-Ser-Ala$_2$)
Lactobacillus confusus	*confusa*	Lys-Ala
halotolerans	*halotolerans*	Lys-Ala-Ser
kandleri	*kandleri*	Lys-Ala-Gly-Ala$_2$
minor	*minor*	Lys-Ser-Ala$_2$
viridescens	*viridescens*	Lys-Ala-Ser
	hellenica sp. nov.[b]	Lys-Ala-Ser

[a] Organisms listed in Bergey's *Manual of Determinative Bacteriology* (Kandler and Weiss, 1986); [b] Newly described species by Collins *et al.* (1993).

food-associated LAB), Hammes and Vogel (1995) (on the genus *Lacto-bacillus*), Dellaglio *et al.* (1995) (on the genus *Leuconostoc*), Schillinger and Holzapfel (1995) (on the genus *Carnobacterium*), and Simpson and Taguchi (1995) (on the genus *Pediococcus*).

Enterococcus and *Lactococcus* are other LAB genera of some commercial significance. *Enterococcus* spp. use the homolactic pathway for energy production, yielding mainly L(+)-lactic acid from glucose at pH < 5.0. At pH >7.0 ethanol, acetic acid and formic acid are the main products of glucose fermentation. In the absence of heme and under aerobic conditions, glucose is converted to acetic acid, acetoin and CO_2. In addition, energy is also obtained by amino acid degradation (DeVriese *et al.*, 1992). The genus *Enterococcus*, formerly the 'group D' or 'faecal' streptococci, is described in detail by DeVriese and Pot (1995); all species differ from the lactococci by their resistance to 40% bile and growth of most species in 6.5% salt. The major representatives typical of the food environment, *E. faecium* and *E. faecalis*, are associated with the gastro-intestinal tract of man and warm-blooded animals and have been suggested as indicators of faecal contamination of meat (Reuter, 1996). Special phenotypic features are their ability to grow at 10 °C and 45 °C and at pH 9.6. Although morphologically different from the rod-shaped carnobacteria, the enterococci show phenotypic similarities to the carnobacteria by virtue of their growth at pH 9.6 and several other phenotypic characteristics, including resistance to azide, antibiotics and heavy metals (Schillinger and Holzapfel, 1995). The genus *Lactococcus* is phylogenetically more closely related to the streptococci than to the enterococci and the genera *Vagococcus*, *Carnobacterium* and *Lactobacillus* (Pot *et al.*, 1994). Its most important representatives, the two subspecies of *Lactococcus lactis*, are typically associated with the dairy environment. Occasionally they may form in addition to *Lactococcus raffinolactis* a minor part of the microflora of fresh meat (Schillinger and Lücke, 1987).

Only distantly related to the 'typical' LAB, the genus *Bifidobacterium* forms part of the so-called '*Actinomycetes*'-branch. Lactic acid and acetic acid are produced in a ratio of 2:3 as major products of glucose fermentation by the so-called 'bifidus-shunt' heterofermentative pathway, and as the Actinomycetes branch constitutes bacterial genera with >55 mol% G+C in the DNA, the bifidobacteria are therefore more closely related to the genera *Brevibacterium*, *Propionibacterium* and *Microbacterium*. These groups are rarely associated with meat. The inclusion of *Bifidobacterium* spp. in meat starter cultures has been suggested recently (Arihara *et al.*, 1996).

2.2.2 *The genera* Brochothrix, Kurthia *and* Listeria

The catalase-positive, non-sporulating Gram-positive genera *Brochothrix*, *Kurthia* and *Listeria* are common to the meat habitat. On account of its strict aerobic growth requirement, *Kurthia* is easily distinguished from the other two species (Table 2.2). *Brochothrix* and *Listeria* may be confused because of their ability to grow under similar (facultatively anaerobic to microaerophilic) conditions and for sharing several morphological and biochemical characteristics. Phenotypic differentiation relies on features such as the inability of *Brochothrix* to grow at 37 °C and differences in the sugar fermentation pattern (Kandler and Weiss, 1986). In addition to phenotypic similarities, these genera are also phylogenetically closely related (see also Fig. 2.1). Some of the properties determining their growth and survival in meat/food systems are compared with those of the LAB in Table 2.1.

Originally allocated to the genus *Microbacterium* as *Microbacterium thermosphactum*, Gram-positive, nonsporulating, nonmotile rod-shaped bacteria isolated from pork sausage meat by Sulzbacher and McClean (1951) were later transferred to the new genus *Brochothrix* (Sneath and Jones, 1986). Whilst *Listeria* apparently only utilizes the glycolysis Embden–Meyerhof–Parnas) pathway for aerobic and anaerobic glucose catabolism, *Brochothrix* possesses enzymes typical for both the HPM- and the EMP-(glycolysis) pathways of glucose catabolism. Under anaerobiosis and glucose limitation, L(+)-lactate and ethanol are produced in the ratio of about 3:1; the ratios of major end products of anaerobic glucose fermentation (lactate, acetate, formate and ethanol) may vary with conditions (Jones, 1992). *Brochothrix thermosphacta* was shown to utilize not only glucose and glutamate during growth on a meat juice medium, but also other compounds including ribose and glycerol (Egan and Roberts, 1987).

2.2.3   *The genera* Micrococcus *and* Staphylococcus

The micrococci and staphylococci share common habitats such as the skin of man and animals, and they exhibit similar (microscopic) morphology. This explains in part why these two phylogenetically distinct genera have have often been grouped together, e.g. as 'micrococci'. With a high mol% G+C content of the DNA (ranging from 66 to 73%) *Micrococcus* spp. form part of the 'Actinomycetes' branch of bacteria – the staphylococci have a mol% G+C of 30–38% (see Table 2.2). A few physiological features, however, allow the relatively reliable differentiation of most strains belonging to these two genera. In contrast to the staphylococci, most micrococci do not ferment (or produce acid) from glucose anaerobically, they are resistant to lysostaphin but sensitive to bacitracin (explained by the differences in their cell wall structure and the interpeptide bridges of the peptidoglycan), and do not contain teichoic acid in the cell wall (Kocur *et al.*, 1992). The ability of staphylococci to produce acid anaerobically from glycerol in the presence of 0.4 µg of erythromycin/ml and their sensitivity to lysostaphin form the basis of a discriminatory test described by Schleifer and Kloos (1975) to distinguish between these two groups.

Mainly micrococci have been reported in earlier publications (e.g. Kitchell, 1962) to be associated with fermented meat products. Correct classification has shown, however, that some of these isolates belonged to the staphylococci, in particular *Staphylococcus carnosus* and *Staph. xylosus*. Some micrococci (mainly *Microcccus varians* and *M. kristinae*) appear to play a desirable role in meat processing (Fischer and Schleifer, 1980).

Being strictly aerobic, most micrococci oxidize carbon sources (e.g. glucose, fructose, glycerol, lactate) to CO_2 and water. They metabolize glucose by the fructose-1,6-diphosphate and hexose monophosphate pathways and with citric acid enzymes. *Microcccus varians* and *M. kristinae* are exceptions in that they may also grow facultatively anaerobically and produce L(+) lactic acid from glucose (Kocur *et al.*, 1992). Being facultative anaerobes, the staphylococci may use both the glycolytic (EMP or Embden–Meyerhof–Parnas) and the hexose-monophosphate (HMP) pathways as main routes of glucose metabolism. The main end product of anaerobic glucose metabolism is (73–94%) L(+)-lactate (with traces of acetate, pyruvate and CO_2). Acetate and CO_2 are the main end products from glucose under aerobic conditions (Kloos *et al.*, 1992).

Of the more than 30 recognized *Staphylococcus* spp., the coagulase-positive representatives *Staph. aureus*, *Staph. intermedius* and *Staph. delphini*, and the coagulase-variable *Staph. hyicus* are generally regarded as potential pathogens. However, a number of coagulase-negative species, e.g. *Staph. haemolyticus*, *Staph. lungdunensis*, *Staph. warneri*, *Staph. hominis*, *Staph. schleiferi*, *Staph. saprophyticus* and *Staph. simulans* have occasionally been isolated from infections in man and animals, hence some of these may be considered as opportunistic pathogens (Kloos *et al.*, 1992). Strains of *Staph. aureus* may produce one or more of five serologically different enterotoxins that cause typical 'staphylococcal' food poisoning on ingestion (Bergdoll, 1979; 1983).

Reports on potential pathogenicity of *Micrococcus* spp. are rare, although *M. kristinae*, *M. luteus* and *M. sedentarius* may have been implicated in some cases of human infections (Kocur *et al.*, 1992).

2.2.4 *The endospore-forming genera* Bacillus *and* Clostridium *(nonpathogenic, nontoxinogenic)*

Bacillus (aerobic to microaerophilic) and *Clostridium* (anaerobic) are the most important endospore-forming genera with respect to meat systems. They are rarely associated with spoilage of meat systems. The one exception is the anaerobic putrefactive (deep-bone) spoilage of meat. Other Gram-positive endospore-forming genera include *Sporolactobacillus* (catalase-negative and facultatively aerobic), Sporosarcina (coccus-shaped with typical 'sarcina' arrangement, and strictly anaerobic), *Thermoactinomycetes* (with typical actinomycete properties) and *Desulfotomaculum*. As the major spore-forming sulphate-reducer in the environment, some strains of the latter may cause H_2S spoilage of canned meats. This anaerobic genus is distinguished from the clostridia by the ability to reduce sulphate. Some key physiological features of the genera *Bacillus* and *Clostridium* are compared in Table 2.2. (Claus *et al.*, 1992; Hippe *et al.*, 1992; Slepecky and Hemphill, 1992; Widdel, 1992). Bacilli isolated from pasteurized, smoked vacuum-packaged Vienna sausages were identified with *Bacillus circulans*, *B. licheniformis*, *B. pumilis*, *B. sphaericus* and *B. subtilis* (Von Holy, 1989).

2.2.5 Other Gram-positive genera

Representatives of the genera *Brevibacterium* (60–67 mol% G+C in the DNA), *Corynebacterium* (51–65 mol% G+C) and *Propionibacterium* (66–67 mol% G+C) together with *Bifidobacterium* all form part of the 'Actinomycete' branch of bacteria. They appear to be rarely associated with meat systems. Reports on 'coryneforms' in or on raw meat may indicate morphological relationship of unidentified meat bacteria with these genera. However, some strains designated 'coryneforms' have been shown to be either identical or related to *Kurthia* and *Brochothrix*. *Bifidobacterium* spp. are typically associated with the gastro-intestinal tract of man and animals. Although the parasitic contamination of meat surfaces must not be ignored, their strict anaerobic and fastidious growth requirements will not generally allow their growth or survival on fresh and processed meats.

Little is known about the genus *Aureobacterium*. With a high GC content (67–70 mol%) it belongs to the Actinomycetes branch of bacteria (Collins and Bradbury, 1992). The dairy environment is considered to be a typical habitat. It was reported to be part of the microflora of dressed lamb carcasses (Sierra *et al.*, 1995). The association of the genus *Microbacterium* with the meat environment has not been confirmed even though it has been isolated from poultry giblets and fresh beef (Kraft *et al.*, 1966). Its GC content ranges from 69 to 75 mol% (Collins and Bradbury, 1992) and phylogenetically it also belongs to the Actinomycetes branch.

2.3 Depots and levels of contamination in the meat environment

The initial microbial population on meat and meat products may range from 10^2 to 10^4 cfu/cm^2 or g of which about 10% may be able to grow at chill temperatures, and a smaller percentage cause spoilage (Borch *et al.*, 1996). Drying of the carcass surface (e.g. through air movement in cold rooms) will reduce the numbers, especially of Gram-negative bacteria such as *Campylobacter* spp. which are sensitive to reduced a_w.

2.3.1 Surface contamination

Fresh/raw meats. Major sources of carcass contamination during and after slaughter originate from the air and soil, and particularly also from the hide, hair and gastro-intestinal tract of the animal. Within the typical whole-sale/retail meat environment, important additional sources of contamination are the hands and clothes of workers, as well as saws and other utensils (Nortjé *et al.*, 1989a). Differential counts on cut meat surfaces revealed that the LAB was the third largest group of contaminants, and served as an indicator of the overall sanitary condition of retail premises (Nortjé *et al.*, 1989b). A study of a non-integrated meat production system of abattoirs, wholesalers and retailers in South Africa, focused on 1265 representative bacteria taken from the highest psychrotrophic plate counts, revealed (Table 2.5) that in descending order of importance *Staphylococcus* spp., 'coryneforms' (excluding *Brochothrix*), *Micrococcus* spp., *Lactobacillus* spp. and *Bacillus* spp. were the major Gram-positive contaminants (Nortjé *et al.*, 1990a). Gram-positive cocci and *Br. thermosphacta* were found on practically all surfaces sampled in the abattoir, wholesale and retail environments, including the hands of staff. Contamination with the micrococci averaged up $\log_{10} 3.0$ cfu/cm^2 in the abattoir, $< \log_{10} 2.9$ cfu/cm^2 in wholesale and $\log_{10} 3.8$/cm^2 in retail situations; corresponding values for *Enterococcus* spp. were 3.0, 3.0 and 5.1, respectively, and for *Br. thermosphacta* 3.0, < 2.9 and 5.3 respectively. The data in Table 2.5 suggest the relative importance of staphylococci and micrococci only as major initial Gram-positive contaminants. Compared to wholesale, an overall increase in retail contamination levels was observed on all surfaces (hands, clothes, saws, mincers and meat) with an especially marked increase in *Brochothrix* numbers, ranging from 5.1 to 6.2 $\log_{10}$/cm^2 or g (Nortjé *et al.*, 1990b). In Japan about 34% of retail pork samples were found to be contaminated with *Erysipelothrix rhusiopathiae*, as compared to 4–54% of pork loins in Sweden (Jay, 1996).

Table 2.5 Proportion (in %) of Gram-positive bacteria, as part of the total microbial population, contaminating meat at different distribution levels (modified from Nortje *et al.*, 1990a).

Genus/Group	Abattoir	Wholesale	Retail	No./% of total isolates
Micrococcus spp.	5.1	13.6	1.9	77/6.1
Staphylococcus spp.	8.8	16.5	12.6	163/12.9
Bacillus spp.	0.7	0.8	1.3	13/1.0
Lactobacillus spp.	0.7	1.9	1.7	19/1.5
'Coryneform' group	7.1	13.6	13.6	153/12.1
Total Gram-positives	21.6	46.4	31.1	425/33.6
Total number of all isolates	295	375	595	1265/100

The most important psychrotrophic Gram-positive contaminants on freshly dressed lamb carcasses were reported to be (in the order of importance): *Br. thermosphacta*, *Caseobacter polymorphus*, coryneforms (*Corynebacterium*, *Kurthia*, *Brevibacterium* and *Cellulomonas*); in addition, *Aureobacterium* and *Listeria* were also isolated (Sierra *et al.*, 1995). *Cellulomonas* (72–76 mol% G+C) is considered to be a typical inhabitant of soil (Stackebrandt and Prauser, 1992). This may well explain its prescence on meat surfaces.

Micrococci (34%), corynebacteria (24%) and lactobacilli (16%), followed by Enterobacteriaceae (16%) and other Gram-negative bacteria (Gallo *et al.*, 1988) were found to be the predominant contaminants on refrigerated fresh broilers.

Assessment of surface contamination at various stages of dressing, chilling and cutting operations in three pig processing plants revealed the domination of Gram-positive bacteria (numbers around $10^3/cm^2$) before and Gram-negative ones (number about $10^4/cm^2$) after scalding. Contamination with lactobacilli and *Br. thermosphacta* occurred during cutting (Gill and Bryant, 1992). In a study on the contamination of 4357 pig hindquarters, 22.7% were found to be contaminated with *St. aureus*, with numbers on the raw, uncured ham ranging from 10 to 10^3 cfu/cm^2 in 89% of the positive samples (Schraft *et al.*, 1992).

Processed products.   Mesophilic and psychrotrophic species of *Lactobacillus* (facultatively heterofermentative) and *Leuconostoc* (including *Weissella*) appear to be the most prominent spoilage bacteria of emulsion-type meat products. Problems may be related to (a) contamination of the raw materials with *Weissella viridescens* which may survive heat processing, and (b) recontamination with other *Lactobacillus* spp. (Borch *et al.*, 1988), probably *Lb. sake* and *Lb. curvatus*. *Lactobacillus sake* in particular, has frequently been reported to predominate in the meat processing environment and in the spoilage association of processed meat products – vacuum packaging and refrigeration may enhance its competitive behaviour (Holzapfel and Gerber, 1986; Von Holy *et al.*, 1992; Dykes and Von Holy, 1994; Dykes *et al.*, 1995; Björkroth and Korkeala, 1996a, b; Björkroth *et al.*, 1996) (see also Table 2.3). The 'leuconostocs' constitute a second major group (Schillinger and Lücke, 1987), with the most prominent species reported to be *Leuc. carnosum*, *Leuc. gelidum* and *W. paramesentoides* (Shaw and Harding, 1989; Von Holy *et al.*, 1991; Dykes *et al.*, 1994) and *Leuc. amelibiosum* (*Leuc. citreum*) (Mäkelä *et al.*, 1992c).

Frozen fresh beef sausages manufactured in Egypt were contaminated with *Br. thermosphacta* and LAB at levels of $8 \times 10^2/g$ and $2 \times 10^2/g$ respectively (Khalafalla and El-Sherif, 1993). In Egypt 35.9% of the raw meat samples from slaughterhouses were contaminated with *S. aureus*, 25% from a mechanical slaughterhouse, 23.3% of cooked meat samples and 27.5% of luncheon meat samples (El-Sherbeeny *et al.*, 1989), indicating recontamination as an important factor. Enterococci were detected in all samples of Egyptian fresh sausages studied by Rashad (1990), whilst low counts (1.3×10^2– 2.6×10^4/g) of lactobacilli were detected throughout storage for 28 days.

Due to their high tolerance to reduced a_w, staphylococci and micrococci were dominant in an intermediate moisture meat product ('basturma') in Greece, and 42% of 120 isolates were identified with *Staphylcoccus epidermidis*, 32% with *Staph. saprophyticus*, 12% with *Staph. simulans*, 2% with *Staph. carnosus*, and 7.5% with *M. varians* (Kotzekidou, 1992).

In a study of contamination sources and lactic spoilage of Finnish cooked ring sausages, the highest LAB counts (10^5 and 10^8 cfu/g respectively) were found in meat trimmings and pork skin emulsion (Mäkelä *et al.*, 1990); indeed LAB were found in all other raw materials except for commercial spice mixtures used in the production of this type of sausage.

Fermented sausages were also implicated as a potential source of contamination of ropy-slime producing LAB (mainly homofermentative lactobacilli) in cooked meat products (Mäkelä, 1992). Using ribotyping, a starter strain of *Lb. sake* used in sausage fermentation was shown to contaminate the processing environment. It could not be implicated however, in the spoilage of vacuum-packed, sliced cooked meat products (Björkroth and Korkeala, 1996a).

2.3.2 Clostridia as contaminants

An intensive study of clostridia as contaminants in the food processing industry was conducted by Eisgruber (1992). Spoilage ('bone taint') of the deep muscle of beef and of cured pork products may be caused by microorganisms that have entered the musculature after slaughter or even as a result of contamination through the bloodstream during the slaughtering process (Egan and Roberts, 1987). The musculature of healthy, rested animals is considered to be essentially germ free. Destruction of the defence mechanisms at slaughter will allow microbes to survive. Although of rare occurrence, *Clostridium* spp. (e.g. *Cl. perfringens* and *Cl. sporogenes*) as well as facultatively aerobic bacteria may be associated occasionally with deep-muscle contamination and anaerobic spoilage. Of 136 samples of fresh sausages studied in Argentina, 110 were found positive for *Cl. perfringens*, with numbers ranging from 10 to 10^9/g (Guzman *et al.*, 1990).

A low redox potential created by vacuum packaging, modified (high CO_2) gas atmospheres and by processing (inner sections of sausages, bacon, etc.) may allow the survival of non-toxinogenic clostridia and other anaerobic to facultative aerobic bacteria in meat and meat products. In the order of frequency of isolation, the following *Clostridium* spp. have been reported in the literature: *Cl. sporogenes, Cl. tertium, Cl. bifermentans, Cl. tyrobu-tyricum, Cl. carnis, Cl. fallax, Cl. butyricum, Cl. plagarum* and *Cl. malenominatum* (Reuter, 1996). Recently psychrotrophic clostridia have been associated with contamination and spoilage, e.g. *Clostridium ester-theticum* in vacuum-packaged meats (Collins *et al.*, 1992), *Cl. laramie* in vacuum-packaged fresh and roasted beef, *Clostridium* spp. (some identified with *Cl. difficile, Cl. beijerinckii* and *Cl. lituseburense*) associated with 'blown pack' spoilage of chilled vacuum-packaged red meats (Broda *et al.*, 1996a), and as yet unidentified *Clostridium* spp. causing cheesy, deep tissue aroma of chilled, vacuum-packaged lamb (Broda, *et al.*, 1996b).

Clostridium thermosaccharolyticum (*Thermoanaerobacterium ther-mosaccharaolyticum*) was implicated in thermophilic, gaseous spoilage of low-acid (pH > 4.6) canned products. *Bacillus coagulans* and *B. stearother-mophilus* may cause flat-sour thermophilic spoilage, and *Cl. nigrificans* and *Cl. bifermentans* may be associated with sulfide spoilage. In medium to low acid foods, proteolytic strains of *Cl. botulinum* may cause putrefactive spoilage. Survival of the mesophilic *Clostridium* spp. (e.g. *Cl. botulinum*) may be due to understerilization of canned meats. On the basis of D_r values, the approximate heat resistance of some mesophilic and thermophilic endosporeformers may be compared as follows (Jay, 1996):

Bacillus stearothermophilus	4.0–5.0
Thermoanaerobacterium thermosaccharolyticum	3.0–4.0
Clostridium nigrificans	2.0–3.0
Cl. botulinum (types A and B)	0.1–0.2
Cl. sporogenes	0.1–1.5
Bacillus coagulans	0.01–0.07

2.4   Growth and metabolism

2.4.1   Factors influencing the growth, survival and persistence of Gram-positive bacteria in the meat environment

Microbial spoilage results directly from the growth and metabolic activity of those bacteria best adapted to a particular meat ecosystem. Most pro-cessing and preservation procedures, ranging from vacuum packaging to curing, will promote a shift in the microflora towards the Gram-positive bac-teria. This is exemplified by the domination of *Lb. curvatus* and *Lb. sake* in cured processed meat ecosystems with desirable contributions to fermented meat products and undesirable ones in vacuum packaged processed prod-ucts (Table 2.3).

Through handling, refrigeration, storage and processing (including pack-aging), a number of factors will influence selectively the growth, survival and eventual persistence of Gram-positive bacteria in or on meat and meat products. Normally, a combination of two or more factors will determine the particular spoilage association of a meat ecosystem. Gram-positive bacteria are generally more resistant to adverse conditions than Gram-negatives ones, especially with respect to pH (low or high), reduced a_w, reduced Eh (e.g. vacuum and reduced oxygen packaging), curing and heat processing. Despite their fastidiousness, this is especially true for the LAB. Information on such parameters for meat-associated Gram-positive bacteria (compared with two Gram-negative representatives) is given in Table 2.1. Variation in resistance/growth ability among representatives of this group is illustrated by:

- differences in ability to grow at reduced $a_w < 0.95$ and in the following approximate order of increased tolerance: *Brochothrix/Kurthia* < *Carnobacterium* < *Lactobacillus* ('atypical') < *Leuconostoc/Lactobacillus* ('typical') < *Pediococcus/Enterocccus/Staphylococcus* (anaerobic) < *Staphylococcus* (aerobic). This partly explains the domination of LAB in 'dry' fermented products, where pH, concomitantly with other factors (reduced Eh, curing salts and competition), also play a part. The safety of many cured products is also related to the fact that endospores of *Clostridium* and *Bacillus* generally do not germinate at $a_w < 0.95$;
- ability of most Gram-positive bacteria to grow at pH levels < 4.7 influences decisively their domination over Gram-negative ones. Putrefactive Gram-positive bacteria such as clostridia and some bacilli are generally inhibited in meat systems with pH < 4.7 and are out-competed by LAB and inhibited by their metabolites. Listeria are inhibited at pH <5.5 whilst some strains may grow up to pH 9.6 (Jones and Seeliger, 1991). The slightly selective effect of the normal pH range of *post mortem* muscle of 5.4 to 5.8 may be amplified in combination with other factors (e.g. vacuum packaging, reduced Eh, and refrigeration). The spoilage potential of refrigerated DFD meat (pH > 6.0) is related mainly to growth of *Brochothrix thermosphacta* in addition to some Enterobacteriaceae and *Pseudomonas*. spp. (see Chapter 1); the last two are inhibited however under vacuum packaging, a practice that favours the growth of the lactobacilli (Gill and Newton, 1979). *Brochothrix* was inhibited and lactobacilli became dominant when high pH meat was stored in 100% CO_2 at -1.5 °C (Jeremiah *et al.*, 1995). *Brochothrix thermosphacta* appears unable to grow under strictly anaerobic conditions on beef pH values < 5.8 (Egan and Roberts, 1987);
- domination at storage temperatures down to -1.5 °C (Gill and Molin, 1991). Gram-positive bacteria (mainly represented by LAB and *Br. thermosphacta*) constitute only a minor proportion of the typical psychrotophic flora on meat stored in air; vacuum packaging and refrigeration at 0–1 °C will however selectively promote an increase in the number of

LAB up to $5 \times 10^7/cm^2$, with *Br. thermosphacta* as a minor component (Egan and Roberts, 1987). The storage temperature also influences the population shift among the LAB, with *Carnobacterium* spp. dominating on pork stored in 100% CO_2 at -1.5 °C, and homofermentative *Lactobacillus* spp. at 4–7 °C (McMullen and Stiles, 1994). *Leuconostoc* spp. (probably *Leuc. carnosum* and *Leuc. gelidum*) have been found to dominate the populations on vacuum packaged meat after 2–4 weeks storage at 2 °C and *Carnobacterium* spp. ('acetic-acid sensitive' bacteria) at 5 °C. Storage at 2 °C (compared to 5 °C) favours the progressive domination of typical psychrotrophic *Leuconostoc* spp. (*Leuc. carnosum* and *Leuc. gelidum*) and meat lactobacilli such as *Lb. sake* in vacuum packaged meat (Holzapfel and Schillinger, 1992; Schillinger and Lücke, 1986; Shaw and Harding, 1989). These organisms are able to grow over the range 1 °C to 37 °C (Shaw and Harding, 1989). *Kurthia* appears not to grow in meat at refrigeration temperatures, but it may constitute a major part of the aerobic population at 16 °C (Gardner *et al.*, 1967; Gardner, 1969).

- lowering of the redox potential (Eh) by vacuum packaging and/or the application of modified atmospheres causes a population shift towards the LAB (mainly *Lactobacillus* spp., *Leuconostoc* spp. and *Carnobacterium* spp.) and away from the Gram-negative groups such as *Pseudomonas* spp. and even Enterobacteriaceae (Borch and Molin, 1988). In the presence of CO_2 and under reduced O_2 conditions the growth rate of LAB and especially that of *Brochothrix* is reduced and the storage life of vacuum-packaged beef may reach 10–12 weeks at 0 °C (Egan, 1983). This trend is illustrated in Table 2.6 in which the dominance of lactobacilli and *Br. thermosphacta* in vacuum packaged meat is shown after 12 days storage at 0–2 °C (Hechelmann, pers. comm.). Other experiments on beef stored in vacuum or 100% CO_2 (at 2 °C or 6 °C) and in different mixtures of CO_2 and N_2 (at -1 °C or 2 °C) showed a strong inhibitory effect of low temperatures and a high concentration of CO_2 on

Table 2.6 Change in the microbial load on the surface of vacuum packaged raw beef during storage at 0–2 °C (Courtesy: H. Hechelmann, Fed. Centre for Meat Research, Kulmbach).

| Micro-organisms/portion | Viable counts/cm² $\bar{x}n = 30$ | | | | | |
| | Directly after portioning | | | After 12 days | | |
	Roast beef	Leg	Neck	Roast beef	Leg	Neck
Total population	6×10^4	3×10^4	2×10^4	3×10^5	1×10^5	6×10^5
Pseudomonadaceae	1×10^4	2×10^3	2×10^3	3×10^4	1×10^3	5×10^4
Enterobacteriaceae	$<10^2$	$<10^2$	1×10^2	6×10^2	4×10^2	1×10^4
Lactic acid bacteria	1×10^2	2×10^3	2×10^2	6×10^5	1×10^4	5×10^4
Bacillus spp.	$<10^2$	$<10^2$	$<10^2$	$<10^2$	$<10^2$	$<10^2$
Brochothrix	$<10^2$	$<10^2$	$<10^2$	2×10^5	3×10^4	8×10^3
Yeasts	3×10^2	6×10^2	5×10^2	6×10^2	2×10^2	9×10^2

the growth of *Br. thermosphacta*, which was inhibited further by high numbers ($\log_{10}$ 5–6) of LAB (Nissen *et al.*, 1996). Leuconostocs were found to dominate the microflora in vacuum and CO_2 packs both at 2 °C and 6 °C, whilst carnobacteria did so in N_2 at -1 °C (Nissen *et al.*, 1996).

• production of antagonistic metabolites and competition contribute to the complexity of 'implicit' factors determining the survival and domination within a microbial population. Relevant aspects are briefly discussed below with respect to beneficial aspects.

2.4.2   Metabolic activities of importance in meat systems

Even in heat-processed, emulsion type meat products, mixed microbial populations are typically encountered. This results in complex interactions and synergies within the microflora on meat substrates. Microbes best adapted to the existing extrinsic and intrinsic conditions and with the highest growth rate will dominate a population, with concomitant effects on the keeping quality and ongoing metabolic activities. Generally, levels of $>10^6$ viable bacteria/g or cm ^{2} are considered to exert detectable (sensory) changes in the substrate (e.g. off-flavours, slime-production).

Most microbial interactions in meat systems are generally considered in the context of spoilage (e.g. souring, proteolysis/putrefaction, etc.) and potential health consequences (e.g. production of toxins, biogenic amines, etc). On the other hand, a number of metabolic activities may be desirable, e.g. the production of lactic acid, nitrate/nitrite reduction during the fermentation of raw sausages and the curing of ham. The presence of LAB may even be desirable in vacuum-packaged meat, their presence leading to the 'natural' control of pathogens and putrefactive spoilage, through the production of lactic acid, and other antimicrobial metabolites, and by competition, etc.

Metabolites with both antimicrobial and sensory effects.   Production of one or more antagonistic metabolites may be part of the complex mechanism by which a micro-organism becomes established in the presence of other competing organisms. Understanding such mechanisms provides a valuable key to our understanding the complexity of microbial interactions in a meat system and hence the basis of 'biological' approaches to food preservation.

The antimicrobial properties of a number of metabolites from LAB are summarized with respect to the inhibition of undesired organisms in Table 2.7. Several of the listed metabolites (e.g. lactic acid) may, of course, cause spoilage or contribute to desirable changes in cured meat, fermented products, etc.

Organic acids.   Fermentation of the carbohydrates, glucose, glycogen, glucose-6-phosphate and small amounts of ribose, in meat and meat

Table 2.7 Metabolic products of lactic acid bacteria with antimicrobial properties (Holzapfel et al., 1995).

Product	Main target organisms
Organic acids	
lactic acid	Putrefactive and Gram-negative bacteria, some fungi
acetic acid	Putrefactive bacteria, clostridia, some yeasts and fungi
Hydrogen peroxide	Pathogens and spoilage organisms, especially in protein-rich foods
Low-molecular-weight metabolites	
reuterin (3-OH-propionaldehyde)	Wide spectrum of bacteria, moulds and yeasts
diacetyl	Gram-negative bacteria
fatty acids	A range of different bacteria
Bacteriocins	
nisin	Some LAB and Gram-positive bacteria, notably endospore-formers
others	Gram-positive bacteria, inhibitory spectrum according to producer strain and bacteriocin type

products, produces organic acids, by glycolysis (EMP-pathway) or the HMP-pathway. Carbohydrate metabolism serves as the main source of energy for most Gram-positive bacteria, and with the LAB it involves substrate-level phosphorylation. It has been shown however that not all LAB utilize all the carbohydrates available in meat. Thus *Cb. piscicola* was found to be unable to catabolize glucose-6-phosphate. Following glucose depletion, it seems, in common with several other LAB, to oxidize indigenous (meat) and microbial L(+)- and D(−)-lactic acid, to acetic acid under atmospheres enriched with CO_2 (Drosinos and Board, 1995). This suggests that acetic acid and D(−)-lactic acid may serve as possible parameters for estimating microbiological quality of packaged meat (Drosinos and Board, 1995).

Lactic acid is a major fermentation end product of LAB and a number of other genera (e.g. *Staphylococcus*, *Brochothrix* and *Listeria*). The LAB in particular are able to reduce the pH to levels where putrefactive (e.g. clostridia and pseudomonads), pathogenic (e.g. salmonellas and *Listeria* spp.) and toxinogenic bacteria (*Staphylococcus aureus*, *Bacillus cereus*, *Clostridium botulinum*) will be either inhibited or killed. In addition, the undissociated acid, on account of its fat solubility, will diffuse into the bacterial cell, thereby reducing the intracellular pH and slowing down metabolic activities, and in the case of Enterobacteriaceae such as *E. coli* inhibiting growth at around pH 5.1. The rapid reduction of the pH below 5.3 during sausage fermentation is sufficient to inhibit growth of salmonellas and *Staph. aureus*. These and other mechanisms related to 'biological preservation' have been reviewed by Holzapfel et al. (1995).

With *Listeria* and *Brochothrix* the redox potential and availability of oxygen influence the amount and ratio of metabolic end products from glucose. *Listeria monocytogenes* produces acetoin (26%) under aerobic but not anaerobic conditions, and in addition 28% lactate and 23% acetate are produced. Under anaerobic conditions 79% lactate, 2% acetate, 5.4% formate, 7.8% ethanol and 2.3% CO_2 are produced (Romick *et al.*, 1996).

Acetic acid (pK_a 4.75) is produced by heterofermentative LAB in equimolar amounts to lactic acid, but it is usually present in small amounts as a consequence of LAB metabolism. It may constitute a vital factor in the establishment of the initial LAB population (notably *Leuconostoc* spp.) during 'spontaneous' fermentation. Under specific conditions of hexose limitation and/or availability of oxygen, homofermentative LAB (e.g. pediococci, lactococci and most *Lactobacillus* spp.) may decompose lactic acid to acetic acid, formic acid and/or CO_2 (Kandler, 1983).

Hydrogen peroxide. Hydrogen peroxide is formed by a number of LAB in the presence of molecular oxygen during the production of lactate, pyruvate and NADH by flavin enzymes (Kandler, 1983). Oxidation of lactate may be by a flavin-containing L(+)-lactate oxidase using either O_2 (examples: *Lb. curvatus* and *Lb. sake*), or methylene blue as the electron acceptor (examples: *Lb. plantarum*, *Lb. casei* and *Lb. coryniformis*). Most undesirable bacteria such as *Pseudomonas* spp. and *Staph. aureus* are 2 to 10 times more sensitive than the LAB to H_2O_2. The bacteriostatic concentration is around 6 $\mu g/ml$ for staphylococci and 23–35 $\mu g/ml$ for pseudomonads (Gudkow, 1987).

H_2O_2 may be formed by microbial reduction of oxygen during food fermentation or as a result of LAB contamination of a food. It may cause detrimental, oxidative changes in a product. One type of greening is typically caused by the accumulation of H_2O_2, produced by some LAB under low Eh conditions, and its reaction with nitrosohemochrome to produce a greenish oxidized porphyrin. Strains, especially of *W. viridescens* and *Leuconostoc* spp., but also of *Lb. fructivorans*, *Ent. faecium* and *Ent. faecalis*, have been associated with this type of spoilage (Grant *et al.*, 1988). As with other metabolic products, the tolerable amount of H_2O_2 is dependent of the product type and situation.

Low-molecular metabolites. Primary metabolites of low molecular weight are known for their relatively potent antimicrobial activities. Diacetyl may be produced by some *Pediococcus* spp. in meat as a result of citric acid degradation. Due to its intense aroma even at low concentrations diacetyl may cause off-flavours. It has little direct impact as an antimicrobial agent in meat systems.

Carbon dioxide, produced by heterofermenters from hexoses, contributes to a reduced Eh and is directly toxic to a number of putrefactive aerobic bacteria, including *Brochothrix*. These attributes may be observed, concomitantly with acid production, in vacuum packaged products.

Reuterin (3-hydroxypropionaldehyde) is produced from glycerin in the presence of coenzyme B_{12} by *Lactobacillus reuteri*. Its broad-spectrum antimicrobial activity may be due to the inhibition of ribonucleotide reductase. It may be used for biopreservation of meat products by using *Lb. reuteri* as a starter culture (Daeschel, 1989; Lindgren and Dobrogosz, 1990).

Bacteriocins.   These highly potent antimicrobial peptides or proteins are produced by a range of bacteria, including the LAB, and are generally active against organisms closely related to the producer. They are ribosomally produced, and are probably inactivated by proteases in the gastro-intestinal tract. The majority of the bacteriocins of LAB are thermostable (100 °C/10 min), hydrophobic and show a tendency towards multimer formation (Klaenhammer, 1988; Schillinger, 1990; Lindgren and Dobrogosz, 1990). Their activity spectrum appears to be determined by the presence of specific receptors on the cell wall, e.g. lipoteichoic acid for pediocin AcH (Ray, 1992). For most bacteriocins, the antimicrobial effect is bactericidal (Schillinger and Lücke, 1989), and, with some exceptions (e.g. those of a glycoprotein nature [Lewus *et al.*, 1992]), bacteriostatic.

The bacteriocins of LAB, which have been the subject of intensive studies in recent years are of special interest with regard to their acceptability and potential for biopreservation. Based on current knowledge, bacteriocins of LAB may be grouped (see Table 2.8) into three classes (Schillinger *et al.*, 1995):

- the lantibiotics (with nisin as a typical example) and non-lanthionine compounds containing small, membrane-active peptides;
- large, heat-sensitive proteins, and
- complex bacteriocins such as glycoproteins.

In the search for potential 'biopreservation' agents, one of the areas that has been focused on is bacteriocinogenic LAB in meat systems. Relative frequencies of bacteriocin producing LAB within typical populations were

Table 2.8 Classification of bacteriocins of lactic acid bacteria (Schillinger *et al.*, 1995).

Class	Properties
I	Small, membrane active, heat-stable peptides (<10 kDa)
Ia	Lantibiotics
Ib	Non-lanthionine-containing peptides
	peptides active against *Listeria* (N-terminal sequence of consensus: -Tyr-Gly-Asn-Gly-Val-Xaa-Cys-)
	Bacteriocins in which the activity depends on the complementary action of 2 peptides ('2-component-bacteriocins')
	thiol active bacteriocins requiring cysteine for their activity
II	Large heat-sensitive proteins (>30 kDa)
III	Complex bacteriocins (requiring a non-protein component, e.g. a carbohydrate or lipid moiety, for the activity)

found to range from 0.6% to 22% (Table 2.9). Some aspects of the potential application of bacteriocins are discussed in the section on 'desirable aspects' above.

Proteolysis. Apart from some *Clostridium* spp. and a few *Bacillus* spp. , most Gram-positive bacteria associated with meat show only weak or no proteolytic activities. The role of proteolytic aerobic *Bacillus* spp. in the spoilage of meat and meat products was discussed by Müller (1995). In contrast very little is known about the enzymatic degradation of meat proteins by LAB. Extracellular protein systems of lactobacilli have been poorly characterized thus far (Hammes *et al.*, 1992), and in general the proteolytic activity of LAB appears to be very weak (Law and Kolstad, 1983). Reuter (1971) observed qualitatively and quantitatively higher extracellular peptidase activity than intracellular activity for lactobacilli isolated from meat products. The proteolytic pathways of LAB associated with dairy products, and especially of *Lactococcus lactis*, have been studied comprehensively (Poolman *et al.*, 1995). No endopeptidase activity was found in strains of *Pediococcus pentosaceus* and *Staph. xylosus* isolated from cured ham; the former however showed strong leucine and valine arylamidase activities (Molin and Toldra, 1992). It has been suggested that the spoilage of meat due to the development of odours or production of desirable flavours in fermented sausages from free amino acids, may not be related to proteolytic activity of LAB.

Although micrococci and staphylococci are known to produce extracellular proteinases, their role in the proteolysis of fermented sausages is not

Table 2.9 Occurrence of bacteriocinogenic lactic acid bacteria in meat systems.

Bacteria	Number of isolates screened	Number (%) of isolates producing bacteriocin-like substances	Origin	References
LAB	1600	40 (2.5%)	Different foods	Harding and Shaw (1990)
LAB	720	119 (16.5%)	Fermented sausages	Sobrino *et al.* (1991)
LAB	168	1 (0.6%)	Italian raw ham	Stecchini *et al.* (1992)
LAB	100	12 (12%)	Fermented sausages	Vignolo *et al.* (1993)
LAB	163	7 (4.3%)	Foods and environment	Arihara *et al.* (1993)
Lactobacilli	221	23 (1%)	Meat products	Schillinger and Lücke (1989)
Lactobacilli	254	56 (22%)	Fermented sausages	Garriga *et al.* (1993)
LAB	1000	43 (4.3%)	Minced beef	Vaughan *et al.* (1994)
LAB	1000	27 (2.7%)	Bacon	Vaughan *et al.* (1994)
LAB	1000	47 (4.74%)	Ham slices	Vaughan *et al.* (1994)
LAB	148*	26 (18%)	Retail meats	Garver and Muriana (1993)

* Number of samples investigated.

yet clear (Nychas and Arkoudelos, 1990). According to Hammes *et al.*
(1995) strains of *Staph. piscifermentans* and *Staph. carnosus* exhibit low but
significant proteolytic activity, although only 20% of the casein-splitting
activity observed for *Bacillus cereus*.

Among the non-sporulating bacteria it is mainly the Gram-negative ones,
and especially *Pseudomonas* spp., that show proteolytic activities which are
mainly confined to the surface or adjacent layers of meat (Egan and
Roberts, 1987). Very little is known about the potential of *Brochothrix* and
Kurthia to degrade proteins. Some strains of *Kurthia* produce H_2S weakly
(Keddie and Jones, 1992).

Lipolysis.   Relatively little is known about the lipolytic activity of LAB in
meat systems. The indications are that such activity is weak. In an extensive
study on mesophilic and thermophilic lactobacilli, esterase and lipase activities were found to be species- and strain-specific. The highest esterase
activity was found with β-naphthyl butyrate and an appreciable one with
caproate and caprylate. In general a lower esterase and lipase activity was
observed with mesophilic strains (Gobbetti *et al.*, 1996).

On the other hand, lipolytic enzymes of micrococci (under aerobic conditions) and staphylococci (under aerobic to anaerobic conditions) appears
to contribute in part to desirable flavour and aroma production in fermented sausages. Some strains of *Staph. piscifermentans* and *Staph.
carnosus* exhibited lipolytic potential on tributyrin agar (Hammes *et al.*,
1995).

Catalase.   Deleterious oxidative effects due to the hydrogen peroxide
formed by different mechanisms in meat products may be counteracted by
catalase or other peroxidases. Indeed catalase may inhibit or reduce rancidity (Nychas and Arkoudelos, 1990). This property seems to be advantageous with respect to starter cultures for fermented meat products
because the accumulation of H_2O_2 may well be prevented by catalases. In
practice suitable strains of micrococci and staphylococci for starter cultures
may be selected on the basis of high catalase activity. Although the biosyn-
thesis of catalase was blocked under anaerobic conditions in the absence of
heme, which is typically present in meat, synthesis was restored on addition
of hematin (Hammes *et al.*, 1995). This indicates that starter strains such as
Staph. carnosus have the ability to produce catalase during fermentation of
raw types of sausage.

A few LAB form a heme-dependent catalase or a non-heme 'pseudocatalase' (Whittenbury, 1964; Wolf and Hammes, 1988). The latter is formed
by some members of the genera *Lactobacillus*, *Leuconostoc*, *Enterococcus*
and *Pediococcus*. In a study of 71 species of LAB, non-heme catalase
activity was found to be restricted to strains of *Lb. plantarum*, *Lb. mali* and
Ped. pentosaceus; heme catalase activity was detected in 21 species,

representing virtually all the genera of LAB (Engesser and Hammes, 1994). Application of catalase-producing *Lb. sake* strains as starters was found to be superior to that of *Lb. curvatus* strains because of improvements in the desired colour of raw fermented sausages (Min, 1994).

Carnobacterium maltaromicus (formerly *Lb. maltaromicus*) may cause spoilage of poultry products. It has been shown to contain functional cytochromes of the b and d types when grown aerobically in the presence of hematin, and enables tenfold higher growth under aerobic conditions than under anaerobic growth, concomitantly with increased acetoin production (Meisel *et al.*, 1994). This property further supports the observed close phylogenetic and physiological relationship between *Carnobacterium* and *Enterococcus*.

Nitrate and nitrite reductase.   Although not a common property of the LAB, several meat-associated strains of *Lactobacillus* and *Weissella* produce either nitrate and nitrite reductase (some strains of *Lb. pentosus* and *Lb. plantarum*) or only nitrite reductases (some strains of *Lb. brevis*, *Lb. farciminis*, *Lb. suebicus*, *Lb. sake* and *W. viridescens*) (Wolf and Hammes, 1988). The reduction of nitrate to nitrite in the traditional slow fermentation of raw sausage is brought about by certain *Micrococcus* and *Staphylococcus* spp. Strains of *Staph. carnosus* and *Staph. piscifermentans*, typical organisms of fermented sausage products, generally exhibit strong nitrate but weak nitrite reductase activities (Hammes *et al.*, 1995). The desired nitrite concentration during fermentation may be reduced because of the chemical oxidation of nitrite or microbial nitrite reduction (Toth, 1982; Min, 1994), thereby emphasizing the role of nitrate reductase-active 'micrococci' in such products.

Biogenic amines.   Biogenic amines may be produced, especially in high-protein foods, by either amino acid decarboxylase activity of microorganisms, or amination and transamination of aldehydes and ketones (Ten Brink *et al.*, 1990; Halász *et al.*, 1994). High levels of biogenic amines in foods may have toxic effects. They may also contribute to food spoilage through production of putrid odours and off-flavours (Taylor *et al.*, 1982; Ten Brink *et al.*, 1990). This property therefore seems especially undesirable for starter cultures, and may well serve as a selection criterion. Amino acid decarboxylases are common for most Enterobacteriaceae, but relatively rare among most Gram-positive bacteria (except for *Bacillus* and *Clostridium* spp.) associated with meat and meat products. Tyramine in particular was found quite commonly in meat systems, and its presence could be related to the metabolic activities of all of the *Carnobacterium* strains and some of the strains of *Lb. curvatus* and *Lb. plantarum* investigated by Masson *et al.* (1996). Micrococcaceae and *Lb. sake* strains did not show this ability. Lactic acid bacteria isolated from dry sausages showed strain-specific production of tyramine and histamine (Maijala, 1993). Strains of *Ped. cerevisiae* and *Lb. plantarum* used as starter cultures in commercial sausage did not exhibit any appreciable tyrosine or histidine decarboxylase activity (Rice and Koehler, 1976). In a study by Hammes *et al.* (1995) strains of *Staph. carnosus* and *Staph. piscifermentans* associated with fermented meat products showed no ability to form histamine, whilst only two strains of *Staph. carnosus* produced tyramine. All strains of *Staph. piscifermentans* showed only a low potential for phenylethylamine formation and six out of 10 for tyramine. None of the strains investigated showed ability to form putrescine or cadaverine. Amino acid decarboxylase activity could not be detected in strains of *Listeria* associated with poultry spoilage (Geornaras *et al.*, 1995). During the ripening of fermented sausages, the concentration of putrescine and cadaverine (produced by *Bacillus* spp.) increased during the first six days of storage, while that of tyramine, a metabolite from LAB, decreased after 24 days (Ramirez *et al.*, 1995).

Production of extracellular polymers.   Slime production, especially as polysaccharide polymer, may occur during the spoilage of vacuum-packaged meat and meat products, and occasionally also in fermented raw sausages. A wide range of lactobacilli associated with meat products produce slime (most notably a dextran) from sucrose (Reuter, 1970b). Slime production by *W. viridescens* in spoiled meat products has been reported by Deibel *et al.* (1961). Korkeala *et al.* (1988) were the first to report ropy slime formation by LAB in vacuum-packaged cooked meat products. The majority of the implicated strains were identified with *Lb. sake* and some with *Leuc. amelibiosum* (Mäkelä *et al.*, 1992a, b, c); the polymer was described as a glucose–galactose containing exopolysaccharide (Mäkelä *et al.*, 1992c).

2.5 Spoilage

The interactive factors and metabolic activities discussed in section 2.4 provide some explanation for the complexity of microbial spoilage by heterogeneous populations as they exist in practice. Due to their higher growth rate under aerobic, chill situations, Gram-negative bacteria (mainly *Pseudomonas* spp.) will generally dominate the spoilage association of refrigerated meat stored in air; *Brochothrix* and *Kurthia* spp. may also form an important part of the competing organisms in aerobic meat ecosystems. Most other processing factors will promote a shift in the population towards the Gram-positive bacteria. These general tendencies with respect to shelf-life and typical spoilage association of different meat product types are shown in Table 2.10. Data on the association of Gram-positive bacterial groups with spoilage of meat and meat products are given in Table 2.11.

Table 2.10 Expected shelf-life and growth ability of different Gram-positive bacterial groups on meat and meat products under refrigerated storage. The information can only be considered as generalized and related to situations of 'natural' contamination (Modified from Borch *et al.*, 1996, and amplified by data from Ahmad and Marchello (1989), Asensio *et al.* (1988), Drosinos and Board (1995), Gallo *et al.* (1988), Kaya and Schmidt (1991), Nissen (1996), Jeremiah *et al.* (1995).

Product	Storage	Expected shelf-life	Growth						
			Leucon./ Weissella	Lactobacillus*	Carnobacterium	Br. thermosphacta	Micro./ Staph.	Coryneforms**	Listeria monocytog.
Meat, normal pH	Air/aerobic	Days	++	+(+)	++	++/+++	++	++	++
	High-O₂-MA	Days	++(+)	++	++(+)	+++	+	++	+
	Vacuum	Weeks-months	+++	+++	++	+	(+)	+	+***
	100% CO₂	Months	+++	+++	+(+)	(+)–	–	–	–
Meat, high pH	Vacuum	Days-week	++/+++	+++	+++	++/+++	+	+(+)	++
	100% CO₂	Weeks-months	++	+++	++	+(–)	–	–	–
Meat products	Air	Days	++	++/+++	+	+++	+	++	+
	Vacuum	Weeks	++	+++	+(+)	+(+)	–	(+)	+***
	CO₂ + N₂	Weeks	++(+)	+++	+	+	–	–	–
Broilers	Air	Days	+	++	++	++	+	++	++

* mainly *Lb. curvatus* and *Lb. sake*; ** mainly *Kurthia* spp.; *** inhibition also due to LAB; +++ extensive growth; ++ good growth; + and (+) moderate to slow growth; – no growth or growth not generally expected.

Table 2.11 Gram-positive non-sporulating bacteria and *Clostridium* associated with spoilage of meat and meat products. The numbers refer to literature references in the foot note. Those in brackets reflect doubt with respect to their importance in spoilage.

Bacterial Group	Fresh	Vacuum-packed	Processed	Fermented	Poultry
Aerococcus viridans			(1)		
Brevibacterium	3		2		
Brochothrix thermosphacta	4	4	3/4		3
Carnobacterium	6, 7, 10		7/8	8	5
Clostridium		9/23/24			
Enterococcus		21			
Kurthia	11, 12		12		
Lactobacillus:	12		15	14	12
heterofermentation	14	6, 15	15, 16	14	15
homofermentation	14	6, 15	15, 16	14	15
Lactococcus		22			
Leuconostoc/Weissella	18	15	16	30	5
Micrococcus	24		(17)		
Pediococcus			(20), 21		

1. Deibel and Niven (1960); 2. Cantoni *et al.* (1969); 3. Gardner (1981); 4. Keddie and Jones (1981); 5. Thornley and Sharpe (1959); 6. Holzapfel and Gerber (1983); 7. Shaw and Harding (1984); 8. Borch and Molin (1988); 9. Broda *et al.* (1996 a,b); 10. Hiu *et al.* (1984); 11. Gardner (1969); 12. Keddie (1981); 13. Hammes *et al.* (1992); 14. Coretti (1958); 15. Von Holy and Holzapfel (1989); 16. Reuter (1975); 17. Alm *et al.* (1961); 18. Sharpe and Pettipher (1983); 19. Holzapfel and Schillinger (1992); 20. Weiss (1992); 21. Grant *et al.* (1988); 22. Borch *et al.* (1996); 23. Kalchayanand *et al.* (1993); 24. Reuter (1996).

2.5.1 Meat (aerobic storage and vacuum or MA packed)

Discoloration. Growth of psychrotrophic 'Micrococcaceae' on meat surfaces may result in a grey-white, dry surface layer and a fruity odour when numbers reach 10^8 cfu/cm^2. Visible slime formation may be caused by *Br. thermosphacta* in numbers of about 10^8/cm^2 (Reuter, 1996). H$_2$S, produced by some strains of *Lb. sake*, for example under oxygen and glucose depletion may convert muscle pigment to green sulphmyoglobin (Egan *et al.*, 1989).

Off-odours and off-flavours. In addition to Enterobacteriaceae, *Br. thermosphacta* and some homofermentative lactobacilli were implicated in the development of cheesy odours (associated with diacetyl and 3-methylbutanol formation) in aerobically stored meat (Borch and Molin, 1989; Dainty and Mackey, 1992). Beef slices (pH 5.5–5.8) inoculated with *Br. thermosphacta* and stored aerobically at 1 °C for 14 days developed off-odours due to the production of acetoin, acetic acid, and isobutyric/isovaleric acids (Dainty and Hibbard, 1980).

'Deep spoilage' of refrigerated beef rounds and quarters is caused by anaerobic (deep muscle) growth of *Clostridium* and *Enterococcus* spp. It is also referred to as 'bone taint' or 'sours' (Jay, 1996). Spoilage of vacuum-packaged beef by a *Clostridium* sp. was associated with gas formation (blowing) and off-odours were described as 'sulphurous', 'fruity', 'solvent-like' and 'strong cheese' (Dainty *et al.*, 1989).

Domination of a spoilage association by LAB results in the production of mainly lactic acid and smaller amounts of acetic acid in vacuum and anaerobic MA-packaged meats. The off-odours are less offensive than those of aerobically stored meat (Dainty and Mackey, 1992).

Off-odours described as 'cheesy' and 'rancid' were associated with spoilage of meat packaged in high oxygen-MA, in which *Br. thermosphacta* and LAB along with *Pseudomonas* spp. were found to dominate (Ordóñez *et al.*, 1991). It is suggested that the presence of oxygen will increase the spoilage potential of both *Br. thermosphacta* and LAB by end-product formation of acetoin and acetic acid (Borch *et al.*, 1996). Under vacuum-packaging and anaerobic conditions glucose depletion leads to the formation of end-products such as acetic acid and H_2S from cysteine by homofermentative lactobacilli (Edwards and Dainty, 1987; Borch *et al.*, 1991). Strains of *Lb. sake* were identified as causative agents of sulphide odour on vacuum-packaged beef (Egan *et al.*, 1989) and (within three weeks storage) on vacuum-packaged pork and beef stored at 2 °C and -1 °C respectively (McMullen and Stiles, 1993; Leisner *et al.*, 1996).

Gas production.   Strains of both homo- and heterofermentative LAB may produce CO_2 in vacuum-packaged meats concomitantly with lactic acid and other non-offensive smelling metabolites. Large amounts of a mixture of H_2 and CO_2, accompanied by foul off-odours, may be produced in vacuum-packaged beef by *Clostridium* spp. (Borch *et al.*, 1996).

2.5.2   Processed meat products

Souring, and off-odours and off-flavours.   Under reduced atmospheric conditions, LAB are generally the major producers of lactic acid, acetic acid and smaller amounts of other organic acids. An appreciable increase in $D(-)$ and $L(+)$-lactic acids may be expected when LAB numbers reach 10^7 cfu/g on the surface of vacuum-packaged cooked sausages (Korkeala *et al.*, 1990). When 3–4 mg of lactic acid/g (pH < 5.8) were exceeded, most sausage samples were considered to be spoiled. Although lactic acid as such does not cause strong off-odours, even in high concentrations, its production is often accompanied by that of other (partly volatile) end-products such as short-chain fatty acids (e.g. acetic and butyric acids) (Dainty, 1982). In addition, souring and prolific growth of LAB is often associated with excessive separation of exudate in vacuum-packaged emulsion-type sausages (Von Holy *et al.*, 1991). Chilled, sliced roast beef stored under CO_2 packaging developed 'slightly acidic' flavours associated with *Lactobacillus*-dominated populations and a 'sweaty' odour or flavour with high numbers of *Br. thermosphacta* (Penney *et al.*, 1993). With high oxygen permeability film vacuum-packaged meat products may develop a slightly sweet, cheesy obnoxious odour due to the formation of acetoin by *Br. thermosphacta, Lactobacillus* spp. and *Carnobacterium* spp. (Borch and Molin, 1989).

Discoloration. Green discoloration, often evident as green spots, is caused by the oxidation of nitrosohaemochrome to choleomyoglobin by H_2O_2. Peroxide-producing bacteria, especially heat-resistant *W. viridescens*, may survive the cooking process in the centre of emulsion type sausages. On exposure to air, the survivors produce H_2O_2 which causes green colouration (Borch *et al.*, 1988). Surface greening may be associated with homo- and heterofermentative *Lactobacillus, Leuconostoc* and *Weissella* spp., *Cb. divergins, Enterococcus* and *Pediococcus* spp. as a result of surface recontamination after cooking (Borch and Molin, 1989; Grant *et al.*, 1988). Discolouration of an Italian ham-style beef product was associated with high numbers (10^8–10^9 cfu/g) of H_2O_2 forming strains of *Cb. divergens, Lb. brevis/buchneri* and *W. viridescens* (Cantoni *et al.*, 1992).

Gas production. Although CO_2 formation is typically associated with the growth of *Leuconostoc* spp. (*Leuc. mesenteroides* spp. *mesenteroides, Leuc. carnosum* and *Leuc. amelibiosum*), lactobacilli and *Weissella* spp. may also contribute to the 'blowing' of packaged meat or meat products (Von Holy *et al.*, 1991; Mäkelä *et al.*, 1992a; Dykes *et al.*, 1994).

Expolysaccharide production. Ropy slime production on vacuum-packaged, cooked meat products has been reported in several countries. It was caused by different LAB, especially by *Lb. sake, Lb. carnosum, Leuc. gelidum, Leuc. mesenteroides* subsp. *dextranicum* and *Leuc. mesenteroides* subsp. *mesenteroides* (Korkeala *et al.*, 1988; Von Holy *et al.*, 1991; Mäkelä *et al.*, 1992a; Dykes *et al.*, 1994). Typical ropy slime-producing strains of *Lb. sake* and *Leuc. amelibiosum* could be recovered directly from the meat-processing environment (Mäkelä *et al.*, 1992b).

2.6 Beneficial associations

It has been shown above that the LAB are so well adapted to the meat substrate that they dominate the populations developing with a variety of conditions of processing, packaging (reduced atmosphere) and refrigerated storage. The species *Lb. sake, Lb. curvatus, Leuc. amelibiosum, Leuc. carnosum* and *Leuc. gelidum* and more rarely the pediococci appear to predominate in different systems, ranging from spoiled, vacuum packaged meat, processed products, raw fermented sausages etc. (compare Table 2.2). Their beneficial association with fermented meat products has been studied intensively in recent years – such studies have also produced valuable information on the role of certain species of staphylococci and micrococci in the fermentation and maturation of meat products. In practice an improved understanding of the underlying antagonistic and competitive mechanisms against spoilage and pathogenic bacteria, has focused interest on the potential of LAB as agents for the 'biological' preservation or safeguarding of food. These two aspects are discussed below.

2.6.1 Meat fermentations

Raw dry sausages (German: 'Rohwurst') are the major product of meat fermentation. The fermentation and maturation of raw meat to which curing salts (nitrite or nitrate mixed with salt) and spices are added, promotes:

- elimination or reduction in the numbers of pathogens and spoilage organisms;
- desirable development of the typical curing colour;
- development of typical aroma and taste attributes;
- development of meat firmness; and
- preservation.

The growth of LAB is favoured by particular conditions prior to and during fermentation, e.g. pH < 6.0, initial a_w around 0.96 (due to 2.5–3.0% salt added), *ca.* 100 mg sodium nitrite/kg, and (quite commonly) the addition of about 0.3% fermentable sugars in order to promote rapid lactic acid production. Furthermore, the redox potential is reduced by firm stuffing of sausage meat into water-permeable casings and a gradual diminution of the temperature, from around 18 °C initially to 12–14 °C. The major physiological attributes of starter cultures are summarized in Table 2.12 (Hammes and Vogel, 1990).

Even with slight technical differences in the manufacturing process in different countries, the microbial populations of spontaneously fermented sausages were shown to be typically dominated (Table 2.3) by *Lb. sake* and

Table 2.12 Physiological attributes of starter cultures (modified after Hammes and Vogel, 1990).

Physiological activity	Influence on food system
Acidification	Improved keeping quality and hygiene; improved taste, texture, colour; improved processing (cooking/baking, etc.) features
Production of different metabolites	Improvement of: aroma, colour, taste; hygienic quality and shelf life; nutritional value/probiosis
Production of enzymes	Improved aroma, texture, colour; improved keeping quality and hygiene

Lb. curvatus (Gürakan *et al.*, 1995; Holzapfel, 1996; Hugas *et al.*, 1993; Samelis *et al.*, 1994). This is explained by the competitive growth ability of these two species at reduced a_w (down to 0.91) and at temperatures around 4 °C. Other psychrotrophic LAB such as *Carnobacterium* spp., *Leuconostoc* spp. and *Weissella* spp. are less salt tolerant (exception *W. halotolerans*) than these two species whilst *Lactobacillus pentosus*, *Lb. plantarum*, *Ped. acidilactici* and *Ped. pentosaceus* grow only poorly at temperatures around 7 °C (Lücke, 1996a).

As traditional fermentations rely on the reduction of nitrate to produce the nitrite needed for curing, slow fermentation and involvement of nitrate reducing 'micrococci' (most strains now classified with either *Staph. carnosus*, *Staph. xylosus* or *Staph. piscifermentans*) are essential for development of the desired sensory attributes. Modern industrial processing of fermented sausages relies increasingly on the application of starter cultures consisting of either single or mixed strains of homofermentative LAB, or (when nitrate is used) a mixture of homofermentative LAB and micrococci (mainly *Staph. carnosus* and *Staph. piscifermentans*) (Hammes *et al.*, 1995; Lücke, 1996a). New developments indicate the possibility of selecting natural strains of *Lb. sake*, *Lb, curvatus* or even *Lb. plantarum* and *Lb. pentosus* with 'additional' properties of technical importance such as the formation of nitrate reductase and non-heme catalase (Wolf and Hammes, 1988; Hammes *et al.*, 1990; Engesser and Hammes, 1994). Strains of *Lb. sake*, especially suited to raw sausage fermentation, have been patented (Hammes, 1993, 1995). It has been claimed that these strains give improved acidification, enhance colour intensity and stability and accelerate ripening.

According to Lücke (1996a) (modified) the following LAB are at present commercially available as starter cultures for sausage fermentation:

• 'Mesophilic' species such as *Lb. pentosus*, *Lb. plantarum* and *Ped. pentosaceus*, which possess nonheme catalase and grow rapidly at 20–25 °C thereby ensuring relatively short fermentation times without impairment of aroma or colour. Some strains of *Lb. pentosus* and *Lb. plantarum* produce nitrate reductase;
• the 'psychrotrophic' species, *Lb. sake* and *Lb. curvatus*, grow well and compete strongly at 18–22 °C. Their smaller share of the market may be related to continuation of acid production at low temperatures leading to 'oversouring'. Strains of *Lb. sake* but not *Lb. curvatus* show nonheme catalase and nitrite reductase activity (Min, 1994);
• in the USA the 'thermophilic' species *Ped. acidilactici* (formerly *Ped. cerevisiae*) is specially selected for the manufacture of summer sausage at 30–38 °C. Slow growth only is achieved at temperatures typically used in Europe;
• separate preparations or mixtures of LAB and 'micrococci' (mainly *Micrococcus varians*, *Staph. carnosus*, *Staph piscifermentans* [Hammes *et*

al., 1995] or *Staph. xylosus*). The micrococci and staphylococci reduce nitrate and evidently protect the product against colour changes and rancidity.

The role and association of staphylococci in fermented sausages are comprehensively reviewed by Nychas and Arkoudelos (1990).

2.6.2 Lactic acid bacteria as protective cultures

Biopreservation aims at the extension of shelf life and enhancement of food safety by using either micro-organisms 'naturally' associated with a food system or their antimicrobial metabolites. Increasing knowledge on the interactions of (a) the LAB in different food systems and (b) their metabolites and their synergistic effect, has opened new opportunities for 'mild' and more acceptable 'biological' food preservation methods. Special attention is also given to the potent antimicrobial polypetides, the bacteriocins, produced by some LAB. Relevant aspects are discussed in section 2.4. 'Natural' bacteriocinogenic LAB in meat systems were found to range between 1 and 24% (Table 2.9), and appear to be a regular part of the population of meat and meat products.

Nisin is virtually the only bacteriocin available commercially; it is licensed for use as a food additive in about 45 countries (Delves-Broughton, 1990). Although studies have shown potential of other bacteriocins for use as food additives and especially as anti-*Listeria* agents, few attempts towards their *in situ* application have been made thus far. Prospects for use of either bacteriocinogenic LAB or their (semi-)purified bacteriocins for preserving meat products are supported by *in situ* research studies in recent years, some of which are summarized in Table 2.13. Practical examples are shown in Figs. 2.5 and 2.6, of (a) the effective inhibition of *Listeria monocytogenes* growth on MA-packed bacon by a bacteriocin-negative (commercially

Table 2.13 Meat systems in which effectiveness of bacteriocins against *Listeria* spp. was tested.

Food System	Bacteriocin	Reference
1. Sterile products		
e.g. ground beef	Pediocin AcH	Motlagh *et al.* (1992)
2. Non-sterile products		
fresh meat	Pediocin PA-1	Nielsen *et al.* (1990)
comminuted, cured raw pork	Sakacin A	Schillinger *et al.* (1991)
minced meat	Sakacin A	Schillinger *et al.* (1991)
fermented sausages	Pediocin	Berry *et al.* (1991)
	Pediocin PA-1	Foegeding *et al.* (1992)
frankfurters	Pediocin	Berry *et al.* (1991)
vacuum-packed wieners	Pediocin AcH	Degnan *et al.* (1991)
fermented sausages	*Lb. sake* (Sakacin K)	Hugas *et al.* (1995)
beef cubes	'*Lb. bavaricus*' (bac+)	Winkowski *et al.* (1993)

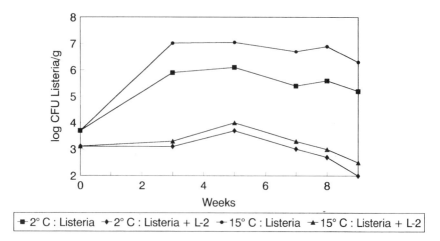

Figure 2.5 Development of *Listeria monocytogenes* (cfu/g) in MA-packed bacon (+0.5% glucose) at 2 °C and 15 °C in presence of *Lactobacillus alimentarius* FloraCarn L-2 (courtesy Chr. Hansen, Denmark).

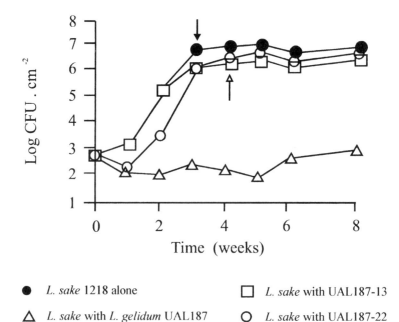

Figure 2.6 Log cfu of *Lactobacillus sake* 1218 showing growth and survival in mixed culture with variants of *Leuconostoc gelidum* per square centimeter of vacuum-packaged beef stored at 2 °C. The solid arrow indicates the sampling time at which a sulphide odour was first detected in samples inoculated with *Lb. sake* 1218. The open arrow indicates the sampling time at which a sulphide odour was first detected in samples inoculated with *Lb. sake* 1218 and *Leuc. gelidum* UAL 187-13 or UAL 187-22. The data represent the means of three trials (Leisner *et al.*, 1996).

available) strain of *Lactobacillus alimentarius*, both at 2 °C and 15 °C (courtesy: Lone Andersen, Chr. Hansen, DK-2970 Hørsholm), and (b) inhibition of the sulphide producing *Lb. sake* strain 1218 by bacteriocinogenic *Leuc. gelidum* UAL187 (Leisner *et al.*, 1996).

The eventual use of bacteriocinogenic 'food-grade' LAB or their bacteriocins will be determined by economic and regulatory factors. Major challenges to scientists would be (amongst others) to develop new ways of extending the activity spectrum of bacteriocins and to overcome 'immunity' of food-relevant pathogens and spoilage bacteria.

2.7 Isolation and cultivation methods

Isolation and cultivation methods for non-sporulating, Gram-positive bacteria have been reviewed with respect to food spoilage bacteria (Holzapfel, 1995) and to *Enterococccus* spp. (Reuter, 1995). Aspects relevant to the major Gram-positive bacterial groups of meat may be summarized as follows.

2.7.1 Lactic acid bacteria

A wide range of elective and selective media is available for the isolation and cultivation of meat-associated LAB. The most important information is summarized in Table 2.14. When using selective media for detection of LAB in meat, account should also be taken of their potential selectivity against LAB. Examples are the weak or failing growth ability of *Carnobacterium* spp. and to some extent of lactococci, on Acetate Agar or Rogosa Agar (pH 5.4). On the other hand, carnobacteria may grow on selective media used for detection of listeria.

For *Carnobacterium* spp. and especially strains of *Cb. divergiens* cresol red thallium acetate agar (CTAS) provides sufficient selectivity against other LAB, except for some strains of *Ent. faecium* and *Ent. faecalis* (Corry *et al.*, 1995).

2.7.2 Brochothrix

Streptomycin-thallous acetate-actidione agar according to Gardner (1966) is still the medium of preference. It inhibits the growth of *Bacillus* spp., coryneforms, lactobacilli and other LAB during incubation at 22 °C for up to five days.

2.7.3 Kurthia

A strictly selective medium for detection of *Kurthia* is not available; however, the unusual cultural properties of *K. zopfii* and *K. gibsonii* allow

Table 2.14 Elective and selective media for isolation of lactic acid bacteria from meat and meat products and generally from foods (modified from Reuter, 1985; Hammes et al., 1992; Holzapfel, 1995).

Medium	pH	Abbreviation	Application(s)*	Recovery**
Elective				
All-purpose Agar with Tween	6.7	APT	Cured meats/general	Lb, Lc, Lt, (Pd), Cb, W
Briggs-Agar	6.8	BRIGGS	Milk products, gut Lb/general	Lb, Lb-t, Lc, Pd, Lt, Et, Cb, W
Modified Homohiochi Medium	5.4	MHM	General/vegetables/sour dough/sake wine	Lb-hetero; lactics, W
Lactobacillus-Agar	6.8	La	General	Lb, Lc, Lt, Pd, (Cb?), Ec, W
deMan, Rogosa, Sharpe-Agar	6.4	MRS	General	Most lactics
Brain-Heart-Infusion-Yeast-Extract-Agar	6.8	BHIYE	Meat lactics/general	Most lactics
Tryptose-Proteose-P-Yeast extract-Eriochrome-Agar	6.8	TPPY	Yoghurt/general	Most lactics
Rogosa-Agar (modified)	6.2	RA 6.2	General	Most lactics
Selective				
Acetate-Agar	5.6	AA	Psychr. meat lactics	Lb, Lc, W
MRS 5.5	5.5	MRSS	General/meats	Lb (Lc?), W
MRS + 0.1 % sorbate	5.7	MRSS	Meats/general	Lb, Lc, W
MRS + 0.1 % thallous acetate	6.5	MRST	General	Most lactics at pH 6.5
Lactobacillus-Agar + 0.04 % sorbate	5.0	LaS	Meats	Lb (Lc), W
Nitrite, Actidione-Polymyxin-Agar	5.5	NAP	Meat lactics	Lb, Lt, (Lc), W
Differential Medium 'Isolation Medium'	?	IM	Salted anchovies, soy sauce,	Tetr. halophilus cured hams
Rogosa-Agar	5.5	RA	General	Lb, Lc, Pd, W

* with respect to food systems; ** Lb = *Lactobacillus*; Lb-t = thermophilic lactobacilli; Lc = *Leuconostoc*; Pd = *Pediococcus*; Lt = *Lactococcus*; Ec = *Enterococcus*; Cb = *Carnobacterium*; W = *Weissella*; Lactics (*Lactobacillus*, *Lactococcus*, *Leuconostoc*, *Carnobacterium*, *Pediococcus*, *Streptococcus*, *Weissella* spp.); hetero = heterofermentative.

the discrimination of typical colonies on yeast-nutrient-gelatin (YNG) or yeast-nutrient-agar (YGA) medium (Keddie and Jones, 1992).

2.7.4   Micrococcus *spp.*

Furazolidone agar may be applied for the direct detection of micrococci and their distinction from staphylococci (Rheinbaben and Hadlok, 1981).

2.7.5   Staphylococcus *spp.*

SK-medium, modified according to Schleifer and Krämer (1980), containing 0.0045% sodium azide may be used for determining viable numbers of staphylococci. Addition of 50 ml egg yolk emulsion to 1 l of medium allows the specific detection of *Staph. aureus*.

2.7.6   *Enterococci*

Several media may be used for the detection of enterococci, although none was found completely selective for all 'group D streptococci' (Reuter, 1995). For *Ent. faecalis* and *Ent. faecium* the medium of preference appears to be citrate azide tween carbonate agar (CATC) (Reuter, 1978; 1995).

2.8   Identification and typing methods

Classical identification methods are comprehensively described in the second edition of the authoritative handbook *The Prokaryotes*, Volume 2 (Springer-Verlag, 1992) and individual chapters are devoted to the respective genera discussed here: *Corynebacterium* – nonmedical (Liebl, 1992); *Micrococcus* (Kocur *et al.*, 1992); *Staphylococcus* (Kloos *et al.*, 1992); *Enterococcus* (DeVriese *et al.*, 1992); *Lactococcus* (Teuber *et al.*, 1992); *Pediococcus* and *Aerococcus* (Weiss, 1992); *Leuconostoc* (Holzapfel and Schillinger, 1992); *Lactobacillus* and *Carnobacterium* (Hammes *et al.*, 1992); *Listeria* (Jones and Seeliger, 1992); *Brochothrix* (Jones, 1992); *Kurthia* (Keddie and Jones, 1992); *Bacillus* – nonmedical (Slepecky and Hemphill, 1992); genera related to the genus *Bacillus* – *Sporolactobacillus*, *Sporosarcina*, *Planococcus*, *Filibacter* and *Caryophanon* (Claus *et al.*, 1992).

Developments in molecular techniques have opened new possibilities for the rapid and specific DNA-based typing of an unknown culture. Although initially developed for the clinical field, these methods are finding increased application for (non-pathogenic) micro-organisms associated with foods. A number of direct colony hybridization methods for *Listeria monocytogenes*, for example, have been developed and are commercially available.

Molecular typing methods can be targeted towards different macromolecules (such as proteins, lipopolysaccharides and the DNA or RNA). Development of the polymerase chain reaction (PCR) technique has especially opened new possibilities for genetic (nucleic acid based) typing methods, such as DNA sequencing, plasmid profiling, restriction endonuclease analysis of (genomic) DNA, oligo nucleotide probes and PCR-based methods.

DNA-based typing offers interesting prospects for the rapid detection of commercially valuable or important food organisms. For LAB these methods have also been applied for characterization of populations in ecosystems such as food and the gastrointestinal tract. Application of plasmid profiling for studying contamination of the meat environment and meat products (Dykes *et al.*, 1993) was found unsatisfactory because of the potential loss or exchange of plasmids (Vogel *et al.*, 1993). Presently, RAPD (random amplification of polymorphic DNA) appears to be the method of choice for studying LAB in food systems, e.g. for different LAB and enterococci (Cocconcelli *et al.*, 1995) or *Lb. plantarum* and *Lb. pentosus* (Van Reenen and Dicks, 1996).

Application of DNA or rRNA probe hybridizations relies on information concerning specific oligonucleotide stretches specific of a species or group. Such oligonucleotides may be labelled and applied as probes in hybridization experiments with unknown isolates; PCR techniques may be used to amplify target sequences and to enhance detection levels. A number of probes have been described for LAB, some of which are of direct relevance to lactic acid bacteria typical of meat systems and the meat environment. Examples are plasmid DNA probes for *Lb. fermentum* (Tannock *et al.*, 1992), randomly cloned DNA probes for *Lb. curvatus* (Petrick *et al.*, 1988), rRNA probes for *Cb. divergens*, *Cb. mobile* and *Cb. piscicola* (Brooks *et al.*, 1992), *Lb. curvatus*, *Lb. pentosus*, *Lb. plantarum* and *Lb. sake* (Hertel *et al.*, 1992), *Lb. plantarum* and *Lb. fermentum* (Hensiek *et al.*, 1992), and *Leuc. mesenteroides* (Klijn *et al.*, 1991).

Interesting and most practical examples of the application of nucleotide based methods are the use of RAPD, PFGE patterns of DNA and rRNA gene restriction patterns as characterization tools for studying *Lb. sake* strains involved in ropy slime spoilage of vacuum-packaged meat products (Björkroth *et al.*, 1996; Björkroth and Korkeala, 1996b). Ropy and non-ropy *Lb. sake* strains and other species of *Lactobacillus*, *Carnobacterium* and *Weissella* could be compared and several ropy and non-ropy *Lb. sake* strains even distinguished below the species level.

References

Aguirre, M. and Collins, M.D. (1992) Phylogenetic analysis of some *Aerococcus*-like organisms from urinary tract infections: description of *Aerococcus urinae* sp. nov. *J. Gen. Microbiol.* **138**, 401–5.

Ahmad, H.A. and Marchello, J.A. (1989) Microbial growth and successions on steaks as influenced by packaging procedures. *J. Food Protect.* **52**, 2236–9.

Alm, F., Erichsen, I. and Molin, N. (1961) The effect of vacuum packaging on some sliced processed meat products as judged by organoleptic and bacteriological analysis. *Food Technol.* **4**, 199–203.

Arihara, K., Cassens, R.G. and Luchansky, J.B. (1993) Characterisation of bacteriocins from *Enterococus faecium* with activity against *Listeria monocytogenes*. *Int. J. Food Microbiol.* **19**, 123–34.

Arihara, K., Ota, H., Itoh, M., Kondo, Y., Sameshina, T., Yamanaka, H., Akimoto, M., Kanai, S. and Miki, T. (1996) Application of intestinal lactobacilli to meat fermentation, Proc. 5th Symp. on Lactic Acid Bacteria, Veldhoven, The Netherlands, September 1996.

Asensio, M.A., Ordóñez, J.A. and Sanz, B. (1988) Effect of carbon dioxide and oxygen enriched atmospheres on the shelf life of refrigerated pork packed in plastic bags. *J. Food Protect.* **51**, 356–60.

Baird-Parker, A.C. (1990) The staphylococci: an introduction. *J. Appl. Bacteriol.*, Symp. Suppl., **69**, 1S–8S.

Bergdoll, M.S. (1979) Staphylococcal intoxications, in *Foodborne Infections and Intoxications*, (eds H. Riemann and F.L. Bryan), Academic Press, New York, pp. 443–93.

Bergdoll, M.S. (1983) Enterotoxin, in *Staphylococci and Staphylococcal Infections*, Vol. 2 (eds C.S.F. Easmon and C. Adlam), Academic Press, London, pp. 559–98.

Berry, E.D., Hutkins, R.W. and Mandigo, R.W. (1991) The use of bacteriocin-producing *Pediococcus acidilactici* to control postprocessing *Listeria monocytogenes* contamination of Frankfurters. *J. Food Protect.* **54**, 681–6.

Björkroth, K.J. and Korkeala, H.J. (1996a) Evaluation of *Lactobacillus sake* contamination in vacuum-packaged sliced cooked meat products by ribotyping. *J. Food Protect.* **59**, 398–401.

Björkroth, K.J. and Korkeala, H.J. (1996b) rRNA restriction patterns as a characterisation tool for *Lactobacillus sake* strains producing ropy slime. *Int. J. Food Microbiol.* **30**, 293–302.

Björkroth, K.J., Ridell, J. and Korkeala, H.J. (1996) Characterisation of *Lactobacillus sake* strains associated with production of ropy slime by randomly amplified polymorphic DNA (RAPD) and pulsed-field gel electrophoresis (PFGE) patterns. *Int. J. Food Microbiol.* **31**, 59–68.

Borch, E. and Agerhem, H. (1992) Chemical, microbiological and sensory changes during the anaerobic cold storage of beef inoculated with a homofermentative *Lactobacillus* sp. or a *Leuconostoc* sp. *Int. J. Food Microbiol.* **15**, 99–108.

Borch, E., Berg, H. and Holst, O. (1991) Heterolactic fermentation by a homofermentative *Lactobacillus* sp. during glucose limitation in anaerobic continuous culture with complete cell cycle. *J. Appl. Bacteriol.* **71**, 265–9.

Borch, E., Kant-Muermans, M.-L. and Blixt, Y. (1996) Bacterial spoilage of meat and cured meat products. *Int. J. Food Microbiol.* **33**, 103–20.

Borch, E. and Molin, G. (1988) Numerical taxonomy of psychrotrophic lactic acid bacteria from prepacked meat and meat products. *Antonie van Leeuwenhoek* **54**, 301–23.

Borch, E. and Molin, G. (1989) The aerobic growth and product formation of *Lactobacillus*, *Leuconostoc*, *Brochothrix* and *Carnobacterium* in batch cultures. *Appl. Microbiol. Biotechnol.* **30**, 81–8.

Borch, E., Nerbrink, E. and Svensson, P. (1988) Identification of major contamination sources during processing of emulsion sausages. *Int. J. Food Microbiol.* **7**, 317–30.

Broda, D.M., DeLacy, K.M., Bell, R.G., Braggins, T.J. and Cook, R.L. (1996a) Psychrotrophic *Clostridium* spp. associated with 'blown pack' spoilage of chilled vacuum-packed red meats and dog rolls in gas-impermeable plastic casings. *Int. J. Food Microbiol.* **29**, 335–52.

Broda, D.M., DeLacy, K.M., Bell, R.G. and Penney, N. (1996b) Association of psychrotrophic *Clostridium* spp. with deep tissue spoilage of chilled vacuum-packed lamb. *Int. J. Food Microbiol.* **29**, 371–8.

Brooks, J.L., Moore, A.S., Patchett, R.A., Collins, M.D. and Kroll, R.G. (1992) Use of polymerase chain reaction and oligonucleotide probes for the rapid detection and identification of *Carnobacterium* species from meat. *J. Appl. Bact.* **72**, 294–301.

Cantoni, C., Bersani, C., D'Aubert, S., Rosa, M. and D'Aubert, S. (1992) Origin and causes of discoloration of bressaola (in Italian). *Ingegneria Alimentare le Conserve Animali* **8**, 20–3.

Cantoni, C., Bianchi, M.A., Renon, P. and D'Autent, S. (1969). Ricerche sulla putrefazione del prosciutto crudo. *Arch. Veterin. Ital.* **20**, 355–70.

Claus, D., Fritze, D. and Kocur, M. (1992) Genera related to the genus *Bacillus – Sporolacto-bacillus, Sporosarcina, Planococcus, Filibacter* and *Caryophanon*, in *The Prokaryotes. A Handbook on Habitats, Isolation and Identification of Bacteria*, Vol. II, 2nd edn, (eds M.P. Starr, H. Stolp, H.G. Trüper, A. Balows and H.G. Schlegel), Springer Verlag, Berlin, pp. 1767–91.

Cocconcelli, P.S., Porro, D., Galandini, S. and Senini, L. (1995) Development of RAPD proto-col for typing of strains of lactic acid bacteria and enterococci. *Lett. Appl. Microbiol.* **21**, 376–9.

Collins, M.D. and Bradbury, J.F. (1992) The genera *Agromyces, Aureobacterium, Clavibacter, Curtobacterium* and *Microbacterium*, in *The Prokaryotes*, Vol. II, 2nd edn, (eds A, Balows, H.G. Trüper, M. Dworkin, W. Harder and K.H. Schleifer), Springer Verlag, Berlin, pp. 1353–68.

Collins, M.D., Farrow, J.A.E., Phillips, B.A., Ferusu, S. and Jones, D. (1987) Classification of *Lactobacillus divergens, Lactobacillus piscicola* and some catalase-negative, asporogenous, rod-shaped bacteria from poultry in a new genus, *Carnobacterium*. *Int. J. Syst. Bacteriol.* **37**, 310–16.

Collins, M.D., Rodriguez, U.M., Ash, C., Aguirre, M., Farrow, J.A.E., Martinez-Murcia, A., Phillips, B.A., Williams, A.M. and Wallbanks, S. (1991) Phylogenetic analysis of the genus *Lactobacillus* and related lactic acid bacteria as determined by reverse transcriptase sequencing of 16S rRNA. *FEMS Microbiol. Lett.* **77**, 5–12.

Collins, M.D., Rodriguez, U.M., Dainty, R.H., Edwards, R.A. and Roberts, T.A. (1992) Taxo-nomic studies on a psychrophilic *Clostridium* from vacuum-packed beef: Description of *Clostridium estertheticum* sp. nov. *FEMS Microbiol. Lett.* **96**, 235–40.

Collins, M.D., Samelis, J., Metaxopoulos, J. and Wallbanks, S. (1993) Taxonomic studies on some leuconostoc-like organisms from fermented sausages: description of a new genus *Weissella* for the *Leuconostoc paramesenteroides* group of species. *J. Appl. Bacteriol.* **75**, 595–603.

Collins, M.D., Williams, A.M. and Wallbanks, S. (1990) The phylogeny of *Aerococcus* and *Pediococcus* as determined by 16S rRNA sequence analysis: description of *Tetragenococcus* gen. nov. *FEMS Microbiol. Lett.* **70**, 255–62.

Comi, G., Manzano, M., Citterio, B., Bersani, C., Cantoni, C. and De Bertoldi, M. (1993) Handwerklich und industriell hergestellte italienische Salami. Physiologische Charakter-isierung und Entwicklung von Laktobazillen. *Fleischwirtschaft* **73**, 1312–18.

Coretti, K. (1958) Rohwurstfehlfabrikate durch Laktobazillen. *Fleischwirtschaft* **4**, 218–25.

Corry, J.E.L., Curtis, G.D.W. and Baird, R.M. (eds) (1995) Culture media for food microbiol-ogy, in *Progress in Industrial Microbiology*, Vol. 34, Elsevier, Amsterdam, pp. 293–5.

Daeschel, M.A. (1989) Antimicrobial substances from lactic acid bacteria for use as food preservatives. *Food Technol.* Jan. 1989, **43**, 164–6.

Dainty, R.H. (1982) Biochemistry of undesirable effects attributed to microbial growth on pro-teinaceous foods at chill temperatures. *Food Chem.* **9**, 103–13.

Dainty, R.H., Edwards, R.A. and Hibbard, C.M. (1989) Spoilage of vacuum-packed beef by a *Clostridium* species. *J. Sci. Food Agric.* **49**, 473–86.

Dainty, R.H. and Hibbard, C.M. (1980) Aerobic metabolism of *Brochothrix thermosphacta* growing on meat surfaces and in laboratory media. *J. Appl. Bacteriol.* **48**, 387–96.

Dainty, R.H. and Mackey, B.M. (1992) The relationship between the phenotypic properties of bacteria from chill-stored meat and spoilage processes. *J. Appl. Bacteriol., Symp. Suppl.*, **73**, 103S–14S.

De Bruyn, I.N., Holzapfel, W.H., Visser, L. and Louw, A.I. (1988) Glucose metabolism by *Lac-tobacillus divergens*. *J. Gen Microbiol.* **134**, 2109–30.

De Bruyn, I.N., Louw, A.I., Visser, L. and Holzapfel, W.H. (1987) *Lactobacillus divergens* is a homofermentative organism. *System. Appl. Microbiol.* **9**, 173–5.

Degnan, A.J., Yousef, A.E. and Luchansky, J.B. (1992) Use of *Pediococcus acidilactici* to control *Listeria monocytogenes* in temperature-abused vacuum-packaged wieners. *J. Food Protect.* **55**, 98–103.

Deibel, R.H. and Niven, C.F. (1960). Comparative study of *Gaffkya homari, Aerococcus viri-dans*, tetrad-forming cocci from meat curing brines, and the genus *Pediococcus. J. Bacteriol.*, **79**, 175–80.

Deibel, R.H., Niven, C.F. and WIlson, G.D. (1961) Microbiology of meat curing. III. Some microbiological and related technical aspects in the manufacture of fermented sausage. *Appl. Microbiol.* **23**, 130–5.

Dellaglio, F., Dicks, L.M.T. and Torriani, S. (1995) The genus *Leuconostoc*, in *The Genera of Lactic Acid Bacteria*, (eds B.J.B. Wood and W.H. Holzapfel), Blackie Academic and Professional, London, pp. 235–78.

Delves-Broughton, J. (1990) Nisin and its uses as a food preservative. *Food Technol.* **44**, 100–17.

DeVriese, L.A. and Pot, B. (1995) The genus *Enterococcus*, in *The Genera of Lactic Acid Bacteria*, (eds B.J.B. Wood and W.H. Holzapfel), Blackie Academic and Professional, London, pp. 327–67.

DeVriese, L.A., Collins, M.D. and Wirth, R. (1992) The genus *Enterococcus*, in *The Prokaryotes. A Handbook on Habitats, Isolation and Identification of Bacteria*, Vol. II, (eds M.P. Starr, H. Stolp, H.G. Trüper, A. Balows and H.G. Schlegel), Springer Verlag, Berlin, pp. 1466–81.

Dicks, L.M.T., Du Plessis, E.M., Dellaglio, F. and Lauer, E. (1996) Reclassification of *Lactobacillus rhamnosus* ATCC 15820 as *Lactobacillus zeae* nom. rev., designation of ATCC 334 as neotype of *L. casei* subsp. *casei*, and rejection of the name *Lactobacillus paracasei. Int. J. System. Bacteriol.* **46**, 337–40.

Drosinos, E.H. and Board, R.G. (1995) Attributes of microbial associations of meat growing as xenic batch cultures in a meat juice at 4 °C. *Int. J. Food Microbiol.* **26**, 279–93.

Dykes, G.A., Burges, A.Y. and Von Holy, A. (1993) Plasmid profiles of lactic acid bacteria associated with vacuum-packaged Vienna sausage manufacture and spoilage. *Lett. Appl. Microbiol.* **17**, 1822–84.

Dykes, G.A., Cloete, T.E. and Von Holy, A. (1994) Identification of *Leuconostoc* species associated with the spoilage of vacuum-packaged Vienna sausages by DNA–DNA hybridisation. *Food Microbiol.* **11**, 271–4.

Dykes, G.A., Cloete, T.E. and Von Holy, A. (1995) Taxonomy of lactic acid bacteria associated with vacuum-packaged processed meat spoilage by multivariate analysis of cellular fatty acids. *Int. J. Food Microbiol.* **28**, 89–100.

Dykes, G.A. and Von Holy, A. (1994) Taxonomic status of atypical *Lactobacillus sake* and *Lactobacillus curvatus* strains associated with vacuum-packaged meat spoilage. *Current Microbiol.* **28**, 197–200.

Edwards, R.A. and Dainty, R.H. (1987) Volatile compounds associated with the spoilage of normal and high pH vacuum-packed pork. *J. Sci. Food. Agric.* **38**, 57–66.

Egan, A.F. (1983) Lactic acid bacteria of meat and meat products. *Antonie van Leeuwenhoek* **49**, 327–36.

Egan, A.F. and Roberts, T.A. (1987) Microbiology of meat and meat products, in *Essays in Agricultural and Food Microbiology*, (eds J.R. Norris and G.L. Pettipher), Wiley, New York, pp. 167–97.

Egan, A.F., Shay, B.J. and Rogers, P.J. (1989) Factors affecting the production of hydrogen sulphide by *Lactobacillus sake* L13 growing on vacuum-packaged beef. *J. Appl. Bacteriol.*, **67**, 255–62.

Eisgruber, H. (1992) Clostridien als Kontaminante im Fleischverarbeitungsprozess: Sicherung der hygienischen Qualität von Fleischerzeugnissen. *Fleischerei* **43**, 548–53.

El-Sherbeeny, M.R., Saddik, M.F., El-Hossany, M.M. and Abd-el-Kadir, M. (1989) *Staphylococcus aureus* associated with meat and meat products. *Bull. Nutr. Inst. Arab Rep. Egypt* **9**, 27–37.

Engesser, D.M. and Hammes, W.P. (1994) Non-heme catalase activity of lactic acid bacteria. *System. Appl. Microbiol*, **17**, 11–19.

Fischer, S. and Schleifer, K.-H. (1980) Zum Vorkommen der Gram-positiven, katalasepositiven Kokken in Rohwurst. *Fleischwirtschaft* **60**, 1046–51.

Foegeding, P.M., Thomas, A.B., Pilkington, D.H. and Klaenhammer, T.R. (1992) Enhanced control of *Listeria monocytogenes* by *in situ*-produced pediocin during dry fermented sausage production. *Appl. Environ. Microbiol.* **58**, 884–90.

Fritze, D. and Claus, D. (1995) Spore-forming lactic acid producing bacteria of the genera *Bacillus* and *Sporolactobacillus*, in *The Genera of Lactic Acid Bacteria*, (eds B.J.B. Wood and W.H. Holzapfel), Blackie Academic and Professional, London, pp. 368–91.

Gallo, L., Schmitt, R.E. and Schmidt-Lorenz, W. (1988) Microbial spoilage of refrigerated fresh broilers. I. Bacterial flora and growth during storage. *Lebensm.-Wiss. u.-Technol.* **21**, 216–23.

Gardner, G.A. (1966) A selective medium for the enumeration of *Microbacterium thermosphactum* in meat and meat products. *J. Appl. Bacteriol.* **29**, 455–60.

Gardner, G.A. (1969) Physiological and morphological characteristics of *Kurthia zopfii* isolated from meat products. *J. Appl. Bacteriol.* **32**, 371–80.

Gardner, G.A. (1981) *Brochothrix thermosphacta* (*Microbacterium thermosphactum*) in the spoilage of meats: a review, in *Psychrotrophic Microorganisms in Spoilage and Pathogenicity* (eds T.A. Roberts, G.A. Hobbs, J.H.B. Christian and N. Skovgaard), Academic Press, London, pp. 139–73.

Gardner, G.A., Carson, A.W. and Patton, J. (1967) Bacteriology of prepacked pork with reference to the gas composition within the pack. *J. Appl. Bacteriol.* **30**, 321–33.

Garriga, M., Hugas, M., Aymerich, T. and Montfort, J.M. (1993) Bacteriocinogenic activity of lactobacilli from fermented sausages. *J. Appl. Bacteriol.* **75**, 142–8.

Garver, K.I. and Muriana, P.M. (1993) Detection, identification and characterisation of bacteriocin-producing lactic acid bacteria from retail food products. *Int. J. Food Microbiol.* **19**, 241–58.

Geornaras, I., Dykes, G.A. and Von Holy, A. (1995) Biogenic amine formation by poultry-associated spoilage and pathogenic bacteria. *Lett. Appl. Microbiol.* **21**, 164–6.

Gill, C.O. and Bryant, J. (1992) The contamination of pork with spoilage bacteria during commercial dressing, chilling and cutting of pig carcasses. *Int. J. Food Microbiol.* **16**, 51–62.

Gill, C.O. and Molin, G. (1991) Modified atmospheres and vacuum packaging, in *Food Preservatives* (eds. N.J. Russel and G.W. Gould), Blackie, Glasgow, pp. 172–99.

Gill, C.O. and Newton, K.G. (1979) Spoilage of vacuum-packaged dark, firm and dry meat at chill temperatures. *Appl. Environ. Microbiol.* **37**, 362–4.

Gobbetti, M., Fox, P.F. and Stepaniak, L. (1996) Esterolytic and lipolytic activities of mesophilic and thermophilic lactobacilli. *Ital. J. Food Sci.* **7**, 127–44.

Grant, G.F., McCurdy, A.R. and Osborne, A.D. (1988) Bacterial greening in cured meats: A review. *Can. Inst. Food Sci. Technol.* **21**, 50–6.

Gudkow, A.V. (1987) Starters: As a means of controlling contaminating organisms. *Milk – The vital Force*, pp. 83–93.

Gürakan, G.C., Bozoglu, T.F. and Weiss, N. (1995) Identification of *Lactobacillus* strains from Turkish fermented sausages. *Lebensm.-Wiss. u. -Technol.* **28**, 139–44.

Guzman, A.M.S., Pagano, C.E. and Gimenez, D.F. (1990) Incidence of *Clostridium perfringens* in fresh sausages in Argentina. *J. Food Protect.* **53**, 173–5.

Halász, A., Baráth, A., Simon-Sarkadi, L. and Holzapfel, W.H. (1994) Biogenic amines and their production by microorganisms in food. *Trends Food Sci. Technol.* **5**, 42–9.

Hammes, W.P. (1993) Zum Reifen von Rohwurst geeignete Mikroorganismen der Art *Lactobacillis sake. German Fed. Rep. Pat.* DE 42 01 050 C1.

Hammes, W.P. (1995) Zum Reifen von Rohwurst geeignete Mikroorganismen vom Stamm *Lactobacillis sake. German Fed. Rep. Pat.* EP 0 641 857 A1.

Hammes, W.P., Bantleon, A. and Min, S. (1990) Lactic acid bacteria in meat fermentation. *FEMS Microbiol. Rev.* **87**, 165–74.

Hammes, W.P., Bosch, I. and Wolf, G. (1995) Contribution of *Staphylococcus carnosus* and *Staphylococcus piscifermentans* to the fermentation of protein foods. *J. Appl. Bacteriol.*, Symp. Suppl., 1995, **79**, 76S–83S.

Hammes, W.P. and Vogel, R.F. (1990) Gentechnik zur Modifizierung von Starterorganismen. *Lebensmitteltechnik* **1–2**, 24–32.

Hammes, W.P. and Vogel, R.F. (1995) The genus *Lactobacillus*, in *The Genera of Lactic Acid Bacteria*, (eds B.J.B. Wood and W.H. Holzapfel), Blackie Academic and Professional, London, pp. 19–54.

Hammes, W.P., Weiss, N. and Holzapfel, W.H. (1992) *Lactobacillus* and *Carnobacterium*, in *The Prokaryotes. A Handbook on Habitats, Isolation and Identification of Bacteria*, 2nd edn, Vol. II, (eds A. Balows, H.G. Trüper, M. Dworkin, W. Harder and K.H. Schleifer), Springer Verlag, New York, pp. 1535–94.

Harding, C.D. and Shaw, B.G. (1990) Antimicrobial activity of *Leuconostoc gelidum* against closely related species and *Listeria monocytogenes. J. Appl. Bacteriol.* **69**, 648–54.

Hensiek, R., Krupp, G. and Stackebrandt, E. (1992) Development of diagnostic oligonucleotide probes for four *Lactobacillus* species occurring in the intestinal tract. *System. Appl. Microbiol.* **15**, 123–8.

Schillinger, U., Holzapfel, W. and Kandler, O. (1989) Nucleic acid hybridization studies on *Leuconostoc* and heterofermentative lactobacilli and description of *Leuconostoc amelibiosum* sp. nov. *System Appl. Microbiol.* **12**, 48–55.

Schillinger, U., Kaya, M. and Lücke, F.-K. (1991) Behavior of *Listeria monocytogenes* in meat and its control by a bacteriocin-producing strain of *Lactobacillus sake. J. Appl. Bacteriol.* **70**, 473–8.

Schillinger, U. and Lücke, F.-K. (1986) Milchsäurebakterien-Flora auf vakuumverpacktem Fleisch und ihr Einfluß auf die Haltbarkeit. *Fleischwirtschaft* **66**, 1515–1520.

Schillinger, U. and Lücke, F.K. (1987) Lactic acid bacteria on vacuum-packaged meat and their influence on shelf life. *Fleischwirtschaft*, **67**, 1244–8.

Schillinger, U. and Lücke, F.K. (1989) Antibacterial activity of *Lactobacillus sake* isolated from meat. *Appl. Environ. Microbiol.* **55**, 1901–6.

Schleifer, K.-H. and Kloos, W.E. (1975) A simple test system for the separation of staphylococci from micrococci. *J. Clinical Microbiol.* **1**, 337–8.

Schleifer. K.-H. and Krämer, E. (1980) Selective medium for isolating staphylococci. *Zbl. Bakt. Hyg., I. Abt. Orig.* **C1**, 270–80.

Schleifer, K.H. and Ludwig, W. (1995) Phylogeny of the genus *Lactobacillus* and related genera. *System. Appl. Microbiol.* **18**, 461–7.

Schraft, H., Kleinlein, N. and Untermann, F. (1992) Contamination of pig hindquarters with *Staphylococcus aureus. Int. J. Food Microbiol.* **15**, 191–4.

Seeliger, H.P.R. and Jones, D. (1986) Genus *Listeria* in *Bergey's Manual of Determinative Bacteriology*, Vol. II, (eds P.H.A. Sneath, N.S. Mair, M.E. Sharpe and J.G. Holt), Williams and Wilkins, Baltimore, London, pp. 1235–45.

Sharpe, M.E. and Pettipher, G.L. (1983) Food spoilage by lactic acid bacteria. *Economic Microbiology* **8**, 199–223.

Shaw, B.G. and Harding, C.D. (1984) A numerical taxonomic study of lactic acid bacteria from vacuum-packed beef, pork, lamb and bacon. *J. Appl. Bacteriol.* **56**, 25–40.

Shaw, B.G. and Harding, C.D. (1985) Atypical lactobacilli from vacuum-packaged meats: comparison by DNA hybridization, cell composition and biochemical tests with a description of *Lactobacillus carnis* sp. nov. *System. Appl. Microbiol.* **6**, 291–7.

Shaw, B.G. and Harding, C.D. (1989) *Leuconostoc gelidum* sp. nov. and *Leuconostoc carnosum* sp. nov. from chill-stored meats. *Int. J. System. Bacteriol.* **39**, 217–23.

Sierra, M., Garcia, M.L., Gonzalez, E., Garcia, M.C. and Otero, A. (1995) Species of psychrotrophic bacteria on freshly dressed lamb carcasses. *Arch. f. Lebensmittelhyg.* **46**, 1–24.

Simpson, W.J. and Taguchi, H. (1995) The genus *Pediococcus*, with notes on the genera *Tetragenococcus* and *Aerococcus*, in *The Genera of Lactic Acid Bacteria*, (eds B.J.B. Wood and W.H. Holzapfel), Blackie Academic and Professional, London, pp. 125–72.

Slepecky, R.A. and Hemphill, H.E. (1992) The genus *Bacillus* – Nonmedical, in *The Prokaryotes. A Handbook on Habitats, Isolation and Identification of Bacteria*, 2nd edn, Vol. II, (eds M.P. Starr, H. Stolp, H.G. Trüper, A. Balows and H.G. Schlegel), Springer Verlag, Berlin, pp. 1663–96.

Sneath, P.H.A. and Jones, D. (1986) Genus *Brochothrix*, in *Bergey's Manual of Determinative Bacteriology*, Vol. II, (eds P.H.A. Sneath, N.S. Mair, M.E. Sharpe and J.G. Holt), Williams and Wilkins, Baltimore, London, pp. 1249–53.

Sobrino, O.J., Rodriguez, J.M., Moreira, W.L., Fernández, M.F., Sanz, B. and Hernández, P.E. (1991) Antibacterial activity of *Lactobacillus sake* isolated from dry fermented sausages. *Int. J. Food Microbiol.* **13**, 1–10.

Stackebrandt, E. and Prauser, H. (1992) The family *Cellulomonadaceae*, in *The Prokaryotes. A Handbook on Habitats, Isolation and Identification of Bacteria*, 2nd edn, Vol. II, (eds A. Balows, H.G. Trüper, M. Dworkin, W. Harder and K.H. Schleifer), Springer Verlag, Berlin, pp. 1323–45.

Stecchini, M.L., Aquili, V., Sarais, J. and Pitotti, A. (1992) Inhibition of *Listeria monocytogenes* by *Lactococcus lactis* subsp. *lactis* isolated from Italian raw ham. *J. Food Safety* **12**, 295–302.

Stiles, M.E. and Holzapfel, W.H. (1997) Lactic acid bacteria of foods and their current taxonomy. *Int. Food Microbiol.* **36**, 1–29.

Sulzbacher, W.L. and McClean, R.A. (1951) The bacterial flora of fresh pork sausage. *Food Technol.* **5**, 7–8.

Tannock, G.W. (1989) Biotin-labelled plasmid DNA probes for detection of epithelium-associated strains of lactobacilli. *Appl. Environ. Microbiol.* **55**, 461–4.

Taylor, S.L., Keefe, T.J., Windham, E.S. and Howell, J.F. (1982) Outbreak of histamine poisoning associated with consumption of Swiss cheese. *J. Food Protect.* **45**, 455–7.

Ten Brink, B., Damink, C., Joosten, H.M.L.J. and Huis in't Veld, J.H.J. (1990) Occurrence and formation of biological active amines in foods. *Int. J. Food Microbiol.* **11**, 73–84.

Teuber, M. Geis, A. and Neve, H. (1992) The genus *Lactococcus*, in *The Prokaryotes. A Handbook on Habitats, Isolation and Identification of Bacteria*, 2nd edn, Vol. II, (eds A. Balows, H.G. Trüper, M. Dworkin, W. Harder and K.H. Schleifer), Springer Verlag, Berlin, pp. 1482–1501.

Thornley, M.J. and Sharpe, M.E. (1959) Microorganisms from chicken meat related to both lactobacilli and aerobic sporeformers. *J. Appl. Bacteriol.* **22**, 368–76.

Toth, L. (1982) Reaktionen des Nitrits beim Pökeln von Fleischwaren. *Fleischwirtsch.* **62**, 1256–63.

Van Reenen, C.A. and Dicks, L.M.T. (1996) Evaluation of numerical analysis of random amplified polymorphic DNA (RAPD)-PCR as a method to differentiate *Lactobacillus plantarum* and *Lactobacillus pentosus*. *Curr. Microbiol.* **32**, 183–7.

Vaughan, E.E., Caplice, E., Looney, R., O'Rourke, N., Coveney, H., Daly, C. and Fitzgerald, G.F. (1994) Isolation from food sources of lactic acid bacteria that produced antimicrobials. *J. Appl. Bacteriol.* **76**, 118–23.

Vignolo, G.M., Suriani, F., de Ruiz Holgado, A.P. and Oliver, G. (1993) Antibacterial activity of *Lactobacillus* strains isolated from dry fermented sausages. *J. Appl. Bacteriol.* **75**, 344–9.

Vogel, R.F., Lohmann, M., Nguyen, M. Weller, A.N. and Hammes, W.P. (1993) Molecular characterisation of *Lactobacillus curvatus* and *Lactobacillus sake* isolated from sauerkraut and their application in sausage fermentations. *J. Appl. Bacteriol.* **74**, 295–300.

Von Holy, A. (1989) Microbial ecology of vacuum-packaged Vienna sausage spoilage. Doctorate Thesis, University of Pretoria, Oct. 1989.

Von Holy, A., Cloete, T.E. and Holzapfel, W.H. (1991) Quantification and characterisation of microbial populations association with spoiled, vacuum-packaged Vienna sausages. *Food Microbiol.* **8**, 95–104.

Von Holy, A. and Holzapfel, W.H. (1989) Spoilage of vacuum packaged processed meats by lactic acid bacteria, and economic consequences. Proceedings of 10th International WAVFH Symposium. July 1989, Stockholm, Sweden, pp. 185–90.

Von Holy, A., Holzapfel, W.H. and Dykes, G.A. (1992) Bacterial populations associated with Vienna sausages packaging. *Food Microbiol.* **9**, 45–53.

Wallbanks, S., Martinez-Murcia, A.J., Fryer, J.L., Phillips, B.A. and Collins, M.D. (1990) 16S rRNA sequence determination for members of the genus *Carnobacterium* and related lactic acid bacteria and description of *Vagococcus salmoninarum* sp. nov. *Int. J. System. Bacteriol.* **40**, 224–30.

Weiss, N. (1992) The genera *Pediococcus* and *Aerococcus*, in *The Prokaryotes. A Handbook on Habitats, Isolation and Identification of Bacteria* 2nd edn, Vol. II, (eds A. Balows, H.G. Trüper, M. Dworkin, W. Harder and K.H. Schleifer), Springer Verlag, New York, pp. 1502–7.

Whittenbury, R. (1964) Hydrogen peroxide formation and catalase activity in the lactic acid bacteria. *J. Gen. Microbiol.* **35**, 12–26.

Widdel, F. (1992) The genus *Desulfotomaculum*, in *The Prokaryotes. A Handbook on Habitats, Isolation and Identification of Bacteria* 2nd edn, Vol. II, (eds A. Balows, H.G. Trüper, M. Dworkin, W. Harder and K.H. Schleifer), Springer Verlag, New York, pp. 1792–9.

Winkowski, K., Crandall, A.D. and Montville, T.J. (1993) Inhibition of *Listeria monocytogenes* by *Lb. bavaricus* MN in beef systems at refrigeration temperatures. *Appl. Environ. Microbiol.* **59**, 2552–7.

Wolf, G. and Hammes, W.P. (1988) Effect of hematin on the activites of nitrite reductase and catalase in lactobacilli. *Arch. Mikrobiol.* **149**, 220–4.

3 Yeasts and moulds associated with meat and meat products

VIVIAN M. DILLON

3.1 Introduction

Yeasts and moulds are important opportunistic spoilage organisms of red meats and meat products, but only cause concern when conditions are such that bacterial competition is reduced (Walker and Ayres, 1970). They occur initially in low numbers and their slow growth rates make them poor competitors of psychrotrophic bacteria at chill temperatures (Ingram, 1958; Walker and Ayres, 1970; Walker, 1977). Fungal contamination of beef and lamb carcasses stored at −1 °C or −5 °C, however, was noted by Haines (1931), Lea (1931a, b) and Empey and Scott (1939) and was probably due not only to the reduced temperature but also to the associated lowered water activity on the carcass surface. In meat products, yeast and mould numbers increase at the expense of Gram-negative bacteria with such treatments as ionizing irradiation, antibiotic treatments (oxytetracyclines or chlortetracyclines), reduced water activity by drying, salting or freezing, or preservation methods (lactic acid or sulfur dioxide). The off-odour of sausages, however, stored at room temperature (Dyett and Shelley, 1962) and the surface yellowish slime observed on sulfited British fresh sausages (Dowdell and Board, 1968, 1971) were both caused by yeast spoilage. Further work has shown that not only does sulfite select for yeast in comminuted meat products, but it also induces acetaldehyde production and subsequent sulfite-binding (Dalton *et al.*, 1984; Dillon and Board, 1989a, 1990).

The accuracy of much of the earlier work in identifying moulds and yeasts isolated from meat, was compromised by the morphological diversity displayed in cultures of different ages and the different pH of the medium. The difficulties are compounded because yeasts are isolated in the imperfect state from environmental samples (soil, vegetation) or meat products so that other morphological characteristics evident in their life cycles are not observed (Davenport, 1981).

The study of food yeasts was also hampered by the diversity of methods used to isolate these micro-organisms. Each method imposed certain selective parameters that isolated only the species most tolerant of these conditions. Hence comparisons between studies are almost impossible. For example, different incubation temperatures were used, which influenced the type and number of yeasts isolated. Indeed, di Menna (1955a) noted that *Rhodotorula mucilaginosa* and *Candida parapsilosis* were the major yeasts isolated from soil with incubation at 37 °C whereas *Cryptococcus albidus*, *Cr. terreus* and *Trichosporon pullulans* dominated the flora at 18 °C. Lower yeast counts were recovered at 32 °C than at 12 °C, 17 °C, 22 °C and 27 °C by Koburger (1970, 1973). Higher yeast counts were recorded at 5 °C than at 15 °C (some yeast species only being isolated at 5 °C) from orchards and vineyards in the United Kingdom (Davenport, 1980a, 1980b). Banks and Board (1987) noted that a higher number of yeasts were recovered at 25 °C compared to 5 °C, except in specific meat products where there was a large population of psychrotrophic yeasts.

Different media have been utilized for isolating yeasts and moulds, using various methods of inhibiting bacteria and restricting the spread of mucoraceous moulds. Initially, acidified agars such as malt extract (adjusted to pH 3.5 with 10% tartaric acid) agar (Holwerda, 1952; Jarvis, 1973) and plate count (adjusted to pH 3.5 with 10% citric acid) agar (Dowdell and Board, 1968) or acidified potato dextrose agar were used to inhibit non-aciduric bacteria. These, however, only selected acid-tolerant fungal species and therefore yielded low yeast counts. The yeasts that did grow produced only very small colonies (Hup and Stadhouders, 1972). Antibiotic media were subsequently used and were found to be superior to acidified ones in controlling bacteria and enabling a larger number of yeasts to be recovered (Koburger, 1970; Jarvis, 1973). In many cases, oxytetracycline glucose yeast extract agar was used with or without the supplementation of gentamicin (Mossel *et al.*, 1970, 1975). Alternatively, combinations of dyes, antibiotics and herbicides, such as Dichloran (2,6-dichloro-4-nitroaniline), Rose Bengal chlortetracycline or chloramphenicol agar based on the medium by King *et al.* (1979) were used to suppress the overgrowth of spreading moulds such as species of *Rhizopus* and *Mucor*. In some cases, Dichloran 18% glycerol agar was used for isolating moulds (Hocking and Pitt, 1980). When the Rose Bengal dye was incorporated in the medium, however, the recovery of some yeast species was reduced through photodynamic death (Banks and Board, 1987) but the advantage was that bacterial and yeast colonies were easily distinguished (Dijkmann *et al.*, 1980). Although oxytetracycline glucose yeast extract agar was superior in many cases, Banks and Board (1987) found that more yeasts were recovered on Rose Bengal chloramphenicol agar from such products as skinless sausages, bacon burgers, and fresh and chilled chicken. The various pressures imposed by the media select specific yeasts and thus hinder comparison between studies.

Progressive changes in yeast taxonomy have also contributed to the problems of trying to compare different studies. Characterization was mainly based on morphological and physiological criteria such as shape of cells, modes of sexual and asexual reproduction, anaerobic fermentation, aerobic assimilation of sugars and specific growth requirements (van der Walt and Yarrow, 1984; Deák and Beuchat, 1987; Fung and Liang, 1990). Davenport (1980a, 1981) suggested that in addition to these conventional tests, environmental features should also be noted. Thus morphological characteristics that enable yeasts to survive and reproduce in certain habitats would be included to aid comparative studies (Davenport, 1973, 1980a, 1981). Biochemical and genetical methods are increasingly being used to aid the study of yeast taxonomy (Miller, 1979; Deák and Beuchat, 1987), including protein, polysaccharide and long chain fatty acid analysis (Deák and Beuchat, 1987; Viljoen *et al.*, 1993), molecular criteria such as differences in guanine and cytosine (G+C) base composition (Price *et al.*, 1978), DNA sequence homology and ribosomal RNA relatedness (Price *et al.*, 1978; Kurtzman, 1984, 1988).

Many of the yeasts quoted in the literature have been reclassified continuously (Kreger-van Rij, 1984; Barnett *et al.*, 1990). The yeasts cited in this chapter are classified according to Barnett *et al.* (1990) but where appropriate the imperfect state is used and not the perfect state by which the yeast is catalogued. Several species occurring in the original articles have been reclassified as one species, e.g. *Debaryomyces hansenii. Deb. kloeckeri* and *Deb. nicotianae* have been reclassified as *Deb. hansenii* var. *hansenii* whereas *Deb. subglobosus* has been reclassified as *Deb. hansenii* var. *fabryi* (Kreger-van Rij, 1984; Barnett *et al.*, 1990). Yeasts such as *Rhodotorula* spp. indicate an imperfect state of their life cycle and, where mating types have been found, were reclassified as basidiomycetous yeasts such as *Rhodosporidium* spp. *Rhodotorula glutinis*, for example, is the imperfect state of *Rhodosporidium diobavatum*, *Rhodosporidium sphaerocarpum* or *Rhodosporidium toruloides* (Kreger-van Rij, 1984; Barnett *et al.*, 1990). *Debaryomyces hansenii* is the ascomycetous form of *Candida famata* (Kreger-van Rij, 1984). Many of the yeast species in the literature were identified with reference to the scheme outlined by Lodder and Kreger-van Rij (1952) and Lodder (1970) which was subsequently revised by Kreger-van Rij (1984). Other notable changes were that *Rhodotorula rubra* was reclassified as *Rh. mucilaginosa*, *Candida curvata* as *Cryptococcus curvatus*, *Candida humicola* as *Cryptococcus humicolus*, *Torulopsis candida* as *Candida saitoana* and *Trichosporon cutaneum* as *Trichosporon beigelii* by Barnett *et al.* (1990).

Similarly, the taxonomy of the important meat moulds such as species of *Penicillium* and *Aspergillus* is problematic (Samson and Frisvad, 1993). Morphological and phenotypic differentiation were used conventionally to identify these moulds (Samson and Frisvad, 1993). Identification was improved by the use of isoenzyme analysis, immunological techniques, DNA–DNA hybridizations and profiles of mycotoxins and other secondary metabolites (Samson *et al.*, 1991; Samson and Frisvad, 1993). One method for differentiating *Penicillium* spp. is by the detection of indole metabolites (Lund, 1995, 1996). The classification of *Aspergillus* spp. was modified by investigation of the occurrence of 9 or 10 isoprene units of ubiquinones (Kuraishi *et al.*, 1990). The moulds in this chapter are classified with reference to Samson and Pitt (1990), Samson *et al.* (1991) and Samson and Frisvad (1993). Notably, *Sporotrichum carnis* has been reclassified as *Chryosporium pannorum* and *Cephalosporium* spp. as *Acremonium* spp. More specifically, *Penicillium cycloplium* has been reclassified as *Pen. aurantiogriseum, Pen. frequentans* as *Pen. glabrum, Pen. janthinellum* as *Pen. simplicissimum, Pen. notatum* as *Pen. chrysogenum* and *Pen. palitans* as *Pen. commune* (Samson *et al.*, 1991).

Reviews or studies of yeasts or moulds in meat are relatively sparse when compared to those concerned with bacteria (Jay, 1978, 1979, 1987; Deák and Beuchat, 1987; Dillon and Board, 1991). In this chapter it will be shown that yeasts and moulds have the potential to be transferred from the environment (Tables 3.1 and 3.3) via the fleece or hide to the carcass (Tables 3.2 and 3.3) and subsequently to a meat product (Tables 3.4 and 3.5). Also, the spoilage potential of yeasts and moulds in processed meat products will be discussed in the context of factors which reduce bacterial competition.

3.2   Yeast and mould contamination of meat animals

3.2.1   The field

Yeasts and moulds are ubiquitous in the environment of the meat animal, occurring on or in plants, air, water and soil (Walker, 1977; Smith and Anderson, 1992). Many of these yeasts (Table 3.1) are psychrotrophic (do Carmo-Sousa, 1969) and are therefore potential spoilage organisms of meat under chill storage. *Bullera alba, Cryptococcus laurentii, Rhodotorula ingeniosa, Rh. glutinis, Rh. graminis, Rh. minuta, Rh. mucilaginosa* and *Sporobolomyces roseus* were the yeasts most frequently isolated from rye grass and white clover (pasture plants) in New Zealand (di Menna, 1959). The yeast population increased from 3.1×10^4 in December to 1×10^8/g (wet weight) in March (di Menna, 1959). The red-pigmented yeasts (*Rhodotorula* spp. and *Sporobolomyces* spp.) decreased from 84–92% during the period from February to March to 4–8% in August (di Menna, 1959). Seasonal variations within the yeast flora from hay and soil samples were also noted in the United Kingdom (Dillon *et al.*, 1991). Carotenoid-pigmented yeasts represented only 5–7.5% in December but as much as 80–87% of the yeast flora in March (Dillon *et al.*, 1991).

Candida famata, Candida sake, Cryptococcus albidus var. *albidus*, and the carotenoid *Cr. infirmo-miniatus* (the imperfect state of *Cystofilobasidium infirmo-minatum*) and *Rh. mucilaginosa* were isolated from grass, turnips and soil samples from sheep pastures in the United Kingdom (Dillon *et al.*, 1991). The majority of the yeasts isolated were in their imperfect state in

Table 3.1 Yeast species isolated from soil, plants and air

	Soil (1, 2, 3)	Plants (4, 5, 6)	Air (7)
Bullera alba		+	
Candida albicans	+	+	
dattila			+
famata	+	+	+
inconspicua	+		
lipolytica	+		
melinii			+
parapsilosis	+		
pintolopesii	+		
rugosa	+		
saitoana	+		+
sake		+	
zeylanoides			+
Cryptococcus albidus	+	+	+
albidus var. albidus	+	+	+
albidus var. aerius		+	
curvatus	+	+	
flavus	+	+	
gastricus	+		
humicolus	+	+	
infirmo-miniatus		+	
laurentii	+	+	+
luteolus	+	+	+
macerans		+	
terreus	+	+	
Debaryomyces hansenii			
hansenii var. hansenii	+		+
hansenii var. fabryi	+		+
Debaryomyces marama			+
Hanseniaspora vineae		+	
Hansenula candensis	+		
Leucosporidium scottii		+	+
Pichia fermentans	+		
Rhodotorula aurantiaca	|		
glutinis	+	+	
graminis	+	+	
ingeniosa	+		
minuta	+	+	+
mucilaginosa	+	+	+
spp.	+	+	
Saccharomyces cerevisiae	+		
Schizoblastosporion starkeyi-henricii	+	+	
Sporobolomyces shibatanus		+	
roseus		+	+
salmonicolor		+	+
spp.		+	
Trichosporon beigelii	+	+	+
spp.			+

1 Stockyard soils, New Zealand; di Menna (1955a).
2 Soil samples from sheep and cattle stockyards and paddocks, New Zealand; Baxter and Illston (1977).
3 Soil under pasture, New Zealand; di Menna (1957, 1960).
4 Leaves and roots of pasture grasses and herbs, New Zealand; di Menna (1957).
5 Pasture plants, New Zealand; di Menna (1958a, b, c; 1959).
6 Field samples, England; Dillon *et al.* (1991).
7 Airborne yeasts, New Zealand; di Menna (1955b).

Table 3.2 Yeasts from fleece, hides, carcasses and processing plants

	Fleece (1, 2)	Carcass (3–7)	Process area (1, 8, 9)
Candida spp.	+	+	
albicans		+	+
castellii			+
famata		+	
glabrata		+	
guilliermondii			+
intermedia			+
mesenterica		+	
parapsilosis			+
pararugosa			+
rugosa		+	
saitoana	+	+	+
silvae		+	
vini		+	
zeylanoides		+	+
Cryptococcus spp.	+	+	
albidus var. *albidus*		+	+
curvatus		+	
gastricus		+	
humicolus		+	
laurentii	+	+	+
luteolus	+		
Debaryomyces hansenii		+	+
marama			+
vanrijiae			+
Galactomyces geotrichum			+
Geotrichum spp.		+	+
Leucosporidium scottii		+	
Pichia angusta			+
farinosa			+
guilliermondii			+
membranaefaciens		+	
Rhodotorula spp.		+	
aurantiaca			+
glutinis		+	
graminis		+	
minuta		+	
mucilaginosa	+	+	+
Sporothrix catenata		+	
Torulaspora delbrueckii			+
Trichosporon beigelii		+	+
Yarrowia lipolytica			+

1 Fleece, processing plant New Zealand; Baxter and Illston (1976; 1977).
2 Fleece, England; Dillon *et al.* (1991).
3 Lamb carcasses, New Zealand; Baxter and Illston (1976; 1977).
4 Lamb carcasses, England; Dillon *et al.* (1991).
5 Lamb carcasses, New Zealand; Lowry (1984).
6 Pig carcasses, New Zealand; Baxter and Illston (1976).
7 Pig carcasses, England; Dalton *et al.* (1984).
8 Abattoir; Refai *et al.*, (1993).
9 Vienna sausage processing plant; Viljoen *et al.*, (1993).

Table 3.3 Moulds associated with meat processing

Moulds isolated from carcasses (1, 2, 4)	Stock yards (2, 3, 4)	Process areas (2, 4)	Fleece, cowhair (1, 2, 4)
Acremonium spp.		+	
Alternaria spp.		+	
alternata	+	+	+
Aspergillus spp.		+	
flavus		+	
fumigatus		+	
niger		+	
ochraceus		+	
parasiticus		+	
terreus		+	
Cladosporium spp.	+	+	+
herbarum	+	+	+
cladosporioides	+	+	+
Curvularia spp.		+	
Epicococcum purpurascens	+	+	+
Fusarium spp.	+	+	
Helminthosporium spp.		+	
Mucor spp.	+	+	+
Paecilomyces spp.		+	
Penicillium spp.	+	+	+
commune			
Rhizoctonia spp.	+	+	+
Rhizopus spp.		+	
Scopuloriopsis spp.		+	
Thamnidium spp.			
elegans			

1 Abattoir; Empey and Scott, 1939.
2 Processing plant; Baxter and Illston, 1976.
3 Processing plant; Baxter and Illston, 1977.
4 Abattoir; Refai *et al.*, 1993.
Process area = Slaughter areas and processing halls.

the United Kingdom field samples (Dillon *et al.*, 1991), soil samples in New Zealand (di Menna, 1955a), on plants (Miller, 1979) and in cooler climates (Kunkee and Amerine, 1970).

Cryptococcus albidus, Cr. curvatus, Cr. terreus and *Schizoblastosporion starkeyi-henricii* dominated the yeast flora from soil samples from pastures in New Zealand (di Menna, 1957, 1960). The yeast flora ranged from 6×10^3 to 2.4×10^5 yeasts/g soil wet weight (di Menna, 1957). In contrast, soil samples taken from sheep yards in New Zealand (Baxter and Illston, 1977), were dominated by *C. saitoana* (3.4×10^3 cfu/g) in association with *Cryptococcus luteolus* (1×10^2 cfu/g) and *Rh. mucilaginosa* (1×10^2 cfu/g wet weight). *Cryptococcus laurentii* also occurred in the soil samples from cattle stockyards (Baxter and Illston, 1977). Additionally, *Rh. mucilaginosa* and *C. parapsilosis* were commonly isolated from stockyard soil in New Zealand by di Menna (1959).

Table 3.4 Species of yeasts isolated from red meats and meat products

	References (see footnote to table)
Brettanomyces spp.	14
Bullera alba	10, 23
Candida spp.	5, 7, 14, 19, 22, 27
albicans	13, 23
apis	25, 29
apicola	25
blankii	20
catenulata	5, 9, 10, 23
dattila	25
diddensiae	20
diversa	20
domercqii	23, 25
ernobii	25
etchellsii	10
famata	6, 9, 10, 21, 26, 30
glabrata	6
glaebosa	20, 25
gropengiesseri	5, 25
guilliermondii	10, 30
haemulonii	25
hellenica	20
holmii	25
inconspicua	6, 23, 26
insectamans	20
intermedia	6, 20, 25
kefyr	25
kruisii	6, 30
lambica	20
lipolytica	5, 9, 10, 12, 13, 16, 20
melinii	12
mesenterica	23, 26
molischiana	25
norvegica	23, 26
parapsilosis	10, 13, 16, 17, 18, 21, 28
pelliculosa	6, 10
pintolopesii	25
pinus	10
reukaufii	10
rugosa	6, 9, 10, 12, 23
saitoana	5, 10, 12, 15, 16, 17, 20, 23
sake	21, 23, 26
scottii	6
silvae	23
silvatica	20
silvicultrix	20
stellata	25
tenuis	25
tropicalis	13, 18, 25
valida	23
vanderwaltii	23
versatilis	12, 20, 23, 25
vini	23, 26
wickerhamii	25
zeylanoides	5, 11, 12, 13, 16, 20, 21, 23, 24, 25, 26, 29, 30

Table 3.4 Continued

	References (see footnote to table)
Citromyces matritensis .	25
Cryptococcus spp.	14, 22
albidus	16, 25, 28
albidus var. *albidus*	21, 23, 26, 30
albidus var. *aerius*	10, 12, 23
aquaticus	25
curiosus	25
curvatus	10, 20, 23
dimennae	25
flavus	25
gastricus	25
humicolus	23, 25, 26, 30
hungaricus	23
infirmo-miniatus	21, 24, 25, 26
laurentii	16, 20, 21, 23, 24, 26
magnus	23
macerans	23
skinneri	23, 30
uniguttulatus	23
Debaryomyces spp.	4, 14, 17, 19, 22
castellii	27
hansenii	1, 2, 5, 9, 10, 15, 17, 23, 27, 28, 29, 30
hansenii var. *fabryi*	5, 9, 10, 12
hansenii var. *hansenii*	1, 3, 5, 9, 10, 12, 17
marama	23, 30
polymorphus	30
vanrijiae	29
Dipodascus capitatus	25
Filobasidium capsuligenum	23
Geotrichum candidum	17
fermentans	25
ingens	23
klebahnii	25
Hyphopichia burtonii	25
Kloeckera apiculata	27
Leucosporidium scottii	23
Metschnikowia pulcherrima	27
Pichia spp.	22
anomala	13
angusta	23
carsonii	23
ciferrii	28
etchellsii	23
farinosa	13
haplophila	10
holstii	28
media	23
membranaefaciens	23
subpelliculosa	12
sydowiorum	28
Rhodotorula spp.	7, 8, 14, 17, 22, 24
aurantiaca	6
buffonii	20
foliorum	23

Table 3.4 Continued

	References (see footnote to table)
glutinis	12, 16, 20, 23, 25, 28, 29
graminis	23, 25
minuta	10, 20, 21, 23, 25, 26
mucilaginosa	6, 10, 12, 13, 16, 17, 20, 21, 23, 25, 26, 29
Saccharomyces cerevisiae	10
Sporobolomyces spp.	14
roseus	10
shibatanus	10
tsugae	23
Sporothrix catenata	23, 25
Torulaspora globosa	17
Trichosporon spp.	14, 19, 22
beigelii	16, 17, 20, 21, 23, 25, 26, 29
pullulans	5, 6, 10, 12, 20, 21, 24, 25, 30
Zygosaccharomyces bailii	12
rouxii	25, 29

1 Sausages; Césari and Guilliermondii (1920).
2 Weiner sausages; Mrak and Bonar (1938).
3 Meat brines; Costilow *et al.* (1954).
4 Luncheon meats; Wickerham (1957).
5 Frankfurters; Drake *et al.* (1959).
6 Antibiotic treated poultry; Walker and Ayres (1959).
7 Beef steaks; Ayres (1960).
8 Dried beef; Frazier (1967).
9 Cured meats; Bem and Leistner (1970).
10 Cured meat, ham, sausages; Leistner and Bem (1970).
11 Spanish sausage; Ramírez and González (1972).
12 Sausages, savoloy, meat, black puddings, pies; Aboukheir and Kilbertus (1974).
13 Tea sausages; Zivanović and Ristić (1974).
14 Serwolatka (fresh pork sausages); Szczepaniak *et al.* (1975).
15 Cured meats; Smith and Hadlok (1976).
16 Sausages; Abbiss (1978).
17 Dry sausages; Comi and Cantoni (1980a,b).
18 Sausages, ham; Staib *et al.* (1980).
19 Sausages; Banks (1983).
20 Ground beef; Hseih and Jay (1984).
21 Minced beef radurized; Johannsen *et al.* (1984).
22 Minced beef; Nychas (1984).
23 Sausages and minced beef; Dalton (1984); Dalton *et al.* (1984).
24 Lamb; Lowry (1984), Lowry and Gill (1984a).
25 Ground beef; Comi and Cantoni (1985).
26 Minced lamb and minced lamb products; Dillon (1988); Dillon *et al.* (1991).
27 Italian Salami; Grazia *et al.* (1989).
28 Dry cured ham; Molina *et al.* (1990).
29 Italian fermented sausages; Buzzini and Haznedari (1995).
30 Greek Dry Salami; Metaxopoulos *et al.* (1996).

Where both perfect and imperfect states have been isolated: *Deb. hansenii* (IMP = *C. famata*); *Pichia fermentans* (IMP = *C. lambica*); *Pichia guilliermondii* (IMP = *C. guilliermondii*); *Pichia anomala* (IMP = *C. pelliculosa*); *Pichia membranaefaciens* (IMP = *C. vallida*); *Metschnikowia pulcherrima* (IMP = *C. dattila*).

Table 3.5 Moulds associated with meat and meat products

	References (see footnote to table)
Acremonium spp.	7, 12
strictum	16
Actinomucor elegans	16
Alternaria spp.	1, 2, 3, 4, 5, 6, 14
Alternaria alternata	16
Aspergillus spp.	1, 2, 3, 4, 5, 6, 7, 13, 15
aureolatus	16
candidus	10
clavatus	10
flavus	10, 14, 16
flavipes	14
fumigatus	8, 10, 14
glaucus	5, 10, 14
nidulans	14
niger	5, 10, 14, 16
ochraceus	10, 14
penicilloides	10
repens	5, 10
restrictus	5, 10, 14
ruber	5, 10
sydowi	9, 10, 16
tamarii	8, 10
terreus	14, 16
ustus	16
versicolor	9, 10, 14, 16
viridi-nutans	10
wentii	5, 10, 14
Aurobasidium pullulans	10, 11
Botryotrichum atrogriseum	16
Botrytis spp.	1, 2
cinerea	16
Chrysosporium pannorum	10, 12
Cladosporium spp.	2, 3, 4, 5, 6, 13, 14, 15
herbarum	2, 10, 11
cladosporioides	10, 11
sphaerospermum	16
Emericella nidulans	16
Epicoccum spp.	5
nigrum	16
Eurotium spp.	7, 15
amstelodami	5, 9
chevalierii	5, 9
repens	5, 8, 9
rubrum	5, 8, 9
Fusarium spp.	1, 5
oxysporum	16
Monilia spp.	1, 2, 4, 10, 14
Mortierella spp.	5, 10
Mucor spp.	1, 2, 3, 4, 5, 7, 13
mucedo	10
racemosus	10, 12, 16
Neurospora sitophila	10
Paecilomyces spp.	5, 14
lilacinus	16

Table 3.5 Continued

	References (see footnote to table)
Paecilomyces variotti	16
Penicillium spp.	1, 2, 3, 4, 5, 6, 9, 13
aurantiogriseum	7, 8, 10, 16
chrysogenum	5, 8, 15, 16
commune	8, 15
corylophilum	12
duclauxii	16
expansum	5, 10
glabrum	10
hirsutum	10, 11
islandicum	16
miczynskii	7
nalgiovense	15
oxalicum	10, 15, 16
pinophilum	16
simplicissimum	5
spinulosum	10
variable	10, 16
verrucosum	15
waksmanii	16
Phoma herbarum	16
Phycomyces spp.	2
Pullularia spp.	14
Rhizopus spp.	1, 2, 4, 5, 13
nigricans	3, 10
stolonifer	16
Scopulariopsis spp.	5, 10
candida	16
Scytalidium lignicola	16
Syncephalastrum spp.	5
Thamnidium spp.	2, 4, 13
elegans	10, 12
Trichoderma spp.	3
Ulocladium chartarum	16
Verticillium spp.	2
Wallemia spp.	15

1 Refrigerated meats; Jensen (1954).
2 Refrigerated meats; Ayres (1955).
3 Chicken meat; Njoku-Obi *et al.* (1957).
4 Beef; Ayres (1960).
5 Cured hams, fermented sausages; Leistner and Ayres (1968).
6 Country cured hams; Sutic *et al.* (1972).
7 Sausages; Takatori *et al.* (1975).
8 Meat products; Hadlok *et al.* (1976).
9 African Biltong; van der Riet (1976).
10 Meat and meat products; see Jay (1978, 1979, 1987).
11 Lamb; Gill and Lowry (1981).
12 Lamb; Lowry and Gill (1984b).
13 Refrigerated meats; see Pestka (1986).
14 Spanish dry-cured hams; Rojas *et al.* (1991).
15 Naturally moulded sausages; Anderson (1993).
16 Beef; Nassar and Ismail (1994).

Yeasts were a minor component of the microbial population present in field samples; they represented only a small proportion of the total microflora of grass and turnips (0.02–2.35%), soil (0.01–2.6%) and faecal material (0.004–0.14%) sampled from sheep pastures in the United Kingdom (Dillon and Board, 1989b). Similarly, it can be calculated from the results of Empey and Scott (1939) in Australia, that yeasts represented only 0.045–0.22% with incubation at 20 °C and 0.35–0.5% at −1 °C of the microbial population of soil and faecal samples.

Yeast cells and mould spores are easily dispersed by air (do Carmo-Sousa, 1969; Smith and Anderson, 1992; Refai *et al.*, 1993) and the types and prevalence depend on season, prevailing weather and geographical location (Smith and Anderson, 1992). *Cryptococcus* spp. represented 42% of the yeasts isolated from air in New Zealand; 26.2% were *Debaryomyces* spp. and 18.6% were pigmented yeast species belonging to the *Sporobolomyces-Rhodotorula* group (di Menna, 1955b). The mould species most common in the air in New Zealand were species of *Cladosporium* and *Penicillium* in association with other meat-borne moulds such as *Aspergillus*, *Alternaria* and *Epicoccum* (di Menna, 1955b). The yeast flora from air samples from animal paddocks in New Zealand was dominated by *C. saitoana* and *Rh. mucilaginosa* (Baxter and Illston, 1977).

3.2.2   The abattoir

The hide, hair or fleece of the live animal is the major route of microbial contamination in the abattoir (Empey and Scott, 1939; Ayres, 1955). The micro-organisms that are present will have originated from the field environment. Psychrotrophic yeasts such as *Cr. laurentii*, *Cr. luteolus*, *Rh. mucilaginosa* and *C. saitoana* were isolated from pastures (Table 3.1), from soil and air samples from paddocks (Baxter and Illston, 1977), and from fleece samples and carcass surfaces (Baxter and Illston, 1976, 1977). Lowry (1984) recorded that *Candida* spp. in association with *Cryptococcus* spp. were dominant on lamb carcasses in winter whereas *Rh. glutinis* was the major yeast species in the summer. These observations were correlated with the seasonal changes noted with the yeast flora on pasture plants in New Zealand by di Menna (1959).

If yeasts are transferred to the carcass from the soil-contaminated animal coats then the low percentage (< 5%) of yeasts evident in the microflora isolated from the field samples will also be reflected in microbial populations from the fleece and hides and subsequently from the carcass surfaces. The investigations by Empey and Scott (1939) supported this hypothesis. They recorded that yeasts accounted for only 0.26% of the microbial population from hides with incubation at −1 °C or 0.018% of the microflora with incubation at 20 °C. Similarly, yeasts represented < 5% of the microflora from the surface of the lamb carcass (Lowry, 1984; Dillon and Board, 1989b) and remained at this consistently low level throughout the processing procedure (Dillon and Board, 1989b). The carcass surface was probably contaminated by contact with the fleece when the latter was removed. Fleece samples from the field also showed that the yeast population was a minor component (0.002–12.67%) of the microbial load (Dillon and Board, 1989b). Carotenoid-pigmented yeasts, such as *Rh. mucilaginosa*, accounted for only 5% of the yeast flora of fleece samples from January to July and as much as 40% in October but the nonpigmented yeasts (*Candida* and *Cryptococcus* spp.) dominated the yeast flora throughout the year (Dillon *et al.*, 1991). Empey and Scott (1939), noted however, that although pigmented yeasts such as *Rhodotorula* spp. accounted for as much as 50% of the yeast flora on hides they were rarely isolated from meat and hence were unimportant in meat spoilage.

Candida, Cryptococcus and *Rhodotorula* spp. have been recovered from the surface of lamb carcasses. *Cryptococcus curvatus, C. famata, Candida glabrata, C. mesenterica, Cr. albidus* var. *albidus, Cr. laurentii, Rh. minuta* and *Rh. mucilaginosa* were isolated from the surface of the lamb carcasses in a slaughterhouse (Dillon *et al.*, 1991). The yeast flora, however, was dominated by the nonpigmented yeasts and the carotenoid yeasts were only present in small numbers (Dillon *et al.*, 1991). *Rhodotorula mucilaginosa* and *C. saitoana* were isolated from fleece samples and the surface of lamb carcasses by Baxter and Illston (1976).

Various items of abattoir equipment such as knives, work surfaces, cutting boards, boning tables, conveyors as well as carcass-washing water, slaughtermen's hands and aprons became contaminated directly from fleece, particularly when wet, or by aerosols (Patterson, 1968; Leach, 1971; Miller, 1979; Nottingham, 1982; Dainty *et al.*, 1983). Yeasts, however, were only a minor component of the total microbial population on slaughtermen's hands and clothing (Dillon and Board, 1989b; Nortjé *et al.*, 1990). *Candida mesenterica, Cr. albidus* var. *albidus* and *Rh. mucilaginosa* were present on slaughtermen's aprons (Dillon *et al.*, 1991). These three species were also isolated from lamb carcasses, suggesting that a possible route of cross-contamination exists via hands and equipment. *Candida guilliermondii* (the imperfect state of *Pichia guilliermondii), Rh. mucilaginosa* and *C. saitoana* were recovered from walls, floors and benches in a meat processing plant (Baxter and Illston, 1976). Species of *Cryptococcus, Rhodotorula, Candida* and *Trichosporon* were isolated from the slaughter area and the lairage in a pig processing plant (Dalton, 1984). The major yeasts recovered from a Vienna sausage processing plant were species of *Candida* and *Debaryomyces* in association with *Yarrowia lipolytica* and species of *Rhodotorula Pichia, Galactomyces, Cryptococcus, Trichosporon* and *Torulaspora* (Viljoen *et al.*, 1993).

Moulds are ubiquitous in nature and are easily transferred to the meat surface. Moulds that are commonly associated with meat and meat products

(Table 3.5), such as species of *Acremonium, Alternaria, Aspergillus, Cladosporium, Epicoccum* and *Penicillium* have been isolated from air in New Zealand by di Menna (1955b), from the abattoir environment and from carcass surfaces (Table 3.3). Empey and Scott (1939) isolated species of *Alternaria, Cladosporium, Mucor, Penicillium* and *Thamnidium* from the surface of beef carcasses. *Penicillium commune, Cladosporium herbarum, Thamnidium elegans, Thamnidium chaetocladioides* and *Chry. pannorum* were identified (Empey and Scott, 1939). *Cladosporium herbarum* and *Cladosporium cladosporioides* were isolated from air and soil in stockyards, the slaughter area, chiller and freezer rooms in a processing plant in New Zealand and from cowhair, fleece and carcass surfaces (Table 3.3; Baxter and Illston, 1976, 1977) as well as meat products (Table 3.5). The floors, walls, utensils and particularly air in the abattoir in Egypt were major sources of meat contamination by moulds (Refai *et al.*, 1993). Moulds, such as *Aspergillus* spp. were isolated from the air and surroundings of the slaughter area and processing halls of an abattoir, from carcasses (Table 3.3) and meat (Table 3.5; Refai *et al.*, 1993). The mould population of fresh and chilled carcasses was dominated by species of *Aspergillus* (*Asp. niger, Asp. flavus, Asp. ochraceus, Asp. terreus* and *Asp. parasiticus*) and *Penicilliu* spp. (Refai *et al.*, 1993).

3.3 Mycoflora of meat and meat products

Meat provides an ideal environment for microbial growth as it has an optimum pH (5.8–6.8), a high water content (a_w = 0.99), a rich supply of nitrogenous substances and a source of carbohydrates and essential growth factors, such as minerals and vitamins (Lawrie, 1985). Bacteria are usually the major spoilage organisms of meat under normal conditions of processing and storage particularly at chill temperatures. Thus in fresh ground beef there is usually a large population of Gram-negative bacteria (2×10^3–6.6 $\times 10^7$/g) associated with low numbers (2×10^1–6.2 $\times 10^4$/g) of yeasts (Jay and Margitic, 1981). This is mainly due to the rapid growth rate of bacteria enabling them to outcompete yeasts.

3.3.1 *Chill storage*

The storage conditions imposed allow only a few of the initial contaminants of meat to proliferate (Mossel, 1971; McMeekin, 1982). Temperature is the most important selective factor influencing the microbial colonization of meat (Ayres 1960; Nottingham, 1982). Spoilage micro-organisms can grow at a wide range of temperatures from −5 °C to +70 °C (Mossel and Ingram, 1955). The widespread use of refrigeration at all stages of meat production, however, selects those micro-organisms that grow well at low temperatures

(0–7 °C). Due to their rapid growth these dominate the microflora (Ingram and Dainty, 1971; Gill and Newton, 1977).

Candida, Torulopsis (mainly reclassified as *Candida* spp.) and *Rhodotorula* species were recovered from refrigerated beef by Ayres (1960). In addition to these Hseih and Jay (1984) also found *Trichosporon* and *Cryptococcus* spp. *Candida* spp. however, were dominant accounting for 82% of the yeast flora with *C. lipolytica* and *C. lambica* (the imperfect state of *Pichia fermentans*) being predominant on fresh beef whereas *C. lipolytica* and *C. zeylanoides* dominated at spoilage (Hseih and Jay, 1984). Notably, with the exception of one sample, only *Candida* spp. were isolated from spoiled beef (Hseih and Jay, 1984). In another study, *Candida* spp. accounted for 60% of the yeast flora in minced beef association with *Cryptococcus* spp. (10%), *Rhodotorula* spp. (3%) and a low incidence of *Trichosporon* spp., *Debaryomyces* spp. and *Pichia* spp. (Nychas, 1984). Similarly, Comi and Cantoni (1985) found that the yeasts that dominated on fresh and refrigerated ground beef were species of *Candida* (including species of *Torulopsis*), *Cryptococcus*, *Rhodotorula* and *Trichosporon*. Moulds such as species of *Rhizopus*, *Mucor*, *Thamnidium*, *Monilia*, *Aspergillus*, *Penicillium*, *Cladosporium* and *Alternaria* were also isolated from beef by Ayres (1960).

Candida spp. accounted for 73% of the yeast flora of minced lamb; 21% were *Cryptococcus* spp. and 6% were *Rhodotorula* spp. (Dillon *et al.*, 1991). *Candida famata*, *C. inconspicua*, *C. lipolytica* (the imperfect state of *Y. lipolytica*), *C. mesenterica*, *C. norvegica*, *C. sake*, *C. vini*, *C. zeylanoides*, *Cr. albidus* var. *albidus*, *Cr. humicolus*, *Cr. infirmo-miniatus*, *Cr. laurentii*, *Rh. minuta* and *Rh. mucilaginosa*, were also isolated (Dillon *et al.*, 1991). *Trichosporon beigelii*, however, was only recovered from one sample of minced lamb obtained from a supermarket (Dillon *et al.*, 1991). Notably, *C. famata*, *Cr. albidus* var. *albidus* and *Rh. mucilaginosa* were also isolated from field samples and the carcass surface at the abattoir (Dillon *et al.*, 1991).

Cold temperature storage (−5 °C) may be a factor that selects a flora dominated by yeasts or moulds. Initially, yeasts represented less than 0.1% (10 organisms/cm^2) of the microflora of lamb loins wrapped in an oxygen-permeable plastic film, but after 20 weeks storage at −5 °C they reached a density of 10^6/cm^2 (Lowry, 1984; Lowry and Gill, 1984a). Considering that 10^6 yeasts/cm^2 is equivalent to the biomass of 10^8 bacteria/cm^2, the level required to initiate superficial spoilage in chilled meat (Gill and Newton, 1978), it is surprising that no off-odours or flavour defects were detectable (Winger and Lowry, 1983). *Cryptococcus laurentii* represented 90% of the yeast flora in association with *Cr. infirmo-miniatus*, *Tr. pullulans* and *C. zeylanoides* (Lowry, 1984; Lowry and Gill, 1984a).

Yeasts can alter their intracellular fatty acid composition in response to the incubation temperature, resulting in a higher level of polyunsaturated fatty acid residues at lower temperatures particularly in psychrotrophic yeasts. This may protect them from membrane damage (Watson *et al.*, 1976; Viljoen *et al.*, 1993). This characteristic, together with tolerance of low and intermediate a_w and resistance to preservatives, may enable yeasts to compete well at low temperatures (Guerzoni *et al.*, 1993). *Debaryomyces hansenii*, for example, exhibits a high growth rate at low temperatures and at intermediate a_w levels (Guerzoni *et al.*, 1993).

After 40 weeks of storage at -5 °C mould growth in the form of black spot or white spot was only barely visible on the surface of lamb (Lowry and Gill, 1984a). The growth of moulds on the meat surface is only evident when the temperature exceeds 0 °C but is restricted at -5 °C (Lowry and Gill, 1984b). Mould spoilage of frozen meat occurs because the growth of bacteria is curtailed by desiccation (Lowry and Gill, 1984b). Spoilage by black spot was believed to be caused by a single species, *Clad. herbarum* (Brooks and Hansford, 1923). Gill and Lowry (1981, 1982), however, noted that spoilage by black spot was caused by *Clad. cladosporioides*, *Clad. herbarum*, *Penicillium hirsutum* and *Aureobasidium pullulans*. Of these *Pen. hirsutum* was on the surface only whereas the other moulds penetrated into the meat tissue (Gill and Lowry, 1981). Mould spoilage is also manifested as white spot caused by *Chry. pannorum* or *Acremonium* sp., whisker colonies caused by *Tham. elegans* or *Mucor racemosus* and blue-green colonies caused by *Penicillium corylophilum* (Lowry and Gill, 1984b). These xerotolerant moulds can grow at low temperatures in association with low a_w. *Thamnidium elegans*, for example, grows at -10 °C with an a_w of 0.94 whereas *Clad. herbarum* grows at -5.5 °C with an a_w of 0.88 and *Pen. corylophilum* grows at -2 °C with an a_w of 0.88 (Gill and Lowry, 1982; Lowry and Gill, 1984b; Gill, 1986). The main psychrotrophic moulds found on imported frozen beef in Egypt were species of *Cladosporium* and *Penicillium* and the mesophilic mycoflora was dominated by species of *Aspergillus*, *Cladosporium* and *Penicillium* (Nassar and Ismail, 1994). Similarly, species of *Aspergillus, Penicillium* and *Mucor* were isolated from raw beef from a Nigerian market (Okodugha and Aligba, 1991).

3.3.2 Processed meat products

The processing of meat selects a restricted range of yeast species (Deák and Beuchat, 1987; Jay, 1987; Fleet, 1990). Thus, *Candida, Cryptococcus, Rhodotorula* and *Trichosporon* spp. (Table 3.4) occur most frequently on processed meat products (Deák and Beuchat, 1987; Jay, 1978, 1987).

Kühl (1910) isolated yeasts from the surface slime of dried sausages and Cary (1916) also noted their presence on sausages. In 1920 unnamed yeasts were described from slimy sausages by Césari and Guilliermond. These were later identified with *Deb. hansenii* and *Deb. kloeckeri* by Lodder and Kreger-van Rij (1952); the latter species was reclassified subsequently as *Deb. hansenii* var. *hansenii* (Lodder, 1970; Barnett *et al.*, 1990). Similarly,

Debaryomyces guilliermondii var. *nova zeelandicus* was isolated from slimy Wiener sausages by Mrak and Bonar (1938). This organism was also identified with *Deb. hansenii* by Lodder and Kreger-van Rij (1952). In another study, species of *Debaryomyces* able to assimilate nitrite were isolated from luncheon meat (Wickerham, 1957). *Candida* and *Debaryomyces* spp. (Table 3.4) are frequently isolated from fermented sausage, country cured ham (Leistner and Bem, 1970), salami and dry sausage (Comi and Cantoni, 1980a, b). More specifically, *Debaryomyces hansenii* and *C. saitoana* were isolated from cured meat products by Smith and Hadlok (1976). *Debaryomyces membranefaciens* var. *hollandicus* (film-forming) and non-film forming *Deb. kloeckeri* (both now reclassified as *Deb. hansenii* var. *hansenii*) were isolated from meat brines (Costilow *et al.*, 1954).

Živanović and Ristić (1974) found that *Candida* spp. dominated the microflora of tea sausages. *Candida iberica* (reclassified as *C. zeylanoides*, Barnett *et al.*, 1990) was first isolated from 'salchichón' (Spanish sausage) by Ramírez and González (1972). Another study showed that *Candida melini* and *Rh. mucilaginosa* were the most commonly isolated yeast species from sausages (Aboukheir and Kilbertus, 1974). The yeast flora isolated from 'Serwolatka' (Polish fresh pork sausages) was mainly composed of *Candida* and *Debaryomyces* spp., although *Brettanomyces* (the imperfect state of *Dekkera* spp.), *Cryptococcus*, *Rhodotorula*, *Sporobolomyces* and *Trichosporon* spp. (Table 3.4) were also present (Szczepaniak *et al.*, 1975). Staib *et al.* (1980) recovered pathogenic *Candida* spp., such as *C. parapsilosis* and *C. tropicalis* from Bologna type sausage, salami sausage and smoked ham.

The reduced water activity of processed meats (Jay, 1978, 1979; Beuchat, 1983) and cured meats (Bem and Leistner, 1970) suppressed the growth of Gram-negative bacteria that can tolerate an a_w of 0.94–0.97 (Scott, 1957; Troller and Christian, 1978). Thus yeasts flourish as they can grow under conditions of lower water content down to an a_w of 0.62 (Corry, 1978). Yeast domination of cured meat flora was by those species that tolerated the lowest water activity (Bem and Leistner, 1970). Of the yeasts isolated from dry cured ham 43% were *Pichia ciferrii*, 36% *P. holstii*, 21% *P. sydowiorum*, 19% *Rh. glutinis*, 9% *Cr. albidus* and 5% *Deb. hansenii* (Molina *et al.*, 1990). Yeasts associated with African Biltong (dried meat) were *C. zeylanoides*, *Deb. hansenii* and *Tr. beigelii* (van der Riet, 1976).

Filamentous moulds can also thrive in conditions of low water content and grow at a_w levels ranging from 0.65 to 0.96 (Mossel and Ingram, 1955; do Carmo Sousa, 1969; Jay, 1992; Corry, 1978). The moulds that are the most xerotolerant are species of *Aspergillus*, *Penicillium* and *Eurotium*. Notably these are the three genera that are important in meat spoilage. *Aspergillus* and *Penicillium* species tolerate an a_w of 0.80 whereas *Eurotium* spp. tolerate an a_w of 0.62–0.70 (Leistner and Rodel, 1976). The low a_w that permits mould growth, however, is not high enough for spore formation and germination or for mycotoxin production (Beuchat, 1983). The common moulds isolated from American country cured hams by Leistner and Ayres (1968) were species of *Aspergillus*, *Penicillium* in association with *Cladosporium*, *Rhizopus* and *Alternaria* spp. In the early stages of ripening the mycoflora was dominated by species of *Penicillium* but *Aspergillus* spp. predominated on older hams probably because of their greater tolerance of low a_w (Leistner and Ayres, 1968; Gardner, 1983). In fact, *Aspergillus* spp. grew better in country cured hams with a low a_w than in the higher a_w fermented sausages where *Penicillium* spp. predominated (Leistner and Ayres, 1968). The dominant moulds on the surface of Spanish country cured hams were also species of *Aspergillus* (*Asp. glaucus*, *Asp. fumigatus*, *Asp. niger* and *Asp. flavus*) and *Penicillium* (Rojas *et al.*, 1991). African Biltong (dried meat) yielded species of *Eurotium* (*Eurotium amstelodami*, *Eurotium chevalieri*, *Eurotium repens* and *Eurotium rubrum*), *Asp. versicolor*, *Asp. sydowii* and *Penicillium* spp. (van der Riet, 1976).

Bacteria are inhibited at a pH below 5.5 but the growth of acid tolerant yeasts is enhanced (Ingram, 1958; Walker and Ayres, 1970; Jay, 1978, 1979, 1987). Thus the resulting low pH of meat products fermented by lactic acid bacteria, such as thuringer, summer sausage, pepperoni, cervelat and Genoa salami (Johnston and Elliot, 1976) favours the growth of yeasts. Isolation of yeasts from Italian fermented sausages revealed (Buzzini and Haznedari, 1995) that 50% were *Deb. hansenii*, 12.1% *Deb. vanrijiae*, 11.2% *C. zeylanoides*, 6.9% *Candida apis*, 4.3% *Rh. glutinis*, 8.6% *Rh. mucilaginosa*, 4.3% *Zygosaccharomyces rouxii* and 2.6% were *Tr. cutaneum* (reclassified as *Tr. beigelii*, Barnett *et al.*, 1990). Similarly, 82% of the yeasts isolated from Italian salami were *Deb. hansenii* and 2% were *Deb. castellii* (Grazia *et al.*, 1989). In addition to these species Grazia *et al.* (1989) noted that 2% were *Metschnikowia pulcherrima* (the perfect state of *Candida dattila*), 5% were *Kloeckera apiculata*, 8% were *Candida* spp. and 1% others. Traditional Greek dry salami yielded a yeast flora with a similar composition where 66% of the population were species of *Debaryomyces* (48% *Deb. hansenii*, 16% *Deb. marama* and 2% *Deb. polymorphus*) (Metaxopoulos *et al.*, 1996). The remaining flora consisted mainly of *Candida* spp. (7% *C. famata*, 6% *C. zeylanoides*, 6% *C. guilliermondii*, 5% *C. parapsilosis* and 2% *C. kruisii*) in association with 3% *Cryptococcus albidus* var. *albidus*, 2% *Cr. humicolus*, 1% *Cr. skinneri* and 2% *Tr. pullulans* (Metaxopoulos *et al.*, 1996).

The resulting low pH of fermented meat products also promotes the growth of moulds. *Aspergillus* spp. and *Penicillium* spp. are the most common moulds isolated from fermented sausages and cured meats (Leistner and Ayres, 1968; Takatori *et al.*, 1975; Hadlok *et al.*, 1976). Species of *Penicillium*, *Scopulariopsis*, *Aspergillus* and *Rhizopus* were isolated from fermented sausages by Leistner and Ayres (1968). The mycoflora of naturally moulded sausages consists mainly (96%) of *Penicillium* spp. (*Pen.*

nalgiovense, *Pen. chrysogenum, Pen. veriucosum, Pen. oxalicum* and *Pen. commune*) in association with species of *Aspergillus, Eurotium, Cladosporium, Wallemia* and yeasts (Anderson, 1993). *Penicillium nalgiovense*, that represented 50% of the mycoflora, can be used as a starter culture in fermented sausages (Anderson, 1993).

3.3.3 Chemical additives

Additives such as selective antibiotics inhibit bacterial growth in meat and hence may favour the growth of yeasts. Such was the case with poultry meat treated with chlortetracycline, oxytetracycline or tetracycline (Ayres *et al.*, 1956; Njoku-Obi *et al.*, 1957; Wells and Stadelman, 1958; Walker and Ayres, 1959). The yeast population increased from 10^4–10^5/cm^2 on untreated poultry meat to 10^5–10^8/cm^2 on that treated with 10 ppm oxytetracycline or chlortetracycline (Wells and Stadelman, 1958). The dominant yeasts of untreated or antibiotic treated poultry meat were *Rhodotorula* spp. (80%) in association with *Candida* and *Cryptococcus* spp. (Wells and Stadelman, 1958). Species of *Trichosporon, Candida* and *Rhodotorula* were recovered from poultry meat treated with chlortetracycline, oxytetracycline or tetracycline (Walker and Ayres, 1959). *Saccharomyces cerevisiae* and *Saccharomyces dairensis* dominated on chlortetracycline (20 ppm) treated poultry whereas *Rh. minuta* and *Geotrichum candidum* (the imperfect state of *Galactomyces geotrichum*) dominated on untreated chicken meat (Njoku-Obi *et al.*, 1957). Species of *Penicillium, Cladosporium, Trichoderma* and *Rhizopus nigricans* were also present on chlortetracycline (20 ppm) treated chicken meat (Njoku-Obi *et al.*, 1957).

Sulfite is used as an antimicrobial agent at a final concentration of 450 μg SO$_2$/g (Anon., 1974) in uncooked comminuted meat products containing cereal, such as sausages, in the United Kingdom (Kidney, 1974; Wedzicha, 1984). The preservative effect, however, is more efficient at an acid pH of 2.8–4.2 as found in wines (Ough and Crowell, 1987) than at the pH of 5.8–6.8 associated with meat products. At higher pH levels the molecular sulfur dioxide concentration falls and there is a concomitant increase in bisulfite and sulfite ions (King *et al.*, 1981). The molecular sulfur dioxide is probably the only form that enters the yeast cell and causes inhibition of enzyme activity, depletes ATP and disrupts the proton gradient (Dillon and Board, 1990). Sulfite inhibition, however, is overcome by fermentative yeasts through their ability to detoxify the preservative by the production of acetaldehyde (Dillon and Board, 1990).

Sulfite inhibits the growth of wild yeasts, moulds and bacteria, thereby selecting the fermentative yeasts desired in wines and ciders (Cruess, 1912). In the case of meat products, the growth of the Gram-negative pseudomonads, the major spoilage bacteria, is repressed by sulfite (Richardson, 1970; Banks and Board, 1981; Banks *et al.*, 1985). Since yeasts are tolerant of the preservative (Brown, 1977; Banks, 1983) and the bacterial competition is reduced (Dalton *et al.*, 1984), a dominant yeast flora is established. The resulting yeast flora is directly responsible for acetaldehyde production and its associated sulfite binding, similar to the situation that occurs in fermented wines (Dalton, 1984; Dillon and Board, 1989a, 1990). As only free sulfite is antimicrobial (Burroughs and Sparks, 1964) this feature plays an important role in allowing the spoilage bacteria to proliferate.

The sulfited British fresh sausage is commonly contaminated by yeasts (Dowdell and Board, 1968). The yeast population ranges from 10 to 2.4 $\times$ 10^5/g of sausage and probably plays a major role in spoilage (Dowdell and Board, 1968, 1971). The thick yellow–green film evident on skins of stale sausages was found to be caused by yeasts (Dowdell and Board, 1971). When the microflora of sausages was dominated by yeasts, *C. saitoana* was the major species whereas *Tr. beigelii* was the main contaminant on sausages dominated by *Brochothrix thermosphacta* (Abbiss, 1978). Species of *Candida* were the main yeasts identified from sausages by Banks (1983). The most common species on unsulfited and sulfited sausages and minced beef, however, were *Deb. hansensii* followed by *C. zeylanoides* and *Pichia membranaefaciens* (the perfect state of *Candida valida*) (Dalton *et al.*, 1984). Furthermore, the sulfite present in sausages favoured the growth of *Deb. hansenii* and *Candida* spp. thereby reducing the proportion of *Cryptococcus* and *Rhodotorula* spp. (Dalton *et al.*, 1984). Notably, *Deb. hansenii, C. zeylanoides, P. membranaefaciens* and *C. saitoana* were able to produce acetaldehyde that bound the majority of the available sulfite (Dalton, 1984). In contrast, *Cr. albidus* and *Rh. mucilaginosa* did not produce acetaldehyde which may account for their reduced numbers in sulfited sausages (Dalton, 1984).

The growth of the Gram-negative bacteria, particularly pseudomonads and Enterobacteriaceae, was inhibited by sulfite in minced lamb, thus favouring the growth of yeasts to such an extent that they dominated the microflora (Dillon and Board, 1989b). In contrast, yeasts were < 5% of the total microflora on unsulfited minced lamb, reflecting the ratio of yeasts to bacteria that were recovered from field samples and carcass surfaces (Dillon and Board, 1989b). Yeasts, however, were able to dominate the microflora, representing 179.7% compared to bacterial numbers, after 4 d storage at 5 °C when sulfite was added to the minced lamb (Dillon and Board, 1989b).

Sulfited minced lamb products such as grills, burgers or sausages contained yeasts able to produce acetaldehyde which were therefore potential sulfite binders (Dillon and Board, 1989a, 1990). An acetaldehyde producing yeast, *C. norvegica*, was associated with bound sulfite, whereas *C. vini* tolerated the preservative although it was a non-binder (Dillon and Board, 1990). The production of acetaldehyde by *C. norvegica* was sulfite-induced and occurred in the exponential phase of growth in sulfited (500 μg SO_2/ml) lab-lemco broth supplemented with glucose, fructose or ethanol and buffered at pH 5, 6 or 7. *Candida vini* grew in sulfited (500 μg SO_2/ml) glucose or lactate lab-lemco broth buffered at pH 6 or 7 although it did not produce acetaldehyde (Dillon and Board, 1990). This suggests that there are at least two mechanisms by which meat yeasts exhibit resistance to this preservative.

Inclusion of lactic acid as a preservative in cured meats (ham, bacon, corned beef, turkey, salami) also inhibits bacteria and favours the growth of yeasts (Houtsma *et al.*, 1993). In fact, *Deb. hansenii, Candida* spp. and *Rh. mucilaginosa* were resistant to sodium lactate at concentrations of 700–1300 mM (Houtsma *et al.*, 1993). Yeasts such as *P. membranefaciens* are tolerant of acetic acid (Guerzoni *et al.*, 1993) and *Sacch. cerevisiae* tolerates 500 ppm sorbic acid (Neves *et al.*, 1994). Moulds are also resistant to sorbic acid and degrade it to 1,3-pentadiene (Marth *et al.*, 1966).

3.3.4 Physical factors

There are a number of studies indicating that gamma-irradiation reduces bacterial numbers and therefore microbial competition and so promotes yeast and mould proliferation. The microbial cell may be destroyed by irradiation either by damage to cellular DNA (Silverman and Sinskey, 1977), by inactivation of multiple targets within the cell (O'Neill *et al.*, 1991) or by the effect on DNA repair mechanisms (Davies *et al.*, 1973). In gamma-irradiated (2.5 kGy) minced beef the numbers of bacteria were reduced, whereas those of psychrotrophic yeasts were unaffected initially and increased subsequently after 14 days storage at 4 °C (Johannsen *et al.*, 1984). Indeed, the surface flora of irradiated (2–5 kGy) frankfurters consisted mainly of *Candida catenulata, C. zeylanoides, Candida* spp., *Deb. hansenii* var. *fabryi, C. saitoana* and *Tr. pullulans* (Drake *et al.*, 1959). Irradiation, however, eliminated *Debaryomyces* spp. more readily than *Candida* spp. (Drake *et al.*, 1959). *Yarrowia lipolytica* (the perfect state of *C. lipolytica*), *C. zeylanoides* and *Tr. beigelii* were isolated from irradiated poultry meat whereas yeasts were not detected in unirradiated meat (Sinigaglia *et al.*, 1994). The resistance of *Y. lipolytica* was particularly evident as high numbers of this organism were recovered post-irradiation (Sinigaglia *et al.*, 1994). The yeast population on untreated crab meat was low (1×10^3/g) but, again following irradiation (4 kGy), the yeast flora exceeded 1×10^5/g in air-packed samples with an extended shelf-life at 0.5 °C and 5.6 °C (Eklund *et al.*, 1966). *Rhodotorula, Cryptococcus, Candida* and *Trichosporon* species and a yeast-like organism resembling *Aur. pullulans* were isolated from untreated or irradiated crab meat (Eklund *et al.*, 1965, 1966). *Trichosporon beigelii* and *C. zeylanoides* showed the greatest resistance to irradiation and survived in irradiated (3 kGy) sausages whereas the numbers of *Debaryomyces* spp. were reduced by 1.5 kGy irradiation (McCarthy and Damoglou, 1993). *Trichosporon beigelii*, however, did not persist in irradiated sulfited sausages whereas *C. zeylanoides* was resistant to the combined effect of these two preservative methods (McCarthy and Damoglou, 1993). In phosphate buffered saline *Tr. beigelii* exhibited the greatest resistance to irradiation with *C. zeylanoides* being less sensitive than *Deb. hansenii* or *Sp. roseus* (McCarthy and Damoglou, 1996). When these yeasts were irradiated in sausage meat a protective effect was evident at higher irradiation (> 2 kGy) doses possibly due to the meat proteins and polysaccharides competing to interact with the active radiolytic products of water (McCarthy and Damoglou, 1996). Species of *Fusarium* and *Alternaria* that have multicellular spores are more resistant to irradiation than *Aspergillus* and *Penicillium* spp. that have unicellular ones (O'Neil *et al.*, 1991).

Since the studies of Hite (1899) high hydrostatic pressure has been considered as a means to preserve foods. The inactivation of micro-organisms may occur through enzyme denaturation, conformational change of membranes and the damaging effect on genetic material (Isaacs *et al.*, 1995). Gram-positive are more resistant to this process than are Gram-negative bacteria (Earnshaw, 1995). The effect on micro-organisms depends on a multiplicity of factors; pressure, time duration, temperature, pH, a_w and composition of the substrate. The use of high hydrostatic pressure of 300 MPa at 25 °C or 250 MPa at 45 °C for 10 min inactivated *Sacch. cerevisiae* and *Zygosaccharomyces bailii* inoculated into spaghetti sauce with meat (Pandya *et al.*, 1995). In 0.1M citrate broths the lethality of high hydrostatic pressure on these yeasts was enhanced by mild heat treatment and increased acidity (Pandya *et al.*, 1995). A change in cellular fatty acid composition was noted when *Y. lipolytica* was exposed to pressure treatment, and was associated with an increase in extracellular proteases and a decline in viability (Lanciotti *et al.*, 1997).

3.4 Chemistry of meat spoilage

Initially the growth of bacteria on the meat surface occurs at the expense of low molecular weight compounds, such as glucose (Gill and Newton, 1977; McMeekin, 1982; Gill, 1986). The ease of glucose assimilation means that the enzymes required for metabolism of other substrates, such as lactic acid and amino acids, are under catabolite repression (Jacoby, 1964; Ornston, 1971). When bacteria on the meat surface attain a density of 10^8 cells/cm^2, the glucose concentration at the meat surface is exhausted (Gill, 1976). The subsequent breakdown of amino acids produces 'off-odours' consisting of malodorous sulphides, esters and amines (Dainty *et al.*, 1983, 1984; Edwards *et al.*, 1983) with a concomitant increase in pH associated with ammonia production (Jay and Kontou, 1967). Studies of the chemistry of yeast growth on the meat surface are not as detailed as those of bacteria. Collins and

Buick (1989) however, made similar observations with *Rh. glutinis* on the surface of frozen peas. The growth of *Rh. glutinis* on the frozen peas was at the expense of the available carbohydrates and the enzymes required for the breakdown of amino acids and lipids were under catabolite repression (Collins and Buick, 1989).

Notably, the total carbohydrate content decreased rapidly in peas stored > 5 °C compared to a slower decrease at lower temperatures and this corresponded to the relative growth rates of the yeast (Collins and Buick, 1989). Subsequent complete depletion of the carbohydrate content of the peas stored at < 5 °C caused production of the objectionable taint of 2-methyl-furan from leucine. Hexanol was also evident indicating lipid oxidation of C_{18} unsaturated fatty acids by lipoxygenases of *Rh. glutinis* (Collins and Buick, 1989).

The low molecular weight compounds such as free amino acids and nucleotides in red meat are degraded by bacteria in preference to more complex compounds and therefore have a sparing action on lipids and proteins (Jay and Kontou, 1967; Jay, 1972; Jay and Shelef, 1976; Gill and Newton, 1980). Further microbial breakdown of meat resulted in fat lipolysis and fatty acid rancidity (Lea 1931a, b) caused by the enzymic hydrolysis of yeasts (Vickery, 1936a, b; Ingram, 1958). In fact, yeasts may be able to compete successfully in the microflora due to their lipolytic activity (Aboukheir and Kilbertus, 1974) particularly in fatty regions of meat. Such yeasts as *Candida, Cryptococcus* and *Trichosporon* spp., particularly *Tr. pullulans* and *Candida scottii*, were isolated from crab meat and found in association with the spoilage of beef, and were noted to produce lipases (Jensen, 1954; Lodder *et al.*, 1958; Eklund *et al.*, 1965, 1966; Johannsen *et al.*, 1984). Additionally, species of *Rhodotorula* and *Trichosporon* isolated from Italian fermented sausages exhibited significant lipolytic and proteolytic activity, respectively (Buzzini and Haznedari, 1995). *Yarrowia lipolytica* is also strongly lipolytic and proteolytic (Guerzoni *et al.*, 1993; Sinigaglia *et al.*, 1994). Oleic acid resulted from the lipolytic activity of *C. lipolytica* (the imperfect state of *Y. lipolytica*) and that of *Tr. beigelii* resulted in the release of myristic acid when grown with Tween 80, glycerol tributyrate and mixed triglycerides of animal and vegetable origin (Alifax, 1979).

The growth of *C. famata* on snail meat indicated that it was capable of utilizing both the carbohydrate and protein content of the meat (Nwachukwu and Akpata, 1987). A decrease in carbohydrate content from 16% to 7% occurred and the protein content was reduced from 2.5% to 0.43% after 4 days storage at 29 ± 1 °C (Nwachukwu and Akpata, 1987). This resulted in a concomitant decrease in pH from 9.5 to 7.4 when spoilage was evident in the form of slime and off-odours (Nwachukwu and Akpata, 1987). Proteinase activity has been noted with *Trichosporon* spp. isolated from crab meat (Eklund *et al.*, 1965, 1966). *Candida, Debaryomyces, Rhodotorula, Cryptococcus, Trichosporon, Brettanomyces* and *Sporobolomyces* spp.

isolated from 'Serwolatka' (fresh pork sausage) were able to utilize carbohydrates (Szczepaniak *et al.*, 1975).

The improved flavour of moulded cured and aged meats is due partially to the proteolytic and lipolytic changes associated with mould growth. Moulded sausages dry in a slow uniform manner resulting in a reduced weight loss and an improved quality (Pestka, 1986). In fact, when whisker moulds are present on aged meats there is an increase in flavour and tenderness (Pestka, 1986). In addition it was shown that *Thamnidium* spp. release proteases that tenderize beef (Ingram and Dainty, 1971).

3.5 Conclusions

It is apparent that the route of contamination by yeasts and moulds mirrors that of bacteria, originating from the environment and arriving in the processing plant via the live animal. This route is suggested for *Candida*, *Cryptococcus* and *Rhodotorula* species and moulds such as *Aspergillus* and *Penicillium* spp. which are commonly isolated from the field, the abattoir, carcass surfaces and meat products.

Moulds and yeasts are opportunistic spoilage organisms, they may flourish when processing, preservation and storage conditions cause suppression of the major Gram-negative spoilage bacteria. The key role of yeasts in meat spoilage, however, is their indirect effect in permitting the growth of the spoilage bacteria by rendering a preservative inactive. Examples of this are the reduction of antimicrobial efficacy of sulfite via acetaldehyde production with associated sulfite binding and also the utilization of organic acids leading to an increased pH with a concomitant decrease in the preservative. Similarly, yeasts and moulds are also tolerant of sorbic acid, which moulds can degrade to 1,3-pentadiene (Marth *et al.*, 1966).

In mixed populations the proteolytic and lipolytic activities of yeasts and moulds may enable them to compete depending on the amount and the rate of enzyme production and penetration into the meat (Aboukheir and Kilbertus, 1974; Pestka, 1986). Yeasts and moulds such as *Penicillium* and *Aspergillus* spp. will proliferate when competing bacteria are restricted, and are therefore important in meat products with a low pH such as those fermented by lactic acid bacteria and processed or cured meats with a low water activity. Low temperatures can also lead to the selection of a microflora dominated by yeasts or moulds. Mould spoilage, in particular, can be a major cause of meat spoilage at chill temperatures. This occurs mainly at temperatures of 0 °C and above as their growth is restricted at −5 °C (Lowry and Gill, 1984b). Such moulds as *Pen. corylophilum* can grow at −2 °C with a low a_w of 0.88 (Gill and Lowry, 1982; Lowry and Gill, 1984b; Gill, 1986) in conditions that bacteria could not tolerate.

Finally, the ability of the yeasts and moulds to survive higher doses of irradiation compared to bacteria may lead to fungi playing a more direct role in meat spoilage in the future. Multicellular spores of *Fusarium* spp. and *Alternaria* spp. are more resistant to irradiation than are the moulds with unicellular spores such as *Aspergillus* spp. and *Penicillium* spp. (O'Neil *et al.*, 1991). Additionally, the use of high hydrostatic pressure to inhibit Gram-negative spoilage bacteria would reduce bacterial competition and may also favour the growth of yeasts and moulds, particularly those with multicellular fungal spores.

References

Abbiss, J.S. (1978) Enzyme-mediated changes in carbohydrate in the British fresh sausage. Ph.D. Thesis, University of Bath, UK.

Aboukheir, S. and Kilbertus, G. (1974) Fréquence des levures dans les denrées alimentaires a base de viande. *Annales de la Nutrition et de L'Alimentation*, **28**, 539–47.

Alifax, R. (1979) Étude de la lipolyse chez quelques levures isolées de denrées alimentaires variées. *Annales de Technologie Agricole*, **28**, 255–72.

Anderson, S.J. (1993) Potential mycotoxin producing moulds isolated from naturally mould-fermented sausages. *Int. Biodeterior. Biodegrad.*, **32**, 225.

Anonymous (1974) Preservatives in Food Regulations. Statutory Instrument 1119, HMSO, London.

Ayres, J.C. (1955) Microbiological implications in the handling, slaughtering and dressing of meat animals. *Adv. Food Res.*, **6**, 106–61.

Ayres, J.C. (1960) Temperature relationships and some other characteristics of the microbial flora developing on refrigerated beef. *Food Res.*, **25**, 1–18.

Ayres, J.C., Walker, H.W., Fanelli, M.J., King, A.W. and Thomas, F. (1956) Use of antibiotics in prolonging storage life of dressed chickens. *Food Technol.*, **10**, 563–8.

Banks, J.G. (1983) Sulphite preservation of British fresh sausage. Ph.D. Thesis, University of Bath, UK.

Banks, J.G. and Board, R.G. (1981) Sulphite: the elective agent for the microbial association in the British fresh sausage. *J. Appl. Bacteriol.*, **51**, ix.

Banks, J.G. and Board, R.G. (1987) Some factors influencing the recovery of yeasts and moulds from chilled foods. *Int. J. Food Microbiol.*, **4**, 197–206.

Banks, J.G., Dalton, H.K., Nychas, G.J. and Board, R.G. (1985) Review – Sulfite, an elective agent in the microbiological and chemical changes occurring in uncooked comminuted meat products. *J. Appl. Biochem.*, **7**, 161–79.

Barnett, J.A., Payne, R.W. and Yarrow, D. (1990) *Yeasts: Characteristics and Identification*, 2nd edn, Cambridge University Press, Cambridge.

Baxter, M. and Illston, G.M. (1976) Psychrotrophic meat spoilage fungi within a freezing works. *New Zealand Vet. J.*, **24**, 177–80.

Baxter, M. and Illston, G.M. (1977) Environmental reservoirs of psychrotrophic meat spoilage fungi. *New Zealand Vet. J.*, **25**, 165–7.

Bem, Z. and Leistner, L. (1970) Die wasseraktivitätstoleranz der bei Pökelfleischwaren vorkommenden hefen. *Fleischwirtschaft*, **50**, 492–3.

Beuchat, L.R. (1983) Influence of water activity on growth, metabolic activities and survival of yeasts and molds. *J. Food Protect.*, **45**, 135–41.

Brooks, F.T. and Hansford, C.G. (1923) Mould growth upon cold storage meat. *Food Invest. Board, Special Report 17*, HMSO: London.

Brown, M.H. (1977) Microbiology of the British fresh sausage. Ph.D. Thesis, University of Bath, UK.

Burroughs, L.F. and Sparks, A.H. (1964) The determination of the free sulphur dixoide content of ciders. *Analyst*, **89**, 55–60.

Buzzini, P. and Haznedari, S. (1995) Caratterizzazione di lieviti isolati da insaccati fermentati prodotti in Umbria. Valutazione preliminare della loro attività proteolitica e lipolitica. *Industrie Alimentari*, **34**, 620–5.

Cary, W.E. (1916) The bacterial examination of sausages and its sanitary significance. *Am. J. Pub. Health*, **6**, 124–35.

Césari, E.P. and Guilliermond, A. (1920) Les levures des saucissons. *Annales de l'Institut Pasteur*, Paris 34, 229–48.

Collins, M.A. and Buick, R.K. (1989) Effect of temperature on the spoilage of stored peas by *Rhodotorula glutinis*. *Food Microbiol.*, **6**, 135–41.

Comi, G. and Cantoni, C. (1980a) Flora blastomicetica superficiale di insaccati crudi stagionati. *Industrie Alimentari*, **19**, 563–9.

Comi, G. and Cantoni, C. (1980b) I Lieviti in insaccati crudi stagionati. *Industrie Alimentari*, **19**, 857–60.

Comi, G. and Cantoni, C. (1985) Lieviti e carni. *Industrie Alimentari* **24**, 683–7.

Corry, J.E.L. (1978) Relationships of water activity to fungal growth. In *Food and Beverage Mycology* (ed L.R. Beuchat), AVI Publishing Company, USA, pp. 45–82.

Costilow, R.N., Etchells, J.L. and Blumer, T.N. (1954) Yeasts from commercial meat brines. *Appl. Microbiol.*, **2**, 300–2.

Cruess, W.V. (1912) The effect of sulfurous acid on fermentation organisms. *J. Indust. Eng. Chem.*, 581–5.

Dainty, R.H., Edward, R.A. and Hibbard, C.M. (1984) Volatile compounds associated with the aerobic growth of some *Pseudomonas* species on beef. *J. Appl. Bacteriol.*, **57**, 75–81.

Dainty, R.H., Shaw, B.G. and Roberts, T.A. (1983) Microbial and chemical changes in chill-stored red meats. In *Food Microbiology: Advances and Prospects*, (eds T.A. Roberts and F.A. Skinner), Academic Press, London, pp. 151–78.

Dalton, H.K. (1984) The yeasts and their chemical changes in the British fresh sausage. Ph.D. Thesis, University of Bath, UK.

Dalton, H.K., Board, R.G. and Davenport, R.R. (1984) The yeasts of British fresh sausage and minced beef. *Antonie van Leeuwenhoek*, **50**, 227–48.

Davenport, R.R. (1973) Vineyard yeasts – an environmental study. In *Sampling: Microbiological Monitoring of Environments*, (eds R.G. Board and D.H. Lovelock), Academic Press, London, pp. 143–74.

Davenport, R.R. (1980a) An introduction to yeasts and yeast-like organisms. In *Biology and Activity of Yeasts*, (eds F.A. Skinner, S.M. Passmore and R.R. Davenport), Academic Press, London, pp. 1–27.

Davenport, R.R. (1980b) Cold-tolerant yeasts and yeast-like organisms. In *Biology and Activity of Yeasts*, (eds F.A. Skinner, S.M. Passmore and R.R. Davenport), Academic Press, London, pp. 215–30.

Davenport, R.R. (1981) Yeasts and yeast-like organisms. In *Smith's Introduction to Industrial Mycology*, (eds A.H.S. Onions, H.O.W. Eggins and D. Allsopp), Edward Arnold, London, pp. 65–92.

Davies, R., Sinskey, A.J. and Botstein, D. (1973) Deoxyribonucleic acid repair in a highly radiation-resistance strain of *Salmonella typhimurium*. *J. Bacteriol.*, **114**, 357–66.

Deák, T. and Beuchat, L.R. (1987) Identification of foodborne yeasts. *J. Food Protect.*, **50**, 243–64.

Dijkmann, K.E., Koopmans, M. and Mossel, D.A.A. (1980) The recovery and identification of psychrotrophic yeasts from chilled and frozen, comminuted fresh meats. In *Biology and Activity of Yeasts*, (eds F.A. Skinner, S.M. Passmore and R.R. Davenport), Academic Press, London, p. 300.

Dillon, V.M. (1988) Sulphite tolerance of yeasts from comminuted lamb products. Ph.D. Thesis, University of Bath, UK.

Dillon, V.M. and Board, R.G. (1989a) A medium for detecting sulphite-binding yeasts in meat products. *Letts Appl. Microbiol.*, **8**, 165–7.

Dillon, V.M. and Board, R.G. (1989b) The significance of the yeast:bacteria ratio in contamination of lamb products. *Letts Appl. Microbiol.*, **8**, 191–3.

Dillon, V.M. and Board, R.G. (1990) A study of sulfite-tolerant yeasts from comminuted lamb products. *Biotechnol. Appl. Biochem.*, **12**, 99–115.

Dillon, V.M. and Board, R.G. (1991) Yeasts associated with red meats. *J. Appl. Bacteriol.*, **71**, 93–108.

Dillon, V.M., Davenport, R.R. and Board, R.G. (1991) Yeasts associated with lamb. *Mycolog. Res.*, **95**, 57–63.

Di Menna, M.E. (1955a) A search for pathogenic species of yeasts in New Zealand soils. *J. Gen. Microbiol.*, **12**, 54–62.

Di Menna, M.E. (1955b) A quantitative study of air-borne fungus spores in Dunedin, New Zealand. *Trans. Br. Mycolog. Soc.*, **38**, 119–29.

Di Menna, M.E. (1957) The isolation of yeasts from soil. *J. Gen. Microbiol.*, **17**, 678–88.

Di Menna, M.E. (1958a) *Torulopsis ingeniosa* n. sp. from grass leaves. *J. Gen. Microbiol.*, **19**, 581–3.

Di Menna, M.E. (1958b) *Candida albicans* from grass leaves. *Nature*, **181**, 1287–8.

Di Menna, M.E. (1958c) Two new species of yeasts from New Zealand. *J. Gen. Microbiol.*, **18**, 269–72.

Di Menna, M.E. (1959) Yeasts from leaves of pasture plants. *New Zealand J. Agric. Res.*, **2**, 394–405.

Di Menna, M.E. (1960) Yeasts from soils under forest and under pasture. *New Zealand J. Agric. Res.*, **3**, 623–32.

Do Carmo-Sousa, L. (1969) Distribution of yeasts in nature. In *The Yeasts*, Volume 1. *Biology of Yeasts*, (eds A.H. Rose and J.S. Harrison), Academic Press, London, pp. 79–105.

Dowdell, M.J. and Board, R.G. (1968) A microbiological survey of British fresh sausage. *J. Appl. Bacteriol.*, **31**, 378–96.

Dowdell, M.J. and Board, R.G. (1971) The microbial associations in British fresh sausages. *J. Appl. Bacteriol.*, **34**, 317–37.

Drake, S.D., Evans, J.B. and Niven, C.F. Jr. (1959) The identity of yeasts in the surface flora of packaged frankfurters. *Food Res.*, **24**, 243–6.

Dyett, E.J. and Shelley, D. (1962) Microbiology of the British fresh sausage. International Congress of Food Science and Technology, London, pp. 393–403.

Earnshaw, R.G. (1995) Kinetics of high pressure inactivation of micro-organisms. In *High Pressure Processing of Foods*, (eds D.A. Ledward, D.E. Johnston, R.G. Earnshaw and A.P.M. Hasting), Nottingham University Press, Loughborough, Leicestershire, pp. 37–46.

Edwards, R.A., Dainty, R.H. and Hibbard, C.M. (1983) The relationship of bacterial numbers and types to diamines in fresh and aerobically stored beef, pork and lamb. *J. Food Technol.*, **18**, 777–88.

Eklund, M.W., Spinelli, J., Miyauchi, D. and Dassow, J. (1966) Development of yeast on irradiated Pacific crab meat. *J. Food Sci.*, **31**, 424–31.

Eklund, M.W., Spinelli, J., Miyauchi, D. and Groninger, H. (1965) Characteristics of yeasts isolated from Pacific crab meat. *Appl. Microbiol.*, **13**, 985–90.

Empey, W.A. and Scott, W.J. (1939) Investigations on chilled beef. Part I. Microbial contamination acquired in the meatworks. *Australian Council Sci. Indust. Res. Bull.*, **126**, p. 75.

Fleet, G.H. (1990) Food spoilage yeasts. In *Yeast Technology*, (eds J.F.T. Spencer and D.M. Spencer), Springer-Verlag, Berlin, pp. 124–66.

Frazier, W.C. (1967) *Food Microbiology*, McGraw-Hill, New York.

Fung, D.Y.C. and Liang, C. (1990) Critical review of isolation, detection, and identification of yeasts from meat products. *Crit. Rev. Food Sci. Nutrition*, **29**, 341–79.

Gardner, G.A. (1983) Microbial spoilage of cured meats. In *Food Microbiology: Advances and Prospects*, (eds T.A. Roberts and F.A. Skinner), Academic Press, London, pp. 179–202.

Gill, C.O. (1976) Substrate limitation of bacterial growth at meat surfaces. *J. Appl. Bacteriol.*, **41**, 401–10.

Gill, C.O. (1986) The control of microbial spoilage in fresh meats. In *Advances in Meat Research, Meat and Poultry Microbiology*, (eds A.M. Pearson and T.R. Dutson), MacMillan, Basingstoke, pp. 49–81.

Gill, C.O. and Lowry, P.D. (1981) A note on the identities of organisms causing black spot spoilage of meat. *J. Appl. Bacteriol.*, **51**, 183–7.

Gill, C.O. and Lowry, P.D. (1982) Growth at sub-zero temperatures of black spot fungi from meat. *J. Appl. Bacteriol.*, **52**, 245–50.

Gill, C.O. and Newton, K.G. (1977) The development of aerobic spoilage flora on meat stored at chill temperatures. *J. Appl. Bacteriol.*, **43**, 189–95.

Gill, C.O. and Newton, K.G. (1978) The ecology of bacterial spoilage of fresh meat at chill temperatures. *Meat Sci.*, **2**, 207–17.

Gill, C.O. and Newton, K.G. (1980) Development of bacterial spoilage at adipose tissue surfaces of fresh meat. *Appl. Environ. Microbiol.*, **39**, 1076–7.

Grazia, L., Suzzi, G., Romano, P. and Giudici, P. (1989) The yeasts of meat products. *Yeast*, **5**, S495–S499.

Guerzoni, M.E., Lanciotti, R. and Marchetti, R. (1993) Survey of the physiological properties of the most frequent yeasts associated with commercial chilled foods. *Int. J. Food Microbiol.*, **17**, 329–41.

Hadlok, R., Samson, R.A., Stolk, A.C. and Schipper, M.A.A (1976) Mould contamination in meat products. *Fleischwirtschaft*, **56**, 374–6.

Haines, R.B. (1931) The growth of micro-organisms on chilled and frozen meat. *J. Soc. Chem. Ind. London*, **50**, 223T–227T.

Hite, B.H. (1899) The effect of pressure on the preservation of milk. *Bull. West Virginia Univ. Agric. Exp. Station*, **58**, 15–35.

Hocking, A.D. and Pitt, J.I. (1980) Dichloran glycerol medium for enumeration of xerophilic fungi from low moisture foods. *Appl. Environ. Microbiol.*, **39**, 656–60.

Holwerda, K. (1952) The importance of the pH of culture media for the determination of the number of yeasts and bacteria in butter. *Netherlands Milk Dairy J.*, **6**, 36–52.

Houtsma, P.C., de Wit, J.C. and Rombouts, F.M. (1993) Minimum inhibitory concentration (MIC) of sodium lactate for pathogens and spoilage organisms occurring in meat products. *Int. J. Food Microbiol.*, **20**, 247–57.

Hsieh, D.Y. and Jay, J.M. (1984) Characterization and identification of yeasts from fresh and spoiled ground beef. *Int. J. Food Microbiol.*, **1**, 141–7.

Hup, G. and Stadhouders, J. (1972) Comparison of media for the enumeration of yeasts and moulds in dairy products. *Netherlands Milk Dairy J.*, **26**, 131–40.

Ingram, M. (1958) Yeasts in food spoilage. In *The Chemistry and Biology of Yeasts* (ed A.H. Cook), Academic Press, New York, pp. 603–33.

Ingram, M. and Dainty, R.H. (1971) Changes caused by microbes in spoilage of meats. *J. Appl. Bacteriol.*, **34**, 21–39.

Isaacs, N.S., Chilton, P. and Mackey, B. (1995) Studies on the inactivation by high pressure of micro-organisms. In *High Pressure Processing of Foods*, (eds D.A. Ledward, D.E. Johnston, R.G. Earnshaw and A.P.M. Hasting), Nottingham University Press, Loughborough, Leicestershire, pp. 65–79.

Jacoby, G.A. (1964) The induction and repression of amino acid oxidation in *Pseudomonas fluorescens. Biochem. J.*, **92**, 1–8.

Jarvis, B. (1973) Comparison of an improved Rose-Bengal-chloretetracycline agar with other media for the selective isolation and enumeration of moulds and yeasts in foods. *J. Appl. Bacteriol.*, **36**, 723–7.

Jay, J.M. (1972) Mechanism and detection of microbial spoilage in meats at low temperatures: A status report. *J. Milk Food Technol.*, **35**, 467–71.

Jay, J.M. (1978) Meats, poultry, and seafoods. In *Food and Beverage Mycology.* (ed L.R. Beuchat), AVI Publishing, New York, pp. 129–44.

Jay, J.M. (1979) Occurrence and significance of specific yeasts and molds in meats, poultry and seafoods. In *Food Mycology* (ed M.E. Rhodes), G.K. Hall, Boston, pp. 44–55.

Jay, J.M. (1987) Meats, poultry, and seafoods. In *Food and Beverage Mycology, 2nd Edn.* (ed L.R. Beuchat), AVI Publishing, New York, pp. 155–73.

Jay, J.M. (1992) *Modern Food Microbiology*, 4th edn, AVI, New York.

Jay, J.M. and Kontau, K.S. (1967) Fate of free amino acids and nucleotides in spoiling beef. *Appl. Microbiol.*, **15**, 759–64.

Jay, J.M. and Margitic, S. (1981) Incidence of yeasts in fresh ground beef and their ratios to bacteria. *J. Food Sci.*, **46**, 648–9.

Jay, J.M. and Shelef, L.A. (1976) Effect of microorganisms on meat proteins at low temperatures. *J. Agric. Food Chem.*, **24**, 1113–16.

Jensen, L.B. (1954) *Microbiology of Meats*, 3rd edn. Garrard Press, Illinois.

Johannsen, E., Niemand, J.G., Eagle, L. and Bredenhann, G. (1984) Yeast flora of nonradurised and radurised minced beef – a taxonomic study. *Int. J. Food Microbiol.*, **1**, 217–27.

Johnston, R.W. and Elliot, R.P. (1976) Meat and poultry products. In *Compendium Methods for Microbiological Examination of Foods* (ed. M.L. Speck), American Public Health Association, Washington, pp. 541–8.

Kidney, A.J. (1974) The use of sulphite in meat processing. *Chem. Ind. (London)*, 717–18.

King, J.R., Hocking, A.D. and Pitt, J.I. (1979) Dichloran – Rose Bengal medium for enumeration and isolation of molds from foods. *Appl. Environ. Microbiol.*, **37**, 959–64.

King, J.R., Ponting, J.D., Sanshuck, D.W., Jackson, R. and Mihara, K. (1981) Factors affecting death of yeast by sulfur dioxide. *J. Food Protect.*, **44**, 92–7.

Koburger, J.A. (1970) Fungi in foods. I. Effect of inhibitor and incubation temperature on enumeration. *J. Milk Food Technol.*, **33**, 433–4.

Koburger, J.A. (1973) Fungi in foods. V. Response of natural populations to incubation temperatures between 12 and 32 °C. *J. Milk Food Technol.*, **36**, 434–5.

Kreger-van Rij, N.J.W. (ed) (1984) *The Yeasts – a Taxonomic Study*, Elsevier Science, Amsterdam.

Kühl, H. (1910) Uber ein vorkommen van hefe auf schmieriger wursthaut. *Zentrablatt fuer Bakteriologie Mikrobiologie und Hygiene Originale B (Abteilung 1)*, **54**, 5–16.

Kunkee, R.E. and Amerine, M.A. (1970) Yeasts in wine making. In *The Yeasts*, Vol. 3, *Yeast Technology*, (eds A.H. Rose and J.S. Harrison), Academic Press, London, pp. 5–71.

Kuraishi, H., Itoh, M., Tsuzaki, N., Katayama, Y., Yokoyama, T. and Sugiyama, J. (1990) The ubiquinone system as a taxonomic aid in *Aspergillus* and its telemorphs. In *Modern Concepts in Penicillium and Aspergillus Classification*, (eds R.A. Samson and J.I. Pitt), Plenum Press, New York, pp. 407–32.

Kurtzman, C.P. (1984) Synonomy of the yeast genera *Hansenula* and *Pichia* demonstrated through comparisons of deoxyribonucleic acid relatedness. *Antonie van Leeuwenhock*, **50**, 209–17.

Kurtzman, C.P. (1988) Identification and taxonomy. In *Living Resources for Biotechnology. Yeasts*, (eds B.E. Kirsop and C.P. Kurtzman), Cambridge University Press, Cambridge, pp. 99–140.

Lanciotti, R., Gardini, F., Sinigaglia, M. and Guerzoni, M.E. (1997) Physiological responses to sublethal hydrostatic pressure in *Yarrowia lipolytica*. *Letts Appl. Microbiol.*, **24**, 27–32.

Lawrie, R.A. (1985) *Meat Science*, Pergamon Press, Oxford.

Lea, C.H. (1931a) Chemical changes in the fat of frozen and chilled meat. Part 1. Frozen mutton and lamb. *J. Soc. Chem. Ind. London*, **50**, 207T–213T.

Lea, C.H. (1931b) Chemical changes in the fat of frozen and chilled meat. Part 11. Chilled beef. *J. Soc. Chem. Ind. London*, **50**, 215T–220T.

Leach, T.M. (1971) Preliminary observations on chemical defleecing of sheep as a means of reducing carcass contamination. Proceedings of the 17th European Meeting of Meat Research Workers, Bristol, 161–8.

Leistner, L. and Ayres, J.C. (1968) Molds and meat. *Fleischwirtschaft*, **48**, 62–5.

Leistner, L. and Bem, Z. (1970) Vorkommen und Bedeutung von Hefen bei Pökelfleischwaren. *Fleischwirtschaft*, **50**, 350–1.

Leistner, L. and Rödel, W. (1976) Inhibition of micro-organisms in foods by water activity. In *Inhibition and Inactivation of Vegetative Microbes* (eds F.A. Skinner and W.B. Hug.), Academic Press, London, pp. 219–37.

Lodder, J. (ed) (1970) *The Yeasts – a Taxonomic Study*, North-Holland, Amsterdam.

Lodder, J. and Kreger-van Rij, N.J.W. (1952) *The Yeasts – a Taxonomic Study*, North-Holland, Amsterdam.

Lodder, J., Sloof, W.C. and Kreger-van Rij, N.J.W. (1958) The classification of yeasts. In *The Chemistry and Biology of Yeasts*, (ed A.H. Cook), Academic Press, New York, pp. 1–62.

Lowry, P.D. (1984) Limiting conditions for yeast growth on frozen meat. Proceedings of the 30th European Meeting of Meat Research Workers, 1984, pp. 255–6.

Lowry, P.D. and Gill, C.O. (1984a) Development of a yeast microflora on frozen lamb stored at −5 °C. *J. Food Protect.*, **47**, 309–11.

Lowry, P.D. and Gill, C.O. (1984b) Temperature and water activity mimima for growth of spoilage moulds from meat. *J. Appl. Bacteriol.*, **56**, 193–9.

Lund, F. (1995) Differentiating *Penicillium* species by detection of indole metabolites using a filter paper method. *Letts Appl. Microbiol.*, **20**, 228–31.

Lund, F. (1996) Direct identification of the common cheese contaminant *Penicillium commune* in factory air samples as an aid to factory hygiene. *Letts Appl. Microbiol.*, **22**, 339–41.

Marth, E.H., Capp, C.M., Hasenzahl, L., Jackson, H.W. and Hussong, R.V. (1966) Degradation of potassium sorbate by *Penicillium* species. *J. Dairy Sci.*, **49**, 1197–205.

McCarthy, J.A. and Damoglou, A.P. (1993) The effect of low-dose gamma irradiation on the yeasts of British fresh sausage. *Food Microbiol.*, **10**, 439–46.

McCarthy, J.A. and Damoglou, A.P. (1996) The effect of substrate on the radiation resistance of yeasts isolated from sausage meat. *Letts Appl. Microbiol.*, **22**, 80–4.

McMeekin, T.A. (1982) Microbial spoilage of meats, in *Developments in Food Microbiology* 1, (ed R. Davies), Applied Science, London, pp. 1–40.

Metaxopoulos, J., Stavropoulos, S., Kakouri, A. and Samelis, J. (1996) Yeasts isolated from traditional Greek dry salami. *Italian J. Food Sci.*, **8**, 25–32.

Miller, M.W. (1979) Yeasts in food spoilage: an update. *Food Technol.*, **33**, 76–80.

Molina, I., Silla, H. and Flores, J. (1990) Study of the microbial flora in dry-cured ham. 4. Yeasts. *Fleischwirtschaft*, **70**, 74–6.

Mossel, D.A.A. (1971) Physiological and metabolic attributes of microbial groups associated with foods. *J. Appl. Bacteriol.*, **34**, 95–118.

Mossel, D.A.A. and Ingram, M. (1955) The physiology of microbial spoilage of foods. *J. Appl. Bacteriol.*, **18**, 232–68.

Mossel, D.A.A., Kleynen-Semmeling, A.M.C., Vincentie, H.M., Beerens, H. and Catsaras, M. (1970) Oxytetracycline-glucose-yeast extract agar for selective enumeration of moulds and yeasts in foods and clinical material. *J. Appl. Bacteriol.*, **33**, 454–7.

Mossel, D.A.A., Vega, C.L. and Put, H.M.C. (1975) Further studies on the suitability of various media containing antibacterial antibiotics for the enumeration of moulds in food and food environments. *J. Appl. Bacteriol.*, **39**, 15–22.

Mrak, E.M. and Bonar, L. (1938) A note on yeast obtained from slimy sausage. *Food Res.*, **3**, 615–18.

Nassar, A.M. and Ismail, M.A. (1994) Psychrotrophic and mesophilic fungi isolated from imported frozen lean meat in Egypt. *J. Food Safety*, **14**, 289–95.

Neves, L., Pampulha, M.E. and Loureiro-Dias, M.C. (1994) Resistance of food spoilage yeasts to sorbic acid. *Letts Appl. Microbiol.*, **19**, 8–11.

Njoku-Obi, A.N., Spencer, J.V., Sauter, E.A. and Eklund, M.W. (1957) A study of the fungal flora of spoiled chlortetracycline treated chicken meat. *Appl. Microbiol.*, **5**, 319–21.

Nortjé, G.L., Nel, L., Jordann, E., Badenhorst, K., Goedhart, E. and Holzapfel, W.H. (1990) The aerobic psychrotrophic populations on meat and meat contact surfaces in a meat production system and on meat stored at chill temperatures. *J. Appl. Bacteriol.*, **68**, 335–44.

Nottingham, P.M. (1982) Microbiology of carcass meats, in *Meat Microbiology*, (ed M.H. Brown), Applied Science, London, pp. 13–65.

Nwachukwu, S.U. and Akpata, T.V.I. (1987) Utilization of carbohydrate and protein by *Candida famata* during spoilage of snail meat (*Archachatina marginata* Swainson). *J. Food Agric.*, **1**, 27–30.

Nychas, G.J. (1984) Microbial growth in minced meat. Ph.D. Thesis, University of Bath, UK.

Okodugha, S.A. and Aligba, L.E. (1991) Microbiological quality of raw beef from Irrua Nigerian market retail table. *J. Food Sci. Technol. – Mysore*, **28**, 244–5.

O'Neill, K., Damoglou, A.P. and Patterson, M.F. (1991) Sensitivity of some common grain fungi to irradiation on grain and in phosphate-buffered saline. *Letts Appl. Microbiol.*, **12**, 180–3.

Onston, L.N. (1971) Regulation of catabolic pathways in *Pseudomonas*. *Bacteriol. Rev.*, **35**, 87–116.

Ough, C.S. and Crowell, A. (1987) Use of sulfur dioxide in winemaking. *J. Food Sci.*, **2**, 386–8, 393.

Pandya, Y., Jewett, F.F., Jr. and Hoover, D.G. (1995) Concurrent effects of high hydrostatic pressure, acidity and heat on the destruction and injury of yeasts. *J. Food Protect.*, **58**, 301–4.

Patterson, J.T. (1968) Hygiene in meat processing plants. 3. Methods of reducing carcase contamination. *Rec. Agric. Res. Min. Agric. Northern Ireland*, **17**, 7–12.

Pestka, J.J. (1986) Fungi and mycotoxins in meats, in *Advances in Meat Research, Meat and Poultry Microbiology*, (eds A.M. Pearson and T.R. Dutson), MacMillan, Basingstoke, pp. 277–309.

Price, C.W., Fuson, G.B. and Phaff, H.J. (1978) Genome comparison in yeast systematics: delimitation of species within the genera *Scheanniomyces*, *Saccharomyces*, *Debaryomyces* and *Pichia*. *Microbiol. Rev.*, **42**, 161–93.

Ramírez, C. and González, C. (1972) *Candida iberica* sp. n. A new species isolated from Spanish sausages. *Can. J. Microbiol.*, **18**, 1778–80.

Refai, M., Mansour, N., El-Nagger, A. and Abel-Aziz, A. (1993) Die pilzflora in modernen ägyptischen schlachthäusern. *Fleschwirtschaft*, **73**, 191–3.

Richardson, K.C. (1970) Chemical preservatives in foods in Australia. *Food Preserv. Q.*, **30**, 2–8.

Rojas, F.J., Jodral, M., Gosalvez, F. and Pozo, R. (1991) Mycoflora and toxigenic *Aspergillus flavus* in Spanish dry-cured ham. *Int. J. Food Microbiol.*, **13**, 249–56.

Samson, R.A. and Frisvad, J.C. (1993) New taxonomic approaches for identification of food-borne fungi. *Int. Biodet. Biodegrad.*, **32**, 99–116.

Samson, R.A., Frisvad, J.C. and Arora, D.K. (1991) Taxonomy of filamentous fungi in foods and feeds, in *Handbook of Applied Mycology*, Vol. 3. *Foods and Feeds*, (eds D.K. Arora, K.G., Mukerji and E.H. Marth), Marcel Dekker, New York, pp. 1–29.

Samson, R.A. and Pitt, J.I. (eds) (1990) *Modern Concepts in* Penicillium *and* Aspergillus *Classification*, Plenum Press, New York.

Scott, W.J. (1957) Water relations of food spoilage microorganisms. *Adv. Food Res.*, **7**, 83–127.

Silverman, G.J. and Sinskey, A.J. (1977) Sterilization by ionising radiation, in *Disinfection, Sterilization and Preservation*, 2nd edn, (ed S.S. Block), Lea and Febiger, Philadelphia, pp. 542–61.

Singigaglia, M., Lanciotti, R. and Guerzoni, E. (1994) Biochemical and physiological characterstics of *Yarrowia lipolytica* strains in relation to isolation source. *Can. J. Microbiol.*, **40**, 54–9.

Smith, J.E. and Anderson, J.G. (1992) Modes of arrival and establishment of microfungi. *J. Appl. Bacteriol. Suppl.*, **73**, 69S–79S.

Smith, M.TH. and Hadlok, R. (1976) Hefen und fleisch: vorkommer, systematik und differenzierung. *Fleischwirtschaft*, **56**, 379–84.

Staib, F., MIshra, S.K., Tompak, B., Grosse, G., Abel, TH., Blisse, A., Folkens, U. and Fröhilich, B. (1980) Pathogene hefeähnliche pilze bie fleischproducten. *Zentralblatt fuer Bakteriologie Mikrobiologie und Hygiene Originale A (Abteilung 1)*, **248**, 422–9.

Sutic, M.J., Ayres, J.C. and Koehler, P.E. (1972) Identification and aflatoxin production by molds isolated from country cured hams. *Appl. Microbiol.*, **23**, 656–8.

Szczepaniak, B., Rzezniczak, J. and Pezacki, W. (1975) Higiena I technologia żywności zwierzęcego pochodzenia. Zmienność drożdzy W wędlinach surowych II charakterystyka jakościowa. *Medycyna Weterynaryjna*, **31**, 623–6.

Takatori, K., Takahashi, K., Suzuki, T., Udagawa, S. and Kurata, H. (1975) Mycological examination of sausages in retail markets and the potential production of penicillic acid of their isolates. *J. Food Hygiene Soc. Japan*, **16**, 307–12.

Troller, J.A. and Christian, J.H.B. (1978) *Water Activity and Food*, Academic Press, New York.

Van der Riet, W.B. (1976) Studies on the mycoflora of biltong. *South African Food Rev.*, **3**, 105–11.

Van der Walt, J.P. and Yarrow, D. (1984) Methods for the isolation, maintenance, classification and identification of yeasts, in *The Yeasts – a Taxonomic Study*, (ed N.J.W. Kreger-van Rij). Elsevier Science, Amsterdam, pp. 45–104.

Vickery, J.R. (1936a) The action of microorganisms on fat. I. The hydrolysis of beef fat by some bacteria and yeasts tolerating low temperatures. *J. Coun. Sci. Ind. Res. Australia*, **9**, 107–12.

Vickery, J.R. (1936b) The action of microorganisms on fat. II. A note on the lipolytic activities of further strains of microorganisms tolerating low temperatures. *J. Coun. Sci. Ind. Res. Australia*, **9**, 196–202.

Viljoen, B.C., Dykes, G.A., Callis, M. and von Holy, A. (1993) Yeasts associated with Vienna sausage packaging. *Int. J. Food Microbiol.*, **18**, 53–62.

Walker, H.W. (1977) Spoilage of food by yeasts. *Food Technol.*, **31**, 57–61, 65.

Walker, H.W. and Ayres, J.C. (1959) Characteristics of yeasts isolated from processed poultry and the influence of tetracyclines on their growth. *Applied Microbiol.*, **7**, 251–5.

Walker, H.W. and Ayres, J.C. (1970) Yeasts as spoilage organisms, in *The Yeasts,* Vol. 3, *Yeast Technology*, (eds A.H. Rose and J.S. Harrison), Academic Press, London, pp. 463–527.

Watson, K., Arthur, H. and Shipton, W.A. (1976) *Leucosporidium* yeasts: obligate psychrophiles which alter membrane-lipid and cytrochrome composition with temperature. *J. Gen. Microbiol.*, **97**, 11–18.

Wedzicha, B.L. (1984) *Chemistry of Sulphur Dioxide in Foods*, Elsevier Science, London.

Wells, F.E. and Stadelman, W.J. (1958) Development of yeasts on antibiotic-treated poultry. *Appl. Microbiol.*, **6**, 420–2.

Wickerham, L.J. (1957) Presence of nitrite-assimilating species of *Debaryomyces* in lunch meats. *J. Bacteriol.*, **74**, 823–33.

Winger, R.J. and Lowry, P.D. (1983) Sensory evaluation of lamb after growth of yeasts at −5 °C. *J. Food Sci.*, **48**, 1883–5.

Živanović, R. and Ristić, D. (1974) Nalaz kvasaca u čajnoj kobasici u toku tehnološkog procesa. *Technologija Mesa*, **15**, 51–2.

4 Microbiological contamination of meat during slaughter and butchering of cattle, sheep and pigs

C.O. GILL

4.1 Introduction

Regulatory activities aimed at assuring the hygienic adequacy of the meat supply have been implemented in most developed countries since the first decades of the century. These, now traditional, activities involve the veterinary inspection of stock being presented for slaughter and of the carcasses, inspection of the general cleanliness of plant structure and equipment, and the supervision of meat handling processes. The principal objectives of traditional meat inspection were the elimination from the meat supply of product from overtly diseased animals and of product visibly contaminated with filth or other inedible matter (Thornton and Gracey, 1974). Aided by changes in the animal production and the meat packing industries, meat inspection has generally achieved those goals during the past 30 years (Blamire, 1984).

With the establishment of control over gross defects in meat hygiene, the public health problem of visibly clean meat from healthy animals being contaminated with enteric pathogens came to the fore. Regulatory authorities attempted to address this problem by elaborating detailed requirements for the construction and operation of meat packing plants (Goodhand, 1983). Those initiatives have been costly to implement and maintain. Indeed they have met with a surprising lack of success, as the available information suggests that enteric diseases associated with meat consumption rose, and continue to rise, rather than declined during the past 30 years (Marks and Roberts, 1993; Park *et al.*; 1991; Tauxe, 1991).

Paralleling the efforts of regulatory authorities to improve control over the contamination of meat with pathogens, operators of packing plants have been increasingly motivated by economics, to improve control over the spoilage bacteria which inevitably limit the storage of chilled products. Control of meat spoilage involves the use of preservative packaging and good management of product temperatures as well as the packaging of product of superior hygienic condition (Gill, 1995a). For some commercial purposes, adequate control over spoilage has been obtained by packaging and temperature control alone. However, reductions in the initial numbers of spoilage bacteria on product has undoubtedly been achieved by packers who routinely ship chilled meat to distant markets (Gill, 1989).

Where improvements in the general hygienic condition of meat have occurred, these appear to have resulted largely from trial and error development of appropriate production practices. Consequently, such improvements have been far from uniform within the industry and seem to have had little affect on some types of product. For example, vacuum packaged venison prepared in New Zealand commonly has a storage life of about 18 weeks whereas vacuum packaged pork from any source cannot be expected to keep for longer than 8 weeks (Gill and Jones, 1996; Seman et al., 1988). The measures by which some packers have improved the general hygienic condition of their products are unclear, and even those who have achieved improvements may be unclear about which of their practices are in fact crucial for assuring the hygienic condition of their product.

It had become apparent that neither traditional meat inspection nor supposedly good manufacturing practices can reliably assure the attainment and maintenance of high hygienic standards for meat in respect of contamination with pathogenic or spoilage bacteria. Equally, it is apparent that the alternative to such ineffective or uncertain systems for control of microbiological quality of product is the establishment of Hazard Analysis: Critical Control Point (HACCP) and Quality Management (QM) systems to control meat production processes with regard to microbiological safety and microbiological storage stability, respectively. Consequently, regulatory authorities are now moving towards requirement for such systems in the meat industry (USDA, 1995).

Nevertheless there is still a general uncertainty as to how effective systems of these types could be implemented in the raw meat industry (USDA, 1996). This is hardly surprising in the absence of published examples of such control systems having objectively verified performance. This chapter will therefore consider the microbiology of raw meat with a view to identifying possible approaches to the practical implementation of assuredly effective systems for the control of bacteria on raw meats.

4.2   Process control

There is always a need in manufacturing to produce product of known characteristics. This assures the quality of the product with respect to attributes which are considered to be commercially important. The microbiological conditions of red meat with respect to safety and storage stability can be regarded as quality attributes which must be assured. Their control would involve the specific application of the management principles which are generally used to assure the desired qualities for any product.

The past approach to assuring product quality was to inspect samples of product from each batch and determine the proportion of samples which failed to meet the expected quality. Failure of too many samples to reach standards would lead to rejection or reworking of the batch (Papadakis, 1985). In the extreme case, each item produced would be inspected for acceptability. Such a procedure is still employed in the regulatory inspection of red meat carcasses and can, at the discretion of meat inspectors, be applied to any other raw-meat product. With items smaller than carcasses batchwise inspection is more usual.

Although this method of quality assurance is feasible when throughputs are small, it becomes increasingly impractical as volumes increase, because its maintenance requires either that a progressively increasing fraction of total product effort is committed to inspection or, if inspection effort is not increased in direct proportion to production, an acceptance of increasing uncertainty about the fraction of product which does not meet the stipulated quality (Schilling, 1982). Moreover, with inspection of the end product there must be delay between the occurrence of misprocessing and action to prevent its continuation.

Because of such limitations, control of product quality on the basis of end product inspection has been widely superseded by control of quality through control of the production process. The principle underlying process control is that if product of known quality enters a process and is subjected to a series of operations of known effect, then the quality of the end product will be certain without a need for inspection of the end product (Ryan, 1989). However, such an approach will be effective only when a process is controlled with reference to objective and statistically valid data on product quality. The data for control purposes, which can be direct measures of product quality or measurements of processing parameters which have been correlated with product quality, must be collected routinely. The control data should preferably be collected on-line and continuously. If that is not feasible, it should be collected sufficiently frequently to avoid the production of large batches of product which have not met the stipulated quality. The control data must be monitored for indications of loss of control. When such loss becomes apparent, there should be immediate and specific actions to bring the process back in control, and to isolate possibly misprocessed product from that of assured quality. HACCP or QM systems for assuring the microbiological safety or storage stability of raw meat are special cases of process control systems for product quality. As such, they must meet the fundamental requirements for all process control systems or fail in their purposes.

4.3 HACCP and QM systems

A HACCP system is a process control system concerned with assuring the quality of safety in a product. A full HACCP system for raw meat should therefore address chemical and physical hazards as well as microbiological ones to consumers that could arise as a result of misprocessing. However, separation of the microbiological from other aspects of product safety may be warranted, not only for purposes of discussion but also in the practice of control.

The need for such separation arises because of the difference in the degree of control which can be exercised over microbiological and other hazards. Contamination of raw meat with hazardous chemicals or foreign bodies can be prevented entirely. The hazardous chemicals or foreign bodies that exist within a plant can be readily identified, and systems developed to ensure that they are always kept separate from the meat or, if that is not possible, to ensure that they will be detected if they do contaminate the meat. Thus chemical and physical hazards are amenable, in principle and probably in practice, to direct, on-line monitoring and to control at type I critical points in processes, where each hazard can be wholly prevented or eliminated (Bryan, 1990). In contrast, contamination of raw meat with bacteria is unavoidable and on-line, direct monitoring of microbiological contamination is not a practical possibility. Thus, control of microbiological hazards must usually be exercised in the absence of data on the microbiological condition of the product being processed, and at type II critical points in the processes, where the hazards can be contained but not wholly prevented or eliminated (Tompkin, 1990).

Consideration of fundamental differences suggests the separation also of the HACCP system for controlling the microbiological safety of raw meat products from the QM system for controlling other aspects of microbiological quality. For these two types of system, the methods of control will be the same and control is likely to be necessary at some of if not all the same points in any process. However, the safety of a food is a quality which, if not absolute, must not be allowed to fall below a broadly recognized standard. Storage stability, on the other hand, is only one of several commercially desirable qualities, and one which may be relatively unimportant in particular sectors of the trade. Thus, general microbiological quality can be varied to suit commercial convenience, but the safety of a product cannot. The formal separation of HACCP and QM systems for microbiological quality of product then seems essential in order to avoid confusion of the two aspects by regulators and plant managers, which may either restrict legitimate commercial decisions or result in unintended compromise of product safety.

4.4   Current recommendations on HACCP and QM implementation

Specific recommendations on the implementation of QM systems for assuring the microbiological quality of raw meat are largely lacking. Recommendations for HACCP systems are available, and the fundamental recommendations for HACCP systems apply equally to QM systems. It is proposed however that whereas Critical Control Points (CCPs) are designated for a HACCP system the equivalent points for a QM system should be designated Quality Control Points (QCPs). With that provision, the matter of recommendations can be discussed with reference only to HACCP systems.

The principles of HACCP implementation for food production processes have been identified by the National Advisory Committee on Microbiological Criteria for Foods of the Food Safety and Inspection Service of the U.S. Department of Agriculture (NAC, 1992). These principles are:

1. Conduct a hazard analysis;
2. Identify the CCPs;
3. Establish performance criteria for each CCP;
4. Establish procedures for monitoring each CCP and for the adjustment of each critical operation in response to the information from monitoring;
5. Establish corrective actions for each critical operation when the control data indicate that the operation is out of control;
6. Maintain documentation of the HACCP system;
7. Establish procedures for verifying the performance of the HACCP system.

These principles, which have been presented in other forms elsewhere, are the restatement for a specific purpose of most of the fundamental requirements for process control. The matter of quality of the material entering the process, i.e. the condition of the stock being presented for slaughter, is considered likely to have major impact on the hygienic quality of meat thus requiring control systems for animal production, transport and lairage.

The current view of how these principles can be applied to processes in a slaughtering plant is apparent from the description of 'generic' HACCP systems for beef production processes (AAFC, 1993; NAC, 1993). All the slaughtering, fabricating and packing activities for the production of boxed beef at a plant are considered to be a single process (Fig. 4.1). It is assumed that a HACCP team will be able to conduct a realistic hazard analysis and identify CCPs by inspection and discussion of such a process. The CCPs are to be controlled by the establishment of Standard Operating Procedures (SOPs). Performance criteria will relate to data which can be collected on-line, such as visible contamination, physical and chemical conditions of wash waters, rates of product cooling in chillers, etc. Corrective actions are plant

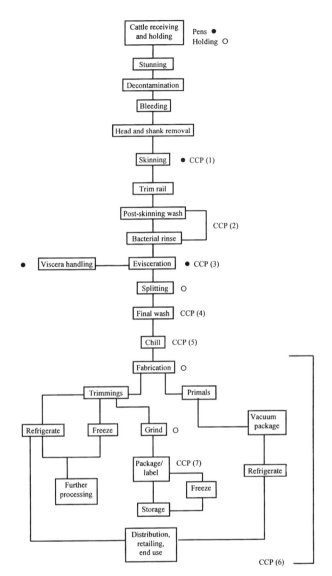

Figure 4.1 Generic HACCP for beef slaughter, fabrication and packaging. Potential site of minor contamination (○); potential site of major contamination (●) (NAC, 1993).

specific. Microbiological data are used for both process verification and for delayed control, as it is recognized that some routine collection of microbiological data is necessary, with corrective responses to such data when a microbiological criterion for the process is exceeded (USDA 1996).

Such a view of HACCP implementation in slaughtering plants presents several problems. Although all activities related to meat production must ultimately be encompassed by a HACCP system, attempting to consider simultaneously the multiplicity of diverse activities in the total process is likely to result in failure to appreciate the importance of some details. That details may well be overlooked is apparent from the designation of areas of multiple activities, such as skinning, as CCPs in the generic HACCP systems. Instead, it can be suggested that, for HACCP implementation purposes, the total process be separated into a series of processes which can be individually considered in detail in an appropriate sequence (Gill *et al.*, 1996d).

It is by no means certain moreover that general perceptions of the hygienic characteristics of meat plant processes will allow a realistic analysis of the microbiological hazards in any particular process, and identification of CCPs by inspection of the process alone. Nor does it seem that limited, routine microbiological sampling of the type currently recommended would provide data which would necessarily correct misconceptions of a process. Obviously, no HACCP or QM system will be effective if the process to be controlled is misunderstood and the control points are wrongly identified. It appears therefore that an approach to HACCP implementation somewhat different from that indicated in current recommendations and regulations may be required.

4.5 Indicator organisms and microbiological criteria

Meat can be contaminated with a large variety of pathogens and spoilage bacteria (see Chapters 1 and 2). The former include *Clostridium perfringens* and *Staphylococcus aureus*, the organism responsible for the majority of cases of food poisoning associated with meat dishes (Foegeding *et al.*, 1994); and *Salmonella*, verotoxigenic *Escherichia coli*, *Campylobacter*, *Yersinia enterocolitica*, *Listeria monocytogenes* and *Aeromonas hydrophila*. All of these organisms cause enteric disease and most can cause systemic disease also (Doyle, 1989). The spoilage organisms include pseudomonads, the bacteria which usually predominate in the spoilage flora of meat stored in air (see Chapter 1), lactic acid bacteria (Chapter 2), which tend to be predominant in the flora of preservatively packaged and comminuted meats, and the psychrotrophic enterobacteria, *Shewanella putrefaciens* and *Brochothrix thermosphacta*, which are facultative anaerobes of high spoilage potential that can precipitate the early spoilage of meat in air or under preservative atmospheres (Gill, 1986).

It would clearly be impossible to monitor meat in any meaningful way for all these organisms. Instead, microbiological monitoring must be concerned with organisms presumed to be indicators of the possible presence of the organisms of concern. Until recently it was perceived that practical considerations limited choices in microbiological monitoring to the enumeration of total aerobic counts (Mackey and Roberts, 1993). However, the general recognition that total counts cannot be adequate indicators of the possible contamination of meat with pathogenic organisms has resulted in proposals for nonpathogenic *E. coli* biotype I to be adopted as the indicator organism for HACCP purposes (Gill, 1995b).

The selection of *E. coli* is justified from its established position as an indicator of faecal contamination and on the assumption that the pathogens associated with meat are largely of faecal origin (USDA, 1996). Discussion in this Chapter of process control then continues with consideration of only total aerobic counts, as the assumed indicator of contamination with spoilage bacteria; *E. coli*, as the assumed indicator of possible contamination with pathogenic bacteria; and coliforms, the count of which is obtained incidentally in some methods for the enumerations of *E. coli* and which can assist in deciding on some of the characteristics of certain processes.

4.6   The condition of stock presenting for slaughter

Although not always entirely true, the meat of unskinned carcasses from healthy animals can, for most practical purposes, be regarded as sterile (Gill, 1979). Exceptions to this assumption are the asymptomatic presence of specific pathogens, such as *Salmonella*, in some organs or lymph nodes of some animals; the presence of clostridial spores in the muscle tissues and organs of some animals; the contamination of tissues remote from the sites of slaughtering wounds when heavily contaminated instruments, such as pithing rods, are used in slaughtering operations; and the contamination of pig carcasses with spores present in water which can enter sticking wounds and the major blood vessels during the scalding of carcasses (Mackey and Derrick, 1979; Sörqvist and Danielsson-Tham, 1986). The incidence of contamination of the deep tissues of carcasses in any of these ways is uncertain. However, it is apparent that the numbers of any organisms which may be involved are small in comparison to those which are deposited on meat during dressing operations. Thus, any bacteria in the meat before dressing can be assumed not to affect significantly the safety or storage stability of the product.

It is thus well recognized that the majority of bacteria occurring on carcasses are deposited on the surface during dressing operations, and that these bacteria originate largely from the hides of animals (Grau, 1986). It would consequently seem obvious that cleaning of the hide before the animals are dressed should reduce the numbers of bacteria available for transfer to the meat. Various studies have failed to confirm this view. With both sheep and cattle it has been shown that dressed carcasses from animals with dirty hides were of no worse hygienic condition than those from animals with relatively clean hides (Biss and Hathaway, 1996; Van Donkers-goed *et al.*, 1997). Indeed, it has been observed that sheep carcasses with long, visually clean fleece may yield dressed carcasses with greater bacterial contamination than those from animals which had short, dirty fleeces (Bell and Hathaway, 1996). Moreover, the dehairing and washing of beef carcasses before dressing did not result in any improvement of the microbiological condition of the dressed carcasses (Schnell *et al.*, 1995).

Although it can be argued that such findings are the result of workers altering their practices to accommodate the state of cleanliness of the hide of each carcass on which they have to operate, these findings show that there is little to be gained in practice by attempting to clean hides before carcasses are dressed. Dressing practices rather than the condition of the hide are the major determinants of the microbiological condition of the dressed carcass. Hence expenditure of effort on providing clean stock for slaughter is not likely to be cost effective.

Another aspect of the microbiological quality of the stock presenting for slaughter has been investigated in recent years, namely the testing of pig herds for *Salmonella* before dispatch to slaughtering plants, with the separate slaughter of animals from infected and non-infected herds and the assignment of carcasses from infected herds to the production of heat-processed meats in which the pathogen *Salmonella* would be destroyed (Bager *et al.*, 1995). Significant reduction in the incidence of *Salmonella* in raw pork has been claimed as a result of such measures.

The production of raw meat only from animals demonstrably free from specific pathogens may thus be a useful approach to assuring the safety of meat in some circumstances. It seems unlikely however that it could be applied simultaneously with respect to several organisms without severely restricting the supply of raw meat on some occasions. Moreover, it would seem to be impossible to apply it effectively to production systems for animals, such as cattle and sheep, which do not involve the rearing of animals in closely controlled groups in controlled environments. Thus, control of asymptomatic infections in stock is unlikely to be a generally practicable substitute for improvement of meat processing hygiene, although it may be a valuable adjunct in some circumstances.

4.7 Sampling of carcasses

When attempting a description of the hygienic condition of carcasses, the method of sampling and the sampling plan which is adopted must be considered. The classical methods of sampling carcass surfaces are by excision of portions of meat surfaces, or by swabbing an area delimited by a template with first a wet then a dry cotton wool swab. Each tissue sample or pair of swabs is then homogenized with a suitable volume of diluent, and pour or spread plates are prepared from portions of each homogenate. The numbers of bacteria recovered by excision sampling will exceed by between 50% and 90% those recovered by cotton wool swabs (Anderson *et al.*, 1987; Lazarus *et al.*, 1977). However, swabbing has often been preferred because of the reduced effort and time required to obtain and process a sample, and because the swab sample is obtained without damage to the meat (Sharpe *et al.*, 1996).

In recent years, there has been a tendency to use swabs made from a mildly abrasive material, such as surgical gauze or cellulose acetate sponges rather than cotton wool. Swabbing with such materials can recover bacteria in numbers approaching those recovered by excision sampling (Table 4.1). Thus swabbing with gauze or sponge would seem to be the optimum method for sampling meat surfaces because convenience is combined with a near maximum recovery of bacteria (Dorsa *et al.*, 1996).

Bacteria are not evenly distributed on meat surfaces. Therefore, the probability of obtaining a representative sample will increase with increasing sample size (Brown and Baird-Parker, 1982). Over the years there has been a tendency to increase the area of surface sampled from 5 cm^2 to, at the extreme, the whole surface of a carcass (Kitchell *et al.*, 1973; Lasta *et al.*, 1992). However, the area to be sampled will usually be a small portion only of the total surface area of a red meat carcass. Due regard must also be paid to the practicability of handling the materials needed for sampling large areas. Consequently, sampling of an area of about 100 cm^2 appears to be optimal.

In recent years hydrophobic grid membrane filtration (HGMF) techniques (Brodsky *et al.*, 1982) have been adopted rather than conventional plate counting methods for the enumeration of bacteria. There are two reasons for the adoption of HGMF. For total counts, which can be recovered from most samples by conventional methods, the advantage of HGMF techniques is mainly the saving of labour in the counting of the bacteria recovered (Jericho *et al.*, 1993). For *E. coli* the advantage is in the level of detection, as all the homogenate from a swab sample can be filtered to give a detection level of 1 *E. coli*/100 cm^2 (Gill *et al.*, 1996b).

Table 4.1 Bacteria recovered from beef by swabbing with cheesecloth pads (A), cellulose acetate sponges (B) or cotton wool swabs (C), as compared with assumed 100% recovery from excised samples of tissue. Values were calculated from the data of Dorsa *et al.* (1996)

Swab type	Bacteria recovered (%)
A	88.5
B	80.8
C	50.0

As regards sampling plans, there have been essentially two approaches to evaluating the numbers of bacteria on carcasses from a processing plant. Both approaches involve the random selection of carcasses.

The first and more usual approach involves the sampling of sites on the carcass which are likely to be heavily contaminated in order to assess the maximum contamination that is likely to be present at any site on the carcass. As no one site will always be the most heavily contaminated on a carcass, sampling of three or more sites on each carcass has been considered necessary. Usually in studies reported in the literature, each sample has been processed separately, in order to obtain information on the distribution of bacteria on carcasses and, on some occasions to relate contamination to particular dressing operations (Hudson et al., 1987; Jericho et al., 1994; Roberts et al., 1984). However, in the requirement for routine microbiological monitoring for E. coli mandated by the USDA, it is stipulated that the samples from three sites on each carcass be combined for analysis (USDA, 1996). That stipulation is explicitly designed to save effort in the evaluation of process performance on the basis of the maximum bacterial contamination likely to be present on any site on any carcass.

The maximum contamination approach permits the setting of a criterion for acceptable processing in the form of a three-point attributes acceptance plan (Bray, 1973). In such a plan, bacterial counts are viewed as quality attributes, the attributes above or below a limit (m) for acceptable contamination. A fraction (c) of n samples is permitted to exceed m. However, no sample may exceed a rejection limit (M) which is greater than m (Table 4.2).

Such a criterion has the merit of simplicity, and has been recommended as the type of criterion appropriate for use with raw meat products (ICMSF, 1974). However, the data obtained with respect to the criterion provide no information about the distribution of bacteria on product units, and hence could be misleading. For example, the data would not distinguish between a situation where the most heavily contaminated site was unique amongst otherwise lightly contaminated sites and one where many other sites on the surface were contaminated similarly to the most heavily contaminated site. Also, the sites designated for sampling might not include the most heavily

Table 4.2 The general form of a three-point attributes acceptance plan for raw meat

Symbol	Definition
n	The number of samples from a lot required for decision on the acceptability of the lot.
m	The maximum numbers (/g or cm^2) of bacteria in wholly acceptable samples.
M	The maximum numbers (/g or cm^2) of bacteria in marginally acceptable samples. Numbers >M are unacceptable.
c	The number of samples in a set on n samples which may have values between m and M in an acceptable lot.

contaminated site on the carcasses produced by some processes. Moreover, in the event of a process failing to meet the criterion, the data would provide no information on where in the process there had been loss of control. An indication of loss of control could instigate only a general inspection of the process to identify and correct the point of process failure. There could then be a prolonged period between loss and re-establishment of control, particularly if inspection unsupported by microbiological data was the only method employed to identify the point of failure.

An alternative approach to devising a sampling plan is based on the assumption that the distribution of bacteria on carcasses, as on or in other forms of meat, approximates the log normal (Kilsby, 1982; Hilderbrandt and Weis, 1994). If that is the case, then estimates of the mean log ($\bar{x}$) and standard deviation (s) for the bacteria on a population of carcasses should be obtainable from the numbers of bacteria in a set of >20 samples obtained from randomly selected sites on randomly selected carcasses in the population. The values for $\bar{x}$ and s can then be used to estimate the log of the arithmetic mean numbers of bacteria on the carcasses (log A) from the equation log $A = \bar{x} + \log_n 10 \, s^2/2$ (Kilsby and Pugh, 1981). This approach permits the formulation of a criterion which simply stipulates a maximum acceptable average load of bacteria on the carcasses leaving a process. If sampling from randomly selected sites is used for routine monitoring and records identifying the counts obtained from each site are maintained then, in the event of failure to meet the criterion, the sites affected by abnormally heavy contamination would be identifiable from the records. Then only operations likely to affect such sites would need to be investigated to identify and correct the loss of control. Thus, if practicable, the randomly selected site approach to sampling would seem to be more compatible with process control systems, and to yield more information for the effort expended, than sampling which involves a few, selected, relatively heavily contaminated sites.

4.8 Carcass dressing processes

A dressing process for beef carcasses will typically involve some 36 operations (Fig. 4.2). Various studies have shown that bacteria are deposited on the carcass during the skinning operations (Grau, 1987; Nottingham, 1982). There are generally relatively few areas of heavy contamination on the carcass after skinning and the distribution of bacteria does not approximate the log normal (Fig. 4.3). During subsequent operations, few additional bacteria may be added to the carcass; handling of the carcass results however in the bacteria deposited during dressing being redistributed to approximate a log normal distribution.

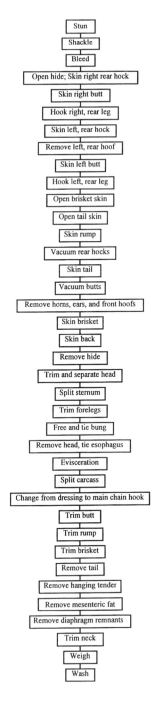

Figure 4.2 Operations in a beef carcass dressing process (Gill *et al.*, 1996c).

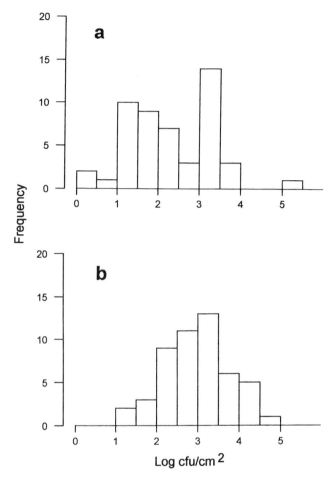

Figure 4.3 Distributions of log total aerobic counts recovered from randomly selected sites on beef carcasses (a) after skinning and (b) after completion of the dressing process.

After splitting, carcasses are trimmed to remove visible contamination and bruised tissue. Traditionally it has been assumed that visible and microbiological contamination are related, so that cutting off visibly contaminated tissue from a carcass will improve its microbiological condition (USDA, 1993). Similarly, it has been assumed that bruised will carry larger bacterial loads than unbruised tissue, and that the removal of bruised tissue would improve the microbiological status of the meat. Laboratory studies of tissue excised from artificially contaminated areas of meat have confirmed that trimming can remove bacterial contamination (Hardin *et al.*, 1995).

In practice however there is no relationship between visible and microbiological contamination on carcasses (Jericho *et al.*, 1993), and the hygienic condition of bruised does not differ from that of unbruised tissue (Rogers *et al.*, 1992). As trimming operations are guided by visible contamination and bruising only, any concurrent removal of bacteria is fortuitous. Consequently, routine trimming of carcasses is apparently without effect on the microbiological condition of the carcass population as a whole (Table 4.3). It is thus questionable whether the effort required for and loss of meat arising from routine trimming are justified by the purely aesthetic improvement that is achieved (Gill *et al.*, 1996a). In contrast the washing of carcasses does appear to have a hygienic effect, in practice approximately halving the numbers of, as well as redistributing, the bacteria on carcasses (Kelly *et al.*, 1981; Ellebroek *et al.*, 1993).

A comparison of beef carcass dressing processes on the basis of the total counts recovered from the equivalent sites on each carcass in several batches from each plant found differences of up to two orders of magnitude between the mean numbers recovered from carcasses at different plants (Roberts *et al.*, 1984). Even so uncertainty persisted about the relative hygienic performances of the plants studied because of the inability to take account of factors extraneous to the plants which might have affected the numbers of bacteria recovered. The observed differences could not be related to factors such as the appearances of plant and equipment, line speed, types of equipment employed, sex and cleanliness of stock, day of slaughter, etc. The lack of effect of upgrading of dressing lines on the hygienic condition of carcasses has also been demonstrated (Hudson *et al.*, 1987). These results indicate that the major factor affecting the hygienic condition of dressed beef carcasses is the skill with which individual dressing operations, particularly ones for skinnings, are performed, rather than details of plant or equipment design or maintenance, or the condition or type of stock slaughtered (Mackey and Roberts, 1993).

A comparison of beef carcass dressing processes on the bases of the mean numbers of total counts and *E. coli* on the carcasses leaving the processing line at each plant revealed large differences in the hygienic performances

Table 4.3 Statistics describing the numbers of bacteria on beef carcasses before and after routine trimming at a packing plant (Gill *et al.*, 1996)

Count	Stage of the process	Statistics			
		$\bar{x}$	s	n	$\log A$
Total aerobic[a]	Before trimming	2.21	1.18	0	3.81
	After trimming	2.51	1.02	0	3.71
Coliforms[b]	Before trimming	0.99	1.13	7	2.46
	After trimming	1.10	1.13	7	2.53
Escherichia coli[b]	Before trimming	0.86	1.11	9	2.28
	After trimming	0.96	1.11	9	2.38

[a] Numbers/cm^2; [b] numbers/100 cm^2; $\bar{x}$ = mean log; s = standard deviation; n = number of samples of 50 in which bacteria were not detected; $\log A$ = estimated log of the arithmetic mean.

of processes, with the mean of both counts differing by up to two orders of magnitude between plants (Table 4.4). It was notable that relatively large numbers of total counts but relatively small numbers of *E. coli* were recovered from carcasses at some plants. This would seemingly confirm that total counts are not a reliable indicator of hygienic performance with respect to safety (Gill *et al.*, 1997b).

As with earlier studies, these findings are open to the objection that a few samples collected during a limited period may not describe adequately the general performance of any dressing process. However, the available data suggest that, at least with sampling for total counts from designated sites, microbiological results from a dressing process may be consistent over long periods (Jericho *et al.*, 1996). Presuming that is so, then log mean numbers of *E. coli* of <1/100 cm^2 and log mean numbers of total aerobic counts about 2/cm^2 would seem to be attainable criteria for beef carcass dressing processes (Gill *et al.*, 1997b).

In addition to preventing contamination, there has been in recent years interest in augmenting dressing processes with novel, decontaminating operations for beef carcasses. Of the proposed decontaminating treatments, treatment of carcasses with solutions of organic acids and the pasteurization of carcasses with hot water have been extensively examined. Pasteurization of carcasses with steam and decontamination of selected areas by application of vacuum nozzles while simultaneously heating the treated areas with hot water or steam are treatments which have been implemented in some North American plants.

Treatment of meat with hot solutions of lactic or acetic acids can reduce the total numbers of bacteria by two or three orders of magnitude (Frederick *et al.*, 1994; Smulders *et al.*, 1986). Species of different bacteria vary

Table 4.4 Estimated log mean numbers (log A) of total aerobic, coliform and *Escherichia coli* counts on beef carcasses leaving the carcass dressing processes at ten beef slaughtering plants (Data from Gill *et al.*, 1997b)

Plant	Log mean numbers		
	Total (log N/cm^2)	Coliforms (log N/100 cm^2)	*E. coli* (log N/100 cm^2)
A	3.42	1.96	2.06
B	3.12	2.03	2.01
C	4.28	3.05	1.98
D	3.62	2.51	1.74
E	4.89	2.99	1.28
F	3.70	1.89	0.79
G	2.78	1.39	0.75
H	2.20	0.77	0.70
I	3.01	1.56	0.58
J	2.04	–	–

– = insufficient data for calculation of the statistic.

widely in their susceptibility to organic acids, and organisms such as *E. coli* and *Salmonella* are notably resistant (Brackett *et al.*, 1994). Moreover, treatments with relatively strong and hot solutions of acids, which damage the appearance of meat, are required to achieve consistently large reductions in numbers (Bell *et al.*, 1986; Woolthuis and Smulders, 1985). Uniform treatment of all surfaces of a carcass by spraying is difficult to achieve unless large volumes of a solution are used on each carcass. Strong organic acids can corrode concrete and metals, thereby damaging plant and equipment. Consequently, treatments with organic acids are unlikely to be generally suitable for reducing the numbers of enteric pathogens on carcasses.

Treatment of both beef and sheep carcasses with hot water has been examined, and an apparatus for the treatment of beef carcasses in commercial circumstances has been designed and tested (Smith and Davey, 1990). A reduction of the natural flora on carcasses by about two orders of magnitude was obtained when carcasses were treated with water of >80 °C for 10 s, although larger reductions were observed in bacteria artificially inoculated onto carcasses (Smith and Graham, 1978; Smith, 1992). Increasing the treatment time or temperature within practical limits does not increase to any significant extent the reduction of bacterial numbers. Without presenting data, various reports indicate that the optimum decontaminating treatment with hot water has no permanent effect on the appearance of carcasses. The examination of the effects of hot water on primal cuts suggests however that cut muscle surfaces, but not fat, membrane-covered or cut bone surfaces would be discoloured by an effective heat treatment (Gill and Badoni, 1997). Such, minor damage to the appearance of carcasses would seem to be commercially tolerable. Pasteurization of beef carcasses with hot water has not been adopted commercially, apparently because large volumes of water would have to be used to assure uniform heating of the carcass surface. For economic reasons, that would require recirculation of the treatment water. Concern about the quality of recirculated water appears to have impeded examination of the routine use of hot water for pasteurization of dressed carcasses under commercial conditions.

Possible problems with the reuse of the hot water are avoided when carcass surfaces are heated by the application of steam. In the commercial apparatus designed for that purpose, steam at a temperature of >100 °C is applied in a chamber in which the pressure rises above atmospheric. The supra-atmospheric pressure is required to assure the uniform condensation of steam onto, and thus the uniform heating of the entire surface of, a carcass (Cygnarowicz-Provost *et al.*, 1994). In commercial practice, the treatment is applied for 6.5 s. This limits degradation of the appearance of cut muscle surfaces while reducing the numbers of *E. coli* by two orders of magnitude and total viable counts by one order of magnitude (Table 4.5). The surviving flora is enriched for Gram-positive species. Thus, the treatment is effective in reducing the numbers of enteric pathogens and

Table 4.5 Statistics describing the numbers of bacteria on beef carcasses before and after their treatment in a routine, commercial operation for pasteurizing carcass surfaces with steam

Count	State of the treatment	Statistics				
		$\bar{x}$	s	n	$\log A$	N
Total aerobic[a]	Before	2.70	0.76	0	3.36	5.23
	After	1.49	0.81	0	2.24	4.19
Coliforms[b]	Before	1.20	1.03	2	2.43	4.06
	After	–	–	37	–	1.69
Escherichia coli[b]	Before	0.84	1.09	8	–	3.84
	After	–	–	45	–	1.11

[a]Numbers/cm^2; [b]numbers/100 cm^2; $\bar{x}$ = mean log; s = standard deviation; n = number of samples of 50 in which bacteria were not detected; $\log A$ = estimated log of the arithmetic mean; N = log of the total numbers recovered from 25 cm^2 or 2500 cm^2; – = insufficient data for calculation of the statistic.

Gram-negative spoilage organisms on carcasses, but probably has a smaller effect on the numbers of bacteria which predominate in the spoilage flora of vacuum packaged meat (Gill, 1995a).

In contrast, hot water–vacuum treatment of selected areas of carcass surfaces with hand-held equipment redistributes but does not greatly reduce the numbers of bacteria on carcasses (Table 4.6). The treatment is relatively ineffective for removing bacteria. In practice, steam or hot water is not applied to any point on the carcass surface for a time sufficient to achieve a pasteurizing effect and only small parts of the surface of any carcass can be treated. None the less the treatment is considered useful for commercial purposes because it is apparently effective for removing visible contamination and so reducing the requirement for trimming carcasses.

Dressing processes for sheep carcasses are similar to those for beef carcass dressing processes in the sense that much of the contamination on dressed carcasses originates from the hide, and is deposited during skinning

Table 4.6 Statistics describing the microbiological effects of routine hot water–vacuum cleaning of the crutch area of beef carcasses passing through a commercial dressing process

Count	State of the treatment	Statistics			
		$\bar{x}$	s	n	$\log A$
Total aerobic[a]	Before	3.76	0.65	0	4.25
	After	3.47	0.46	0	3.71
Coliforms[b]	Before	3.05	1.27	0	4.89
	After	2.30	1.27	0	4.17
Escherichia coli[b]	Before	2.78	1.17	0	4.36
	After	2.06	1.39	1	4.28

[a]Numbers/cm^2; [b]numbers/100 cm^2; $\bar{x}$ = mean log; s = standard deviation; n = number of samples of 50 in which bacteria were not detected; N = log of the total numbers recovered from 25 cm^2 or 2500 cm^2.

operations (Gill, 1987). If the same absolute area of a carcass surface is contaminated during skinning by contact between the meat and the outer surface of the hide, or contaminated hands or equipment, then the fraction of the surface which is heavily contaminated will be larger with the smaller sheep carcasses than with much larger beef carcasses. Thus, in commerce, sheep carcasses may tend to be more evenly and perhaps more heavily contaminated than those of beef (Gill and Baker, 1997). Hygienically superior dressing of sheep carcasses probably does occur but it does not seem to have been reported, except for the case of inverted dressing (Bell and Hathaway, 1996).

Unlike beef carcasses, those of sheep can be and are dressed while suspended by the forelegs (Longdell, 1996). The manning level of a dressing line can be reduced with inverted dressing. The results in a reduced handling of the rumps and legs of carcasses (Fig. 4.4). This apparently reduces not only contamination of the high-value hind quarters but also the total contamination on carcasses.

The dressing of sheep carcasses also differs from that of beef in that, the throat and the floor of the mouth of sheep are usually opened, and the base of the tongue is pulled from the throat whilst the head is still attached to the carcass. The mouth harbours large numbers of bacteria, including *E. coli* (Gill and Baker, 1997). Thus, if other areas of the carcass are handled by a worker immediately after handling the tongue, as is usual in some processes, then the fore parts of the carcass will be heavily contaminated with bacteria from the mouth.

While pig carcasses are skinned in some processes, in most others they are dressed with the skin still on. The dressing processes for most pig carcasses have initial operations – scalding, dehairing, singeing and polishing – which are intended to clean and remove the hair from the skin (Fig. 4.5). Scalding, usually by immersion in a tank of water of *ca.* 60 °C for about 8 min., destroys most of the bacteria on the surface of the carcass (Nickels *et al.*, 1976). However, the carcasses are commonly recontaminated during mechanical dehairing by bacteria which persist and grow on detritus and in the circulating water of the dehairing equipment (Table 4.7). The numbers of such contaminants may be reduced by singeing. As singeing of the surface is usually uneven, large numbers of bacteria can persist in some areas. During polishing of the singed carcass, the surviving bacteria are redistributed over the entire carcass and their numbers augmented by bacteria which contaminate the polishing brushes and flails. Consequently, the visibly clean, polished carcasses may well be contaminated by substantial numbers of bacteria, including both spoilage and pathogenic types (Gill and Bryant, 1992, 1993).

The numbers of bacteria on polished carcasses can be reduced by two orders of magnitude by pasteurizing carcasses before evisceration (Gill *et al.*, 1997a). The bacteria deposited on carcasses during evisceration will then

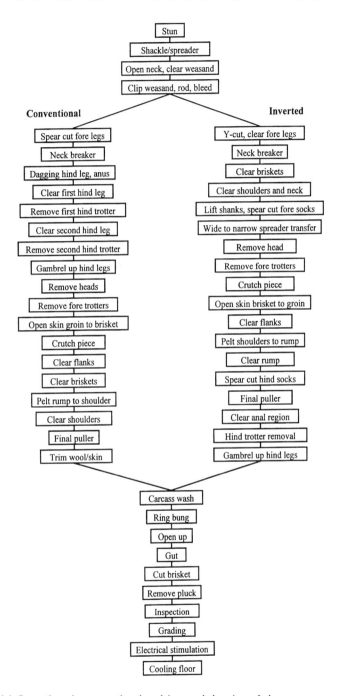

Figure 4.4 Operations in conventional and inverted dressing of sheep carcasses (Bell and Hathaway, 1996).

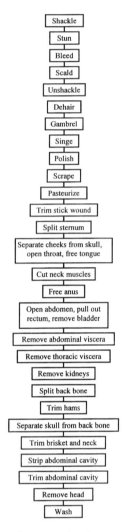

Figure 4.5 Operations in a pig carcass dressing process (Gill and Jones, 1997).

form the majority of those on carcasses at the end of the dressing process. Apart from contaminants originating from improperly cleaned equipment, the bacteria deposited on carcasses during the eviscerating operations originate from the mouth and the bung (Gill and Jones, 1997).

As with sheep carcasses, pig carcasses are usually dressed before the head is removed. Consequently, bacteria are spread from the mouth to the fore parts of the carcass. With carcasses pasteurized after polishing, the mouth is the source of the majority of *E. coli* on carcasses leaving the processing

Table 4.7 Log mean numbers of total aerobic counts recovered from the loins of carcasses after various stages of the dressing processes at two pig slaughtering plants (Gill and Bryant, 1992)

Operating preceding sampling	Total aerobic counts ($\log_{10} N/cm^2$)	
	Plant A	Plant B
Scalding	2.58	2.86
Dehairing	4.00	4.26
Singeing	4.26	3.81
Polishing	3.70	3.20
Washing after dressing	3.11	3.36

line. It is generally also the major source of the *Salmonella* on dressed carcasses (Christensen and Luthje, 1994). It would obviously be desirable to avoid such contamination by removing the head before the mouth and throat are handled. Dressing of the head separated from the carcass would ensure that workers who handle the throat and tongue did not handle the carcass. Unfortunately, regulatory requirements for inspection of the throat while the head is still attached to the carcass prevents the implementation of such an arrangement at most plants.

The carcass may also be contaminated with enteric organisms if the bung or the cut end of the intestine is allowed to make contact with the carcass when the large intestine is pulled from the body cavity. Such contamination is commonly avoided by enclosing the freed bung in a plastic bag when the large intestine is pulled from the body cavity, and retaining the bag in place during the removal of the intestines (Nesbakken *et al.*, 1994).

If polished pig carcasses are pasteurized, and the dressing process were modified to prevent contamination from the mouth and the intestine, then it would seem possible to produce dressed pig carcasses with average log numbers of *E. coli* <1/carcass and log numbers of total counts <2/cm² (Gill and Jones, 1997).

The relatively small amounts of microbiological data that can be collected from a carcass dressing process will reflect the general hygienic performance of the process. However, misprocessing, principally the spillage of gut contents onto some carcasses, will occur on some occasions in all processes. In general, it is the usual practice to continue with the normal processing of a carcass following such a misprocessing event, but with enhanced trimming and washing, sometimes after the carcass has been removed to a detaining rail, in order to remove visible contamination.

Such an approach to misprocessing is contrary to the principles of process control. To conform with such principles, a misprocessing event should immediately precipitate a series of actions to prevent the event affecting following product, and to remove the compromised product from the normal to a separate process, where it may be treated such that its microbiological condition is comparable with that of properly processed product, or diverted for use where the microbiological condition of the raw material will not compromise the safety or storage stability of the final product. Appropriate reactions to a misprocessing event could include: the immediate cleaning of all carcass-contacting equipment which may have been contaminated by the event; the cessation of processing of the carcass on the dressing line, even though it might well have to continue moving on the main line until it arrives at a point designed to allow for diversion of carcasses from the line; and the subjection of the carcass to one or more decontaminating treatments of assured efficacy. Appropriate, defined and automatic reactions to misprocessing events do not appear to be a usual feature of control systems for carcass dressing. However, it must be recognized that without an established procedure for dealing with such events, a control system for a dressing process will be incomplete.

4.9   Product cooling processes

While meat is cooling from body to chiller temperatures there is potential opportunity for the rapid proliferation of both pathogenic and spoilage bacteria. To avoid this, cooling processes must be well controlled and properly targeted at carcasses, offals or hot boned meat.

On hygienic grounds it would in principle be desirable to cool all meat as rapidly as possible. With beef and sheep carcasses, however, the cooling of muscle tissue to chiller temperatures before the development of *rigor* can result in contraction of the muscle fibres with permanent toughening of the tissue, and consequent loss of quality of the meat (Tornberg, 1996). The onset of rigor can be accelerated by the electrical stimulation of carcasses during or immediately after the dressing process. This can allow carcasses to be cooled rapidly without the risk of unacceptable toughening of much of the product (Locker, 1985). Moreover, recent studies have indicated that toughening can be avoided if cooling is very rapid (Joseph, 1996). The problem of cold shortening and toughening does not generally arise with pig carcasses because *rigor* development in pig muscle is intrinsically faster than in that of cattle or sheep (Murray, 1995). In fact, rapid cooling of pig muscle may be advantageous for meat quality. The incidence of Pale, Soft, Exudative (PSE) muscle, a condition arising from the muscle entering rigor while still warm, can be reduced by rapid cooling (Jones *et al.*, 1991).

Despite the possibilities for rapidly cooling carcasses without undue risk to the eating quality of muscle tissue, practical and economic considerations result in most cooling processes for carcasses being operated in order to reduce deep temperatures to 7 °C or below over periods of between 12 h and 24 h (James and Bailey, 1990). In practice with beef carcasses, which are large compared with those of sheep or pigs and hence intrinsically slow to cool, a substantial proportion of those carcasses leaving many cooling processes after 24 h may still be above 7 °C, a temperature that is widely regarded as the maximum temperature for ensuring that mesophilic, enteric pathogens will not proliferate (Smith, 1985). There is no absolute hygienic need to reduce the internal temperatures of all carcasses below 7 °C during carcass cooling. As bacteria are on the surface of the carcass – the deep tissues are essentially sterile – control of conditions at the surface only are required to assure the hygienic adequacy of the carcass cooling process.

It has long been recognized that factors other than temperature may also be of importance in controlling the growth of surface bacterial contaminants during carcass cooling. In a classic study by Scott and Vickery (1939), it was shown that the microbiological condition of cooling beef carcasses was largely dependent upon the extent to which the surface of a carcass dried during cooling. If the speed and humidity of the air passing over the carcass caused the surface to dry during the early stages of cooling, then decreases rather than increases in the numbers of spoilage bacteria could be obtained. However, prevention of drying by shielding the surface from the chilled air resulted in large increases in the numbers of spoilage bacteria. These results were confirmed by studies involving carcasses cooled in experimental situations; decreases, of about 0.5 log units, or increases in total counts could be achieved by varying the speed and humidity as well as the temperature of the air to which the carcasses were exposed (Nottingham, 1982).

Such studies resulted in the belief, which is still widely held and manifested in the regulatory requirements of many countries, that drying of the carcass surface during cooling is essential for assuring the safety and storage stability of the product (Bailey, 1986). However, drying of the surface inevitably results in loss of carcass weight, which is economically undesirable (Gigiel et al., 1989). During the past decade, therefore, most slaughtering plants in North America adopted the practice of intermittently spraying carcasses with water during the first few hours of cooling – the weights of the cooled carcasses are then only slightly little less than those of the warm ones. Contrary to the classic expectation that such treatment must result in large increases in the numbers of bacteria on the carcass, studies under experimental conditions have shown that increases or decreases similar to those observed with air chilling can be obtained by adjustment of air conditions and the frequency, intensity and duration of spraying (Greer et al., 1990; Strydom and Buys, 1995). No explanation seems to have been advanced to account for the observed control of bacterial numbers in spray cooling processes.

Despite the recognition that temperature data alone cannot describe the hygienic performance of a carcass cooling process, such data have been commonly used in attempts to define hygienic performance criteria for these processes. The simplest criterion is to stipulate a maximum time allowable to attain a deep temperature at or below an assumed safe temperature, usually of 10 °C or 7 °C. Lower temperatures have been suggested (USDA, 1970).

Apart from taking no account of the uncertainty of the relationship between surface and deep temperatures, the range of air conditions, and the variations in the rates of cooling which will exist within any batch chiller, such a criterion ignores the possibility that some carcasses may retain a warm surface temperature for several hours thereby allowing substantial growth of bacteria. Subsequent rapid cooling may well allow the process to meet the criterion but the carcass would have a compromised hygienic condition.

Some of these shortcomings have been addressed by the stipulation of an ideal curve for deep temperatures during acceptable cooling or, for a better definition, the stipulation of two cooling curves for deep temperatures which identify an acceptable range of temperatures for any carcass at any time during the cooling process (Armitage, 1989). Such approaches seem to be obvious improvements over the simple type of criterion. Even so any relationship between conformance or otherwise to specified deep temperature cooling curves and the behaviour of pathogenic bacteria, in particular, and other bacteria on the carcass surface remains uncomfortably vague.

A further refinement of the cooling curve approach involves the collection of temperature histories for the warmest point on the surface of each carcass in a group of >20 selected at random from those passing through a process. The warmest point on the surface does not have to be determined for individual carcasses, because the surface within the aitch bone pocket is always amongst the warmest sites on any carcass (Gill *et al.*, 1991a). The surface temperature history recorded for each carcass is integrated with respect to a model which describes the temperature dependency of the aerobic growth rate of *E. coli*, to give a proliferation value for each carcass. The hygienic performance of the process is then described by the distribution of values in the set of *E. coli* proliferation values.

The rationale for such an approach is that, while it will not identify the overall changes in *E. coli* numbers on the carcasses passing through a cooling process, it will identify the range of maximum possible proliferations in the carcass population, as temperature must restrict the growth whenever no other factors are inhibiting bacterial proliferation. The surface temperature history approach aims therefore at identifying the worst possible hygienic performance of a process, with the assumption that processes can be usefully compared on the basis of their estimated, worst-possible performances (Gill and Jones, 1992a). The approach has the merit of reducing any cooling curves, which may be complex, to a readily comprehended, single number of apparent hygienic significance. It also allows the setting of a criterion in the form of a three-point, attributes-acceptance plan. It is notable, however, that the values of *E. coli* proliferation for beef carcass cooling processes, in particular, appear to be far larger than general considerations would suggest could be acceptable, even when processes appear to be operated with due regard to good manufacturing practices (Gill *et al.*, 1991b).

The assessment of the hygienic performances of carcass cooling processes by reference to only time and temperature data, however they be manipulated, is therefore questionable. Until recently, microbiological data which could allow decision as to the validity of assessments based on temperature history were entirely lacking. The application of the random sampling procedure used in the assessment of carcass dressing processes would, however, seem to have resolved the matter (Gill and Bryant, 1997).

In a recent study of two spray-cooling processes for beef carcasses, it was found that, although the two processes appeared to be similar from the temperature history data and estimations of *E. coli* proliferation, in one process the log mean numbers of both total and *E. coli* counts were reduced by <0.5 log units, whereas in the other process log mean numbers of total counts were reduced by >0.5 log units, and those of *E. coli* by about 2 log units (Table 4.8). In a process for cooling sheep carcasses in air, the reductions in numbers of total viable organisms and *E. coli* were similar to those observed for the second beef cooling process, even though surface cooling to <7 °C was slower for the sheep carcasses than it was with the beef carcasses. When pig carcasses were blasted with air of –20 °C while passing through a tunnel for about an hour before entering the chiller, the rate of cooling of carcass surfaces was equal to or more rapid than that in the normal processes used for beef or sheep carcasses. After such coolings the log mean numbers of *E. coli* were little changed while the total counts increased by about 1 log unit.

For these four processes, temperature history data were a wholly inadequate guide to the behaviour of the microflora. Microbiological data alone however would be of limited value for controlling a cooling process because, unlike with carcass dressing, they could yield little information about specific events in a process. Thus, it would seem appropriate to control carcass cooling processes largely on the basis of temperature history data,

Table 4.8 Effects of cooling processes on the log mean numbers of total aerobic counts (N/cm^2) and *Escherichia coli* counts ($N/100\ cm^2$) on carcasses

Carcass type	Cooling process	Log mean numbers			
		Total counts ($\log N/cm^2$)		*E. coli* counts ($\log N/100\ cm^2$)	
		Before cooling	After cooling	Before cooling	After cooling
Beef	Spray chilling	4.03	3.58	0.37	0.06
Beef	Spray chilling with surface freezing	3.12	2.48	2.01	–0.14
Pig	Blast freezing then spray chilling	1.83	2.57	0.35	0.42
Sheep	Air chilling	3.33	2.86	3.57	1.49

which had previously been properly assessed by reference to appropriate microbiological data from the process. The hygienic adequacy of a process would be most appropriately assessed from microbiological data. It can be suggested that appropriate criteria for safety and storage stability would be no increase in the log mean numbers of *E. coli* or total counts respectively, during carcass cooling.

The reduction in bacterial numbers during the air cooling of sheep carcasses is caused by surface drying. The relatively larger decrease in *E. coli* than in the total counts arises from the greater sensitivity to drying of Gram negative as opposed to Gram-positive organisms (Leistner *et al.*, 1981). The reduction in bacterial numbers during spray cooling requires explanation. It can be postulated that small and equal reductions of total and *E. coli* counts are a consequence of the physical removal of bacteria by repeated spraying, while a reduction in *E. coli* counts accompanied by a smaller reduction of total counts is possibly the result of slow freezing of the carcass surface, with Gram-negative organisms being more sensitive than Gram-positive ones (Lowry and Gill, 1985).

In general the cooling of offals presents a simpler, though more often mismanaged hygienic situation than the cooling of carcasses. Some offals, particularly liver, may be cooled in air while suspended from a rack. The hygienic consequences of that type of cooling have not been reported, but it can be reasonably supposed that there would be some similarity to the cooling of carcasses in air. However, most offals are collected into bulk containers or packed into boxes soon after they are separated from the carcass, with some types, notably head meats, commonly being washed with cold water before they are packed.

Most offals present a high pH (>6.0) environment for the microflora. Thus, during cooling, bacterial growth will be restrained only by temperature and the anaerobic conditions which develop within the mass of bulked offals (Gill and DeLacy, 1982). While temperatures remain close to or above 30 °C, the growth of *E. coli* is favoured over that of other organisms. If such temperatures are maintained, a flora dominated by *E. coli* will develop (Gill and Penney, 1984). As temperatures are lowered, the growth of lactic acid bacteria will be favoured.

Offals have traditionally been regarded as being of intrinsically poor microbiological quality. While that may be true of head meats, which will be unavoidably contaminated by bacteria from the mouth and gullet, it is not true of other tissues. The poor microbiological condition often found in these is commonly the result of inadequately controlled cooling processes (Hinson, 1968). Slow cooling of offals is often inevitable in current commercial processes because they are collected into bulk containers, or into large boxes which are assembled in stacks before being subjected to chilling or freezing (Gill and Jones, 1992b). Even in freezers of high refrigerative capacities, product at the centres of large containers or stacks will cool only slowly (Baxter, 1962), affording opportunities for large increases in bacterial numbers.

As temperature is the only variable factor affecting bacterial growth, the hygienic effects of offal cooling processes should be assessable from product temperature history data. It appears that actual bacterial proliferation can be estimated with reasonable accuracy by temperature function integration techniques (Gill and Harrison, 1985). A procedure for evaluating the hygienic performances of processes for offal cooling from temperature history data has been described (Gill *et al.*, 1995b). It involved measuring the rates of cooling at the centres – the slowest cooling point – of randomly selected boxes of offals, and estimation of the *E. coli* growth permitted by each temperature history. A clearer view of the overall bacterial behaviour during cooling would be obtained by collecting temperature histories from randomly selected sites within product units, and by integrating these with respect to the growth of some suitable spoilage organisms as well as *E. coli*. However, such a refinement of the approach has not been attempted.

Assessment of an offal cooling process from temperature history data will allow estimation of bacterial growth during cooling, but it will not take account of reductions in numbers resulting from freezing if the process is designed to yield a frozen product. In that case, microbiological data from random samples of product would be required to assess the overall effects of cooling and freezing. For a better understanding of such a process, it would be preferable to obtain data for both chilled and frozen product during chilling and freezing, so that the microbiological effects of cooling to chiller temperatures and freezing are distinguished for appropriate control of each phase of the process. In general, rapid freezing as done with cryogenic liquids, will result in only small reductions of bacterial numbers, but slow freezing, as will be inevitable with boxed product, can be expected to result in reductions in the numbers of Gram-negative bacteria of, or in excess of, an order of magnitude (Davies and Obafemi, 1985).

In current commercial practice, hot boning involves either the breaking down of hot pig carcasses, with the meat being passed immediately to further processing into comminuted, cured or cooked products, or the collection of hot beef into boxes as manufacturing meat. Muscle which is ground pre-rigor has functional properties superior to those of post-rigor muscle (Cuthbertson, 1980). In the usual mixing of batches of ground meat, to obtain a desired fat content or to prepare a batter, the temperature of the product is easily controlled by the addition of ice or carbon dioxide snow. Thus, the cooling of hot-boned meat is usually well controlled when the meat is processed further immediately after collection.

Processes for the cooling of manufacturing beef in packed boxes, usually for freezing, are similar to processes used for the cooling of offals, and have been assessed from temperature histories obtained from the centres of boxes during cooling (Reichel *et al.*, 1991). If the meat remains at high pH

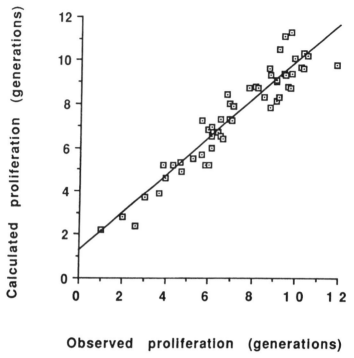

Observed   proliferation   (generations)

Figure 4.6 Growth of *Escherichia coli* during the cooling of hot boned beef. Correlation between growth determined from plate counts or estimated from product temperature histories.

during cooling then, as with offals, estimates of *E. coli* proliferation will approximate the actual increases (Fig. 4.6). If the pH falls below 5.8, then the growth of *E. coli* could be expected to fall below estimated values (Grau, 1983). As yet, however, the effects of the development of *rigor* on the growth of bacteria on hot boned meat has not been investigated. As with offals, the effects of freezing would have to be determined from microbiological data.

4.10 Carcass breaking processes

The majority of the carcasses produced in most slaughtering plants are now broken down to at least primal cuts. It is obvious that the contamination of meat with bacteria from equipment and from personnel should be minimized during carcass breaking.

For many years there have been requirements regarding the clothing and personal hygiene of workers. If properly enforced, these should control contamination from workers' bodies (Kasprowiak and Hechelmann, 1992;

Restaino and Wind, 1990). There are also requirements regarding the cleanliness of fixed equipment, with regulatory (daily) inspection of cleaned equipment before the commencement of work and, often, microbiological sampling of meat contact-surfaces by plant personnel. Recently, the requirements in that area have been expanded by stipulations that meat cutting plants develop, document and adhere to formal, daily cleaning programmes as a prerequisite for the development of HACCP systems (USDA, 1996).

Despite the meat packing industry's experiences with equipment cleaning, the major concerns in cleaning processes still appear to be commonly misunderstood. The designation of cleaning as a 'pre-HACCP requirement' (USDA, 1995), that is as a process controlled by means that do not amount to a HACCP system rather than one for which a HACCP system should be developed, would seem to be a symptom of such misunderstanding.

In the current assessment of equipment cleaning, emphasis is usually given to the visible cleanliness of meat-contact surfaces and the use of sanitizers to reduce bacterial numbers. Microbiological monitoring of cleaned surfaces may be used to establish the efficacy of the cleaning and sanitizing procedures (Ingelfinger, 1994). There is usually little concern about the persistence of pools of rinse water on cleaned equipment, and of small quantities of detritus which may persist on obscured or guarded parts of equipment which do not make direct contact with meat.

Unfortunately, the current emphasis tends to misdirect cleaning practices. Meat-contacting, impervious surfaces which are visibly clean can be expected to harbour bacteria at numbers of $10–10^3/cm^2$; the actual number will depend on the extent to which a surface is pitted and/or scratched (Edelmeyer, 1980). If the surfaces are dried, the numbers of bacteria will diminish rather than increase (Schmidt, 1983). In practice, the surfaces will carry no more and probably less bacteria than the incoming meat, and the microflora will have a composition which does not differ in any respect important for safety or storage stability from the composition of the flora on the incoming meat. Thus, the condition of the surfaces will not compromise the microbiological condition of the first few pieces of incoming meat. Thereafter, the condition of the working surfaces during processing will be determined by the condition of the incoming meat. The use of sanitizers on meat-contact surfaces which are properly dried after cleaning is likely to be of relatively little hygienic benefit.

However, any pools of water left on the equipment will tend to allow the development of large population of Gram-negative organisms, including spoilage types and the cold-tolerant pathogen, *Aeromonas hydrophila* (Gill and Jones, 1995). These bacteria will certainly contaminate the first product that enters the processing line. If there are pools of water on equipment at sites from which it may be sprayed or flow during operation of the equipment, then all the meat which is processed may be contaminated to some extent by a possibly hazardous Gram-negative flora from the equipment. The observed efficacies of some sanitizers for reducing contamination from equipment may then arise from their control of the flora in persisting water; sanitizers would not be necessary if the equipment were dried.

Detritus which tends to dry but which is periodically wetted may accumulate in areas which are not easily accessible for cleaning, such as within guards for drive shafts. Such detritus tends to harbour a large population of predominantly Gram-positive organisms which may include the cold-tolerant pathogen, *Listeria monocytogenes* (Gobat and Jemmi, 1991). Such detritus, or bacteria from biofilms, may be transferred to working surfaces during operation of the equipment and hence contaminate all product moving through the process (Zoltola and Sasahara, 1994).

It is thus apparent that the principal objective in cleaning equipment should be to ensure that water, detritus or biofilms does not persist in equipment from day to day. Unfortunately, fixed equipment in meat cutting facilities has often been constructed without great regard for the cleanability of areas other than meat contacting surfaces. Thus, the presence of a persistent source of contamination might not be readily apparent from inspection.

The common failure to recognize persisting sources of contamination from carcass breaking equipment is evident from the lack of control over the cleaning of articles of workers' personal equipment. Often, the cleaning of such equipment is left to the discretion of workers, and without the provision of facilities specifically designed for the cleaning of items such as mesh gloves and guards, which may harbour large amounts of detritus (Van Klink and Smulders, 1989). Consequently, the cleaning of personal equipment is often variable, with uncertain impact on the microbiological condition of meat (Smeltzer *et al.*, 1980).

The assurance of proper equipment cleaning would seem to require that the cleaning process, which should involve both fixed and personal items of equipment, be treated as all other meat plant processes, with the development of HACCP and QA systems based on appropriate microbiological data. Random sampling of product entering and leaving the carcass breaking process, in a manner similar to that applied to carcass dressing processes, should provide objective information on the extent to which the product is adversely affected by the breaking process. However, estimation of mean numbers of total counts, coliforms and *E. coli* on product might be usefully augmented by examining product for organisms such as aeromonads, listeriae and/or other bacteria which might prove to be suitable indicators of contamination from sources persisting in equipment. If the process results in substantially increased contamination of product, then the major sources of contamination should be identifiable by determining the points in the process at which bacterial numbers markedly increase, with subsequent examination of the equipment in use at each such point, to ascertain the source of the contamination and appropriate actions for its control.

4.11 Storage and transport processes

Most meat is stored and transported at chilled temperatures. At temperature below 7 °C, growth of mesophilic pathogens will not occur (Bogh-Sorensen and Olsson, 1990). Consequently, temperatures <7 °C are generally regarded as safe for the storage and transport of meat although the possible growth of cold-tolerant pathogens at these temperatures remains a matter of concern (Palumbo, 1986). Apart from assuming that temperature abuse does not occur, the control of product temperatures during the storage and transport of chilled meat is then largely related to the assurance of an adequate storage life for the product.

As meat is a complex material, it freezes over a range of sub-zero temperatures, with the ice fraction increasing with decreasing temperature and it tends to supercool (Mazur, 1970). Freezing of meat commences at about –1 °C but, in practice, boxed meat can be held indefinitely at –1.5 ± 0.5 °C without any obvious formation of ice in the tissue (Gill *et al.*, 1988). Spoilage bacteria will grow at these and at somewhat lower temperatures (Rosset, 1982). Thus, spoilage of meat at chiller temperatures cannot be prevented but only delayed, the maximum storage life being obtained when meat is held at the minimum temperature which can be maintained without risk of freezing.

The rates of growth of psychrotrophic spoilage bacteria increase greatly with small increases of temperature within the chill-temperature range. At temperatures of 0 °C, 2 °C and 5 °C, meat will have a storage life of about 70, 50 and 30% of the storage life at –1.5 °C (Fig. 4.7). Thus, close control of product temperatures at or near the optimum –1.5 °C is required when chilled meat must remain unspoiled for extended periods, as with the surface shipment of chilled product to overseas markets. The storage life of chilled meat is greatly reduced by the usual practice of storing and distributing meat in facilities operated with air temperatures of 2 ± 2 °C (Gunvig and Bogh-Sorensen, 1990).

As well as failure to recognize that chilled meat should preferably be stored at sub-zero temperatures, there is often failure to assure that chilled meat is cooled to the intended target temperature. Storage facilities for chilled meat are commonly designed to maintain rather than reduce product temperatures. As boxed product is often ≥5 °C when packed, it will cool only slowly in such facilities and it may well be dispatched while still relatively warm (Gill and Jones, 1992c). The refrigeration units of road trailers which carry meat are now, more often than previously, operated with off-coil air temperatures of about 0 °C. It has become the usual practice to carry chilled meat in sea containers operated at –1.5 ± 0.5 °C. Both types of transport facility are designed to maintain rather than to reduce product temperatures. Consequently, failure to properly cool product before it is loaded for transport can result in some product remaining at

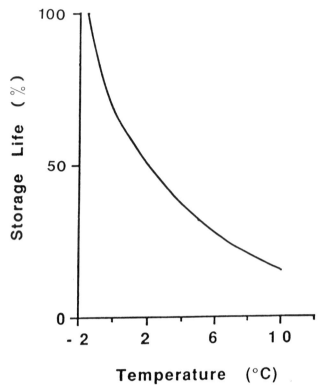

Figure 4.7 The relationship between the storage temperature and the storage life of vacuum packaged beef.

relatively high temperatures for long periods, with a substantial reduction of storage life.

The performances of storage and transport processes can be assessed from temperature histories collected from randomly selected product units moving through each process, with integration of each temperature history with respect to the dependency on temperature of the growth of appropriate spoilage organisms (Gill *et al.*, 1995a). The assessment of the data can be simplified by the calculation of a storage efficiency factor from each temperature history, this factor being the percent ratio of the bacterial proliferation calculated from the temperature history to the proliferation which would be calculated if a temperature of –1.5 °C had been maintained for the duration of the temperature history (Gill and Phillips, 1993). The avoidable loss of storage life in a process is then readily assessable from the distribution of the values for storage efficiency factors (Fig. 4.8).

For the storage of frozen meat, the only microbiological concern is that product temperatures remain below –5 °C, the minimum temperature for growth of moulds and yeasts (Lowry and Gill, 1984a,b). Even if marginal

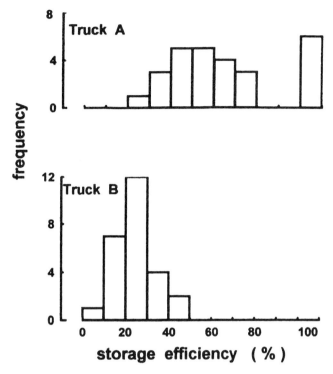

Figure 4.8 The frequency distributions of storage efficiencies for two processes for the distribution of raw meat sausages in refrigerated trucks (Gill *et al.*, 1995a).

freezing temperatures are experienced, they would have to be maintained for long periods before mould colonies became visible. At such temperatures, the growth of yeasts is relatively rapid if desiccation of the meat surface does not occur. However, the spoilage of frozen meat by yeasts is not common. It appears that, in commercial circumstances, mould or yeast spoilage of meat usually occurs when the surface temperature of frozen meat rises to chiller temperatures under conditions where drying of the meat surface inhibits the growth of the spoilage bacteria which would otherwise spoil the meat before visible growth of fungi or yeasts became apparent. Yeasts will predominate when the surfaces have an intermediate water activity, but mould will predominate when the surfaces are desiccated.

4.12 Concluding remarks

Meat packing plant activities inevitably involve a number of processes, which are likely to be performed with differing degrees of hygienic adequacy, within any plant as well as at different plants. In devising HACCP and QA systems for these processes, meat plant personnel should first assess objectively the hygienic effects of each of their processes, to identify and improve the one or two processes which have the greatest adverse hygienic effects on their product. Such an approach to the implementation of control systems is logical as, while the most deleterious processes continue, unaltered efforts expended on less damaging processes will not yield any substantial improvement of the microbiological condition of the final product. For example, a ten-fold reduction in the numbers of *E. coli* on carcasses will be of little consequence for safety if the numbers of *E. coli* are increased a hundred fold during breaking of the carcasses. The approach is practicable, because techniques for assessing process performance from limited amounts of microbiological and/or temperature history data have been identified. Such an approach would seem to be necessary in view of recent developments which show that it is in practice possible to produce carcasses free of detectable *E. coli* by a combination of consistent, but not necessarily superior dressing practices with effective decontaminating treatments. The wide adoption of such practices is obviously a necessary step for improving the safety of the meat supply. However, the safety of the product presented to consumers may in fact be little improved if, as at present, the hazards arising from other sources are regarded as of minor importance largely because of their being poorly characterized. Until the hygiene performances of processes at all stages of meat production are objectively assessed as a matter of course, hazardous processing in some areas will continue unrecognized, to compromise all other efforts to improve the safety of the meat supply.

References

AAFC. (1993) HACCP generic model. Slaughter of cull cows and boneless beef. Agriculture and Agri-Food Canada, Food Protection and Inspection Branch, Ottawa, Canada.

Anderson, M.E., Huff, H.E., Naumann, H.D., Marshall, R.T., Damare, J., Johnson, R. and Pratt, M. (1987) Evaluation of swab and tissue excision methods for recovering microorganisms from washed and sanitized beef carcasses. *J. Food Protect.*, **50**, 741–3.

Armitage, N.H. (1989) Application of temperature function integration to derive a regulatory standard for ovine carcass cooling. Ministry of Agricultural Fisheries, New Zealand.

Bager, F., Emborg, H.D., Sorensen, L.L., Halgaard, C. and Jensen, P.T. (1995) Control of *Salmonella* in Danish pork. *Fleischwirtsch. Int.* (4), 27–8.

Bailey, C. (1986) Current issues affecting meat chilling and distribution, in *Recent Advances and Developments in the Refrigeration of Meat by Chilling*, Int. Inst. Refrig., Paris, pp. 18–20.

Baxter, D.C. (1962) The fusion times of slabs and cylinders. *J. Heat Transfer.*, **84**, 317–20.

Bell, R.G. and Hathaway, S.C. (1996) The hygienic efficiency of conventional and inverted lamb dressing systems. *J. Appl. Bacteriol.*, **81**, 225–34.

Bell, M.F., Marshall, R.T. and Anderson, M.E. (1986) Microbiological and sensory tests of beef treated with acetic and formic acids. *J. Food Protect.*, **49**, 207–10.

Biss, M.E. and Hathaway, S.C. (1996) Effect of pre-slaughter washing of lambs on the microbiological and visible contamination of the carcasses. *Vet. Rec.*, **138**, 82–86.

Blamire, R.V. (1984) Meat hygiene: the changing years. *State Vet. J.*, **39**, 3–13.

Bogh-Sorensen, L. and Olsson, P. (1990) The chill chain, in *Chilled Foods: the State of the Art* (ed. T.R. Gormley), Elsevier, London, pp. 245–67.

Brackett, R.E., Hao, Y.-Y. and Doyle, M.P. (1994) Ineffectiveness of hot acid sprays to decontaminate *Escherichia coli* 0157:H7 on beef. *J. Food Protect.*, **57**, 198–203.

Bray, D.F., Lyon, D.A. and Burr, I.W. (1973) Three class attributes plans in acceptance sampling. *Technomet.*, **14**, 575–85.

Brodsky, M.H., Entis, P., Sharpe, A.N. and Jarvis, G.A. (1982) Enumeration of indicator organisms in foods using the automated hydrophobic grid-membrane filtration technique. *J. Food Protect.*, **45**, 292–6.

Brown, M.H. and Baird-Parker, A.C. (1982) The microbiological examination of meat, in *Meat Microbiology* (ed. M.H. Brown), Applied Science, London, pp. 423–520.

Bryan, F.L. (1990) Application of HACCP to ready-to-eat chilled foods. *Food Technol.*, **44**, 70–7.

Christensen, H. and Luthje, H. (1994) Reduced spread of pathogens as a result of changed pluck removal technique. Proceedings 40th International Congress of Meat Science and Technology, The Hague, The Netherlands, paper No S III.06.

Cuthbertson, A. (1980) Hot processing of meat: a review of the rationale and economic implications, in *Developments in Meat Science*, vol. 1 (ed. R. Lawrie), Elsevier, London, pp. 61–88.

Cygnarowicz-Provost, M., Whiting, R.C. and Craig, J.C. Jr. (1994) Steam surface pasteurization of beef frankfurters. *J. Food Sci.*, **59**, 1–5.

Davies, R. and Obafemi, A. (1985) Responses of micro-organisms to freeze-thaw stress, in *Microbiology of Frozen Foods* (ed. R.K. Robinson), Elsevier, London, pp. 83–107.

Dorsa, W.J., Cutter, C.N. and Siragusa, G.R. (1996) Evaluation of six sampling methods for recovery of bacteria from beef carcass surfaces. *Letts Appl. Microbiol.*, **22**, 39–41.

Doyle, M.P. (1989) *Foodborne Bacterial Pathogens*, Dekker, New York.

Edelmeyer, H. (1980) Cleaning and disinfection in the meat industry: What are the considerations? *Fleischwirtschaft*, **60**, 448–53.

Ellerbroek, L.I., Wegener, J.F. and Messier, S. (1993) Does spray washing of lamb carcasses alter bacterial surface contamination? *J. Food Protect.*, **56**, 432–6.

Foegeding, P.M., Roberts, T., Bennett, J.M., Bryan, F.L., Cliver, D.O., Doyle, M.P., Eden, R.F., Flowers, R., Foreman, C.T., Lorber, B., Madden, J.M., Rose, J.B., Smith, J.L., Todd, E.C.D. and Wekell, M.M. (1994) Foodborne pathogens: Risks and consequences, Task Force Report no. 122. Council for Agricultural Science and Technology (CAST), Ames, Iowa.

Frederick, T.L., Miller, M.F., Thompson, L.D. and Ramsey, C.G. (1994) Microbiological properties of pork cheek meat as affected by acetic acid and temperature. *J. Food Sci.*, **59**, 300–2.

Gigiel, A.J., Collett, P. and James, S.J. (1989) Fast and slow beef chilling in a commercial chiller and the effects of operational factors on weight loss. *Int. J. Refrig.*, **12**, 338–49.

Gill, C.O. (1979) Intrinsic bacteria in meat. *J. Appl. Bacteriol.*, **47**, 367–78.

Gill, C.O. (1986) The control of microbial spoilage in fresh meats, in *Advances in Meat Research*, vol. 2 (eds A.M. Pearson and T.R. Dutson), AVI Publishing, Westport, Connecticut, pp. 49–88.

Gill, C.O. (1987) Prevention of microbial contamination in the lamb processing plant, in *Elimination of Pathogenic Organisms From Meat and Poultry* (ed. J.M. Smulders), Elsevier, Amsterdam, pp. 203–20.

Gill, C.O. (1989) Packaging meat for prolonged chilled storage: the Captech process. *Brit. Food J.*, **91**, 11–15.

Gill, C.O. (1995a) MAP and CAP of fresh, red meats, poultry and offals, in *Principles of Modified-Atmosphere and Sous Vide Product Packaging* (eds J.M. Farber and K.L. Dodds), Technomic Publishing, Lancaster, Pennsylvania, pp. 105–36.

Gill, C.O. (1995b) Current and emerging approaches to assuring the hygienic condition of red meats. *Can. J. Anim. Sci.*, **75**, 1–13.

Gill, C.O. and Badoni, M. (1997) The effects of hot water pasteurizing treatments on the appearances of pork and beef. *Meat Sci.* (in press).

Gill, C.O. and Baker, L.M. (1997) Assessment of the hygienic performance of a sheep carcass dressing process. *J. Food Protect.* (in press).

Gill, C.O. and Bryant, J. (1992) The contamination of pork with spoilage bacteria during commercial dressing, chilling and cutting of pig carcasses. *Int. J. Food Microbiol.*, **16**, 51–62.

Gill, C.O. and Bryant, J. (1993) The presence of *Escherichia coli*, *Salmonella* and *Campylobacter* in pig carcass dehairing equipment. *Food Microbiol.*, **10**, 337–44.

Gill, C.O. and Bryant, J. (1997) Assessment of the hygienic performances of two beef carcass cooling processes from product temperature history data or enumeration of bacteria. *Int. J. Food Microbiol.* (in press).

Gill, C.O. and DeLacy, K.M. (1982) Microbial spoilage of whole sheep livers. *Appl. Environ. Microbiol.*, **43**, 1262–6.

Gill, C.O. and Harrrison, J.C.L. (1985) Evaluation of the hygienic efficiency of offal cooling procedures. *Food Microbiol.*, **2**, 63–9.

Gill, C.O. and Jones, T. (1992a) Assessment of the hygienic efficiencies of two commercial processes for cooling pig carcasses. *Food Microbiol.*, **9**, 335–43.

Gill, C.O. and Jones, T. (1992b) Evaluation of a commercial process for collection and cooling of beef offals by a temperature function integration technique. *Int. J. Food Microbiol.*, **15**, 131–43.

Gill, C.O. and Jones, S.D.M. (1992c) Efficiency of a commercial process for the storage and distribution of vacuum-packaged beef. *J. Food Protect.*, **55**, 880–7.

Gill, C.O. and Jones, T. (1995) The presence of *Aeromonas*, *Listeria* and *Yersinia* in carcass processing equipment at two pig slaughtering plants. *Food Microbiol.*, **12**, 135–41.

Gill, C.O. and Jones, T. (1996) The display life of retail packaged pork chops after their storage in master packs under atmospheres of N_2, CO_2, or $O_2 + CO_2$. *Meat Sci.*, **42**, 203–13.

Gill, C.O. and Jones, T. (1997) Assessment of the hygienic characteristics of a process for dressing pasteurized pig carcasses. *Food Microbiol.*, **14**, 81–91.

Gill, C.O. and Penney, N. (1984) The shelf life of chilled sheep livers packed in closed tubs. *Meat Sci.*, **11**, 73–7.

Gill, C.O. and Phillips, D.M. (1993) The efficiency of storage during distant continental transport of beef sides and quarters. *Food Res. Int.*, **26**, 239–45.

Gill, C.O., Badoni, M. and Jones, T. (1996a) Hygienic effects of trimming and washing operations in a beef-carcass-dressing process. *J. Food Protect.*, **59**, 666–9.

Gill, C.O., Bedard, D. and Jones, T. (1997a) The decontaminating performance of a commercial apparatus for pasteurizing polished pig carcasses. *Food Microbiol.*, **14**, 71–9.

Gill, C.O., Deslandes, B., Rahn, K., Houde, A. and Bryant, J. (1997b) Evaluation of the hygienic performances of ten beef-carcass-dressing processes. *J. Appl. Microbiol.* (in press).

Gill, C.O., Friske, M., Tong, A.K.W. and McGinnis, J.C. (1995a) Assessment of the hygienic characteristics of a process for the distribution of processed meats, and of storage conditions at retail outlets. *Food Res. Int.*, **28**, 131–8.

Gill, C.O., Harrison, J.C.L. and Phillips, D.M. (1991a) Use of a temperature function integration technique to assess the hygienic adequacy of a beef carcass cooling process. *Food Microbiol.*, **8**, 83–94.

Gill, C.O., Jones, S.D.M. and Tong, A.K.W. (1991b) Application of a temperature function integration technique to assess the hygienic adequacy of a process for spray chilling beef carcasses. *J. Food Protect.*, **54**, 731–6.

Gill, C.O., McGinnis, J.C. and Badoni, M. (1996b) Use of total or *Escherichia coli* counts to assess the hygienic characteristics of a beef carcass dressing process. *Int. J. Food Microbiol.*, **31**, 181–96.

Gill, C.O., McGinnis, J.C. and Badoni, M. (1996c) Assessment of the hygienic characteristics of a beef carcass dressing process. *J. Food Protect.*, **59**, 136–40.

Gill, C.O., McGinnis, J.C. and Jericho, K.W.F. (1996d) Implementation of hazard analysis: critical control point (HACCP) systems for beef carcass dressing processes. Agriculture and Agri-Food Canada, Lacombe Research Centre, HACCP Manual No. 1.

Gill, C.O., Phillips, D.M. and Harrison, J.C.L. (1988) Product temperature criteria for shipment of chilled meats to distant markets, in *Refrigeration for Food and People*, Int. Inst. Refrig., Paris, pp. 40–47.

Gill, C.O., Taylor, C.M., Tong, A.K.W. and O'Laney, G.B. (1995b) Use of a temperature function integration technique to assess the maintenance of control over an offal cooling process. *Fleischwirtschaft*, **75**, 682–4.

Gobat, P.F. and Jemmi, T. (1991) Epidemiological studies of *Listeria* spp. in slaughterhouse. *Fleischwirtsch. Int.*, (1), 44–9.

Goodhand, R.H. (1983) The future role of meat inspection in the field of meat hygiene. *R. Soc. Health J.*, **103**, 11–15.

Grau, F.H. (1983) Growth of *Escherichia coli* and *Salmonella typhimurium* on beef tissue at 25 °C. *J. Food Sci.*, **48**, 1700–3.

Grau, F.M. (1986) Microbial ecology of meat and poultry, in *Advances in Meat Research*, Vol. 2 (eds A.M. Pearson and T.R. Dutson), AVI Publishing, Westport, Connecticut, pp. 1–47.

Grau, F.M. (1987) Preventation of microbial contamination in the export beef abattoir, in *Elimination of Pathogenic Organisms from Meat and Poultry* (ed. F.J.M. Smulders), Elsevier, Amsterdam, pp. 221–33.

Greer, G.G., Jones, S.D.M., Dilts, B.D. and Robertson, W.M. (1990) Effects of spray chilling on the quality, bacteriology and case life of aged carcasses and vacuum packaged beef. *Can. Inst. Food Sci. Technol. J.*, **23**, 82–6.

Grunvig, M. and Bogh-Sorensen, L. (1990) Time-temperature in distribution of meat, in *Processing and Quality of Foods*, Vol. 3 (eds P. Zeuthen, J.C. Cheftel, C. Eriksson, T.R. Gormley, P. Linko and K. Paulus), Elsevier, London, pp. 244–7.

Hardin, M.D, Acuff, G.R., Lucia, L.M., Oman, J.S. and Savell, J.W. (1995) Comparison of methods for decontamination from beef carcass surfaces. *J. Food Protect.*, **58**, 368–74.

Hildebrandt, G. and Weis, H. (1994) Sampling plans in microbiological quality control 2. Preview and future prospects. *Fleischwirtsch. Int.*, (3), 49–52.

Hinson, L.E. (1968) Why off-condition offal? *Nat. Provision.*, **159**(22), 14–18.

Hudson, W.R., Roberts, T.A. and Whelehan, O.P. (1987) Bacteriological status of beef carcasses at a commercial abattoir before and after slaughtering improvement. *Epidem. Inf.*, **91**, 81–6.

ICMSF. (1974) International Commission on Microbiological Specifications for Foods. Appropriate sampling plans, in *Microorganisms in Foods 2, Sampling for Microbiological Analysis: Principles and Specific Applications*, University of Toronto Press, Toronto, pp. 19–31.

Ingelfinger, E. (1994) Cleaning and disinfection of wet rooms. *Fleischwirtsch. Int.*, (4), 18–20.

James, S.J. and Bailey, C. (1990) Chilling of beef carcasses, in *Chilled Foods, the State of the Art* (ed. T.R. Gormley), Elsevier, London, pp. 195–81.

Jericho, K.W.F., Bradley, J.A., Gannon, V.P.J. and Kozub, G.C. (1993) Visual demerit and microbiological evaluation of beef carcasses: methodology. *J. Food Protect.*, **56**, 114–19.

Jericho, K.W.F., Bradley, J.A. and Kozub, G.C. (1994) Bacteriological evaluation of groups of beef carcasses before the wash at six Alberta abattoirs. *J. Appl. Bacteriol.*, **77**, 631–4.

Jericho, K.W.F., Kozub, G.C., Bradley, J.A., Gannon, V.P.J., Golsteyn-Thomas, E.J., Gierus, M., Mishiyama, B.J., King, R.K., Tanaka, E.E., D'Souza, S. and Dixon-MacDougall, J.M. (1996) Microbiological verification of the control of the processes of dressing, cooling and processing of beef carcasses at a high line-speed abattoir. *Food Microbiol.*, **13**, 291–302.

Jones, S.D.M., Greer, G.G., Jeremiah, L.E., Murray, A.C. and Robertson, W.M. (1991) Cryogenic chilling of pork carcasses: effects on muscle quality, bacterial populations and palatability. *Meat Sci.*, **29**, 1–16.

Joseph, R.L. (1996) Very fast chilling of beef and tenderness. *Meat Sci.*, **43**, S217–27.

Kasprowiak, R. and Hechelmann, H. (1992) Weak points in the hygiene of slaughtering, cutting and processing firms. *Fleischwirtsch. Int.*, (2), 32–40.

Kelly, C.A., Dempster, J.F. and McLouglin, A.J. (1981) The effect of temperature, pressure and chlorine concentration of spray washing water on numbers of bacteria on lamb carcasses. *J. Appl. Bacteriol.*, **51**, 415–24.

Kilsby, D.C. (1982) Sampling schemes and limits, in *Meat Microbiology* (ed. M.H. Brown), Applied Science, London, pp. 387–421.

Kilsby, D.C. and Pugh, M.E. (1981) The relevance of the distribution of microorganisms within batches of food to the control of microbiological hazards from foods. *J. Appl. Bacteriol.*, **51**, 345–54.

Kitchell, A.G., Ingram. G.C. and Hudson, W.R. (1973) Microbiological sampling in abattoirs, in *Sampling: Microbiological Monitoring of Environments* (eds R.G. Board and D.W. Lovelock), Academic Press, London, pp. 43–61.

Lasta, J.A., Rodriguez, R., Zanelli, M. and Margaria, C.A. (1992) Bacterial count from bovine carcasses as an indicator of hygiene at slaughtering places: a proposal for sampling. *J. Food Protect.*, **54**, 271–8.

Lazarus, C.R., Abu-Bakar, A., West, R.L. and Oblinger, J.L. (1977) Comparison of microbial counts in beef carcasses by using the moist-swab contact method and secondary tissue removal technique. *Appl. Environ. Microbiol.*, **33**, 217–18.

Leistner, L., Rodel, W. and Krispien, K. (1981) Microbiology of meat and meat products in high- and intermediate-moisture ranges, in *Water Activity: Influences on Food Quality* (eds L.B. Rockland and G.F. Stewart), Academic Press, London, pp. 885–916.

Locker, R.H. (1985) Cold induced toughness of meat, in *Advances in Meat Research*, vol. 1, (eds A.M. Pearson and T.R. Dutson), AVI Publishing, Westport, Connecticut, pp. 1–43.

Longdell, G.R. (1996) Recent developments in sheep and beef processing in Australasia. *Meat Sci.*, **43**, S165–74.

Lowry, P.D. and Gill, C.O. (1984a) Temperature and water activity minima for growth of spoilage moulds from meat. *J. Appl. Bacteriol.*, **56**, 193–9.

Lowry, P.D. and Gill, C.O. (1984b) Development of a yeast microflora on frozen lamb stored at –5 °C. *J. Food Protect.*, **47**, 309–11.

Lowry, P.D. and Gill, C.O. (1985) Microbiology of frozen meat and meat products, in *Microbiology of Frozen Foods* (ed. R.K. Robinson), Elsevier, London, pp. 109–68.

Mackey, B.M. and Derrick, C.M. (1979) Contamination of the deep tissues of carcasses by bacteria present on the slaughtering instruments or in the gut. *J. Appl. Bacteriol.*, **46**, 355–66.

Mackey, B.M. and Roberts, T.A. (1993) Improving slaughtering hygiene using HACCP and monitoring. *Fleischwirtsch. Int.*, (2), 40–5.

Marks, S. and Roberts, T. (1993) *E. coli* 0157:H7 ranks as the fourth most costly foodborne disease. *Food Rev.*, **16**, 51–9.

Mazur, P. (1970) Cryobiology: the freezing of biological systems. *Science*, **168**, 939–49.

Murray, A.C. (1995) The evaluation of muscle quality, in *Quality and Grading of Carcasses of Meat Animals* (ed S.D. Morgan Jones), CRC Press, Boca Raton, Florida, pp. 83–107.

NAC. (1992) U.S. Department of Agriculture, National Advisory Committee on Microbiological Criteria for Foods. Hazard analysis and critical control point system. *Int. J. Food Microbiol.*, **16**, 1–23.

NAC. (1993) U.S. Department of Agriculture, National Advisory Committee on Microbiological Criteria for Foods. Generic HACCP for raw beef. *Food Microbiol.*, **10**, 449–88.

Nesbakken, T., Nerbrink, E., Rotterud, O.J. and Borch, E. (1994) Reduction of *Yersinia enterocolitica* and *Listeria* spp. on pig carcasses by enclosure of the rectum during slaughter. *Int. J. Food Microbiol.*, **23**, 197–208.

Nickels, C., Svensson, I., Ternstrom, A. and Wickbery, L. (1976) Hygiene and economy of scalding with condensed water vapour and in tank. Proceedings of the 22nd European Meeting of Meat Research Workers, Malmo, Sweden, vol. C1, pp. 1–10.

Nottingham, P.M. (1982) Microbiology of carcass meats, in *Meat Microbiology* (ed. M.H. Brown), Applied Science, London, pp. 13–66.

Palumbo, S.A. (1986) Is refrigeration enough to restrain foodborne pathogens? *J. Food Protect.*, **49**, 1003–9.

Papadakis, E.P. (1985) The Deming inspection criterion for choosing zero or 100 percent inspection. *J. Qual. Technol.*, **17**, 121–7.

Park, R.W.A., Griffiths, P.L. and Moreno, G.S. (1991) Sources and survival of campylobacters: relevance to enteritis and the food industry. *J.Appl. Bacteriol.*, Symp. Suppl., **70**, 97S–106S.

Reichel, M.P., Phillips, D.M., Jones, R. and Gill. C.O. (1991) Assessment of the hygienic adequacy of a commercial hot boning process for beef by a temperature function integration technique. *Int. J. Food Microbiol.*, **14**, 27–42.

Restaino, L. and Wind, C.K. (1990) Antimicrobial effectiveness of hand washing for food establishments. *Dairy Food Environ. Sanit.*, **10**, 136–41.

Roberts, T.A., Hudson, W.R., Whelehan, O.P., Simonsen, B., Olgaard, K., Labots, H., Snijders, J.M.A., Van Hoof, J., Debevere, J., Dempster, J.F., Devereux, J., Leistner, L., Gehra, M., Gledel, J. and Fournaud, J. (1984) Number and distribution of bacteria on some beef carcasses at selected abattoirs in some member states of the European Communities. *Meat Sci.*, **11**, 191–205.

Rogers, S.A., Hollywood, N.W. and Mitchell, G.E. (1992) The microbiological and technological properties of bruised beef. *Meat Sci.*, **32**, 437–47.

Rosset, R. (1982) Chilling, freezing and thawing, in *Meat Microbiology* (ed. M.H. Brown), Applied Science, London, pp. 265–318.

Ryan, T.P. (1989) *Statistical Methods for Quality Improvement*, Wiley, New York, NY.

Schilling, E.G. (1982) *Acceptance Sampling in Quality Control*, Marcel Dekker, New York.

Schmidt, U. (1983) Cleaning and disinfection of slaughterhouses and meat processing factories. *Fleischwirtschaft*, **63**, 1188–91.

Schnell, T.D., Sofos, J.N., Littlefield, V.G., Morgan, J.B., Gorman, B.M., Clayton, R.P. and Smith, G.C. (1995) Effects of post-exsanguination dehairing on the microbial load and visual cleanliness of beef carcasses. *J.Food Protect.*, **58**, 1297–302.

Scott, W.J. and Vickery, J.R. (1939) Investigations on chilled beef, part II. Cooling and storage at the meat works. *Council Sci. Ind. Res. Australia*, Bull. No. 129.

Seman, D.L., Drew, K.P., Clarken, P.A. and Littlejohn, R.P. (1988) Influence of packaging methods and lengths of chilled storage on microflora, tenderness and colour stability of venison loins. *Meat Sci.*, **22**, 276–82.

Sharpe, A.N., Kingombe, C.I.B., Watney, P., Parrington, L.J., Dudas, I. and Diotte, M.P. (1996) Efficient nondestructive samples for carcasses and other surfaces. *J. Food Protect.*, **59**, 757–63.

Smeltzer, T., Thomas, R. and Collins, G. (1980) The role of equipment having accidental or indirect contact with the carcass in the spread of *Salmonella* in an abattoir. *Austral. Vet. J.*, **56**, 14–17.

Smith, M.G. (1985) The generation time, lag time and minimum temperature of growth of coliform organisms in meat and the implication for codes of practice in abattoirs. *J. Hyg, Camb.*, **94**, 289–300.

Smith, M.G. (1992) Destruction of bacteria on fresh meat by hot water. *Epidemiol. Infect.*, **109**, 491–6.

Smith, M.G. and Davey, K.R. (1990) Destruction of *Escherichia coli* on sides of beef by a hot water decontamination process. *Food Australia*, **42**, 195–8.

Smith, M.G. and Graham, A. (1978) Destruction of *Escherichia coli* and salmonellae on mutton carcasses by treatment with hot water. *Meat Sci.*, **2**, 119–28.

Smulders, F.J.M., Barendsen, P., Van Logtestijn, J.G., Mossel, D.A.A. and van der Marel, G.M. (1986) Lactic acid: considerations in favour of its acceptance as a meat decontaminant. *J. Food Technol.*, **21**, 419–36.

Sorquist, S. and Danielsson-Tham, M.L. (1986) Bacterial contamination of the scalding water during vat scalding of pigs. *Fleischwirtschaft*, **66**, 1745–8.

Strydom, P.E. and Buys, E.M. (1995) The effects of spray-chilling on carcass mass loss and surface associated bacteriology. *Meat Sci.*, **39**, 265–76.

Tauxe, R.V. (1991) Salmonella: a postmodern pathogen. *J. Food Protect.*, **54**, 563–8.

Thornton, H. and Gracey, J.F. (1974) *Textbook of Meat Hygiene*, 6th edn, Balliere Tindall, London.

Tompkin, R.B. (1990) The use of HACCP in the production of meat and poultry products. *J. Food Protect.*, **53**, 795–803.

Tornberg, E. (1996) Biophysical aspects of meat tenderness. *Meat Sci.*, **43**, 5175–91.

USDA. (1970) U.S. Department of Agriculture. Guidelines for chilling, freezing, shipping and packaging meat carcasses and meat byproducts. *Agric. Handbook* No. 412.

USDA. (1993) U.S. Department of Agriculture. Beef carcass trimming versus washing study: Fiscal year 1993. *Fed. Regist.*, **58**, 33925–32.

USDA. (1995) U.S. Department of Agriculture. Pathogen reduction: hazard analysis and critical control point systems; proposed rule. *Fed. Regist.*, **60**, 6774–889.

USDA. (1996) U.S. Department of Agriculture. Pathogen reduction; hazard analysis and critical control point (HACCP) system; Final rule. *Fed. Regist.*, **61**, 38805–989.

Van Donkersgoed, J., Jericho, K.W.F., Grogan, H. and Thorlakson, B. (1997) Preslaughter hide status of cattle and the microbiology of carcasses. *J. Food Protect.* (in press).

Van Klink, E.G.M. and Smulders, F.J.M. (1989) Sanitation by ultrasonic cavitation of steel mesh gloves used in the meat industry. *J. Food Protect.*, **52**, 660–4.

Woolthuis, C.H.J. and Smulders, F.J.M. (1985) Microbial decontamination of calf carcasses by lactic acid sprays. *J. Food Protect.*, **48**., 832–7.

Zottola, E.A. and Sasahara, K.C. (1994) Microbial biofilms in the food processing industry – should they be a concern? *Int. J. Food Microbiol.*, **23**, 125–48.

5 The microbiology of the slaughter and processing of poultry

N.M. BOLDER

5.1 Introduction

Over the past 20 years (Mulder, 1993a) poultry meat production worldwide has increased rapidly with an annual growth rate of 6%. This has led to intensive animal production with an increase in both the number of farms, and in flock size. Both have raised specific problems, such as contamination with human and animal pathogens, and welfare and environmental problems. With poultry meat processing there has been a very rapid transition from the handcraft operations of the 1950s and 1960s to an almost fully automated and mechanized process today (Hupkes, 1996). This development enables poultry processors to slaughter large numbers of animals without much handling of the product. This favours the hygienic quality of the final product. Competition increases the pressure on poultry meat producers. This forces them towards innovation in processing and the development of new, added quality and guaranteed safe products.

Products are perceived to be safe when microbiological and chemical hazards are absent. Live poultry for meat production are normally raised on litter floors. This may lead to contamination of poultry with human pathogens, such as *Salmonella*, *Campylobacter*, *Listeria*, *Escherichia coli*, *Clostridium* and *Staphylococcus aureus*. Additionally spoilage microorganisms, mainly psychrotrophs such as pseudomonads, lactic acid bacteria and yeasts, are present on live animals. Only occasionally do young animals show symptoms of bacterial infection but most are healthy carriers of pathogens, such as *Salmonella* and *Campylobacter*. As long as these pathogens are not excluded from animal husbandry, poultry and poultry products may well be contaminated. In practice, however, only Specific Pathogen Free (SPF) systems are likely to guarantee pathogen-free animals being presented for slaughter. This system is not economically feasible at present. During poultry processing, contamination levels can be controlled by taking hygienic measures, based on the HACCP principle, to avoid cross contamination, both between products and between equipment and product. Even so complete eradication of pathogens from poultry products seems impossible without additional decontamination treatments.

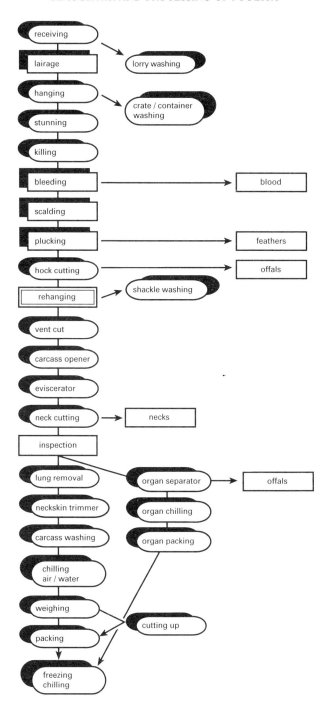

Figure 5.1 Schematic flow sheet of a poultry processing line.

5.2 Poultry processing

Processing of poultry is a complicated operation, which has been mechanised extensively during the past decades. Figure 5.1 gives a simplified flow sheet for a poultry slaughter line.

Manufacturers of equipment for poultry processing have to consider three priorities when designing and producing machines – machines must perform according to specification, be safe and be hygienic during operation (Hupkes, 1996). Equipment is considered to meet the last criterion, when it is easy to clean and disinfect and does not have an adverse influence on products. In this respect, cleaning of machine stations in between contact with products would be the most favourable option. Line speeds do not often allow such an option and will not do so until appropriate cleaning-in-place (CIP) systems have been developed (Stals, 1992). EU Directive 71/118/EEC defines hygienic standards in poultry processing, but in actual practice it appears difficult to meet these standards.

5.3 Microbiology of poultry

Poultry has a very complex microflora, which is partly of intestinal origin, due to the production system, flocks of large numbers of fast growing animals and being reared in climatized houses on litter floors. Kotula and Pandya (1995) found high numbers of human pathogens (Table 5.1) on the feathers and skin of broilers entering a processing plant. Bacteria that come in contact with feathers, skin or meat surfaces, will attach either by physical forces or chemical means such as polysaccharride bonds (Firstenberg-Eden, 1981). The contact time is also an important factor in attachment.

McMeekin and Thomas (1979) have compared the microflora of live poultry in a number of studies. These studies showed differences in microflora of processed carcasses from year to year (Table 5.2).

Table 5.1 Bacterial counts of broiler chickens entering the processing plant (Kotula and Pandya, 1995)

Sample	Log $_{10}$ cfu/g			
	Total count	Escherichia coli	Salmonella	Campylobacter
Feathers on				
breast	7.6	8.0	7.2	7.5
thigh	7.7	7.8	6.5	7.3
drum	7.7	7.9	6.5	7.4
Skin on				
breast	8.3	7.4	6.3	6.9
thigh	8.0	6.3	5.9	6.4
drum	8.3	6.6	5.8	6.4

Table 5.2 Microflora of poultry carcasses immediately after processing (McMeekin and Thomas 1979)

Year of study	1966	1976	1978	1979
Organism				
Micrococcus spp.	50**	–	30	3
Gram-positive rods	14	<10		4
Flavobacteria	14	20	18	8
Enterobacteriaceae	8	–	8	8
Pseudomonads	2	<10	26	5
Acinetobacter/Moraxella	7	50	4	65
Yeasts and others	5	<10	14	7
Ratio of motile/non motile bacteria	26/74	30/70	48/52	21/79

** % isolated; –, not analysed.

As was mentioned above, the heterogeneous microflora on processed poultry consists of both pathogenic and spoilage micro-organisms. It is well established that *Salmonella* and *Campylobacter* are the major cause of poultry borne infections of humans. The isolation rates of *Staphylococcus aureus*, *E. coli*, *Listeria*, *Aeromonas* and *Clostridium* from poultry have increased over time. The incidence of human food-borne infection does not reflect this trend (Mulder, 1993a). Waldroup (1996), who collected data on the contamination of raw poultry with pathogens, concluded that there is very little available information on the 'new' pathogens *Yersinia*, *Hafnia*, *Bacillus* and *E. coli* O157:H7. The spoilage micro-organisms, *Acinetobacter*, *Brochothrix*, *Pseudomonas*, lactic acid bacteria and yeasts, grow relatively rapidly at low temperatures and have a major impact on the shelf life of fresh poultry products (see Chapters 1 and 2).

5.4 Microbiology of poultry processing

5.4.1 Transport of live animals

Poultry have to be caught and transported to a processing plant, an operation that has a major impact on the birds. This is accentuated when birds have been denied feed for a certain period.

Traditional wooden transport boxes have been replaced by plastic transport coops and these have been developed into large transport-container systems which have improved animal welfare by increased automation in the reception area of processing plants (Veerkamp, 1995). The large openings of these modern containers – *vis à vis* those of plastic crates – cause less stress and injury to animals during the loading and unloading operations.

Additionally cleaning and disinfection of such systems are much easier than with the old-fashioned crates. Harvesting machines for loading broilers into these containers are now becoming available.

Stress during transport can change the excretion patterns of birds carrying *Salmonella* through disturbance of intestinal function and even damage of the intestinal tract such that the immune system is adversely affected (Mulder, 1996). Excretion of pathogenic bacteria during transport will cause cross contamination among the animals in a container. Moreover studies on broiler transportation have shown clearly that ineffective cleaning and disinfection of crates leads also to cross contamination. Bolder and Mulder (1983) isolated a larger number of *Salmonella* serotypes from crates than from the caecal contents of broilers from the flock transported in these crates (Table 5.3). This phenomenon was found also in a field study of broilers (Goren *et al.*, 1988) in which the *Salmonella* reduction achieved by competitive exclusion was eliminated by transport of the birds. Rigby *et al.* (1980) found increasing numbers of *Salmonella* serotypes after washing the crates. In practice, however, thorough cleaning of the transport crates, followed by effective disinfection will prevent the *Salmonella* contamination of the following flock. Cleaning of crates calls for a well managed, multistage washing operation – flushing away of the major part of organic material and the soaking of that which remains, a washing step with a detergent at an elevated temperature of 40 °C, followed by flushing with clean water and finally disinfection (Bolder 1988). The disinfectant ought to be sprayed onto the crates from which the rinse water has drained away; the disinfectant will then persist on the crates.

Table 5.3 *Salmonella* isolations from broilers after slaughter and faecal material from transportation crates (Bolder and Mulder, 1983)

Flock	*Salmonella* serotypes isolated from:	
	Caecal samples	Faecal material from crates
1		blockley hadar cerro
2		infantis livingstone typhimurium
3		blockley
4	montevideo	blockley give montevideo senftenberg
5	anatum	anatum

5.4.2   Scalding

After stunning, killing and bleeding (Fig. 5.1), poultry carcasses are submerged in a warm waterbath (the so-called scald tank) at 50–60 °C, for up to 3 min. Scalding loosens feathers and makes feather removal easier. The effect of scalding on poultry skin depends on the water temperature and time combination. At high temperatures of 58–60 °C, the epidermal skin layer will be removed in the pluckers. With this type of 'high scalding', poultry is normally produced for the frozen or fresh, nonfrozen market. Carcasses scalded at 50–52 °C can be air chilled without deterioration of the skin. These are intended mainly for the fresh, nonfrozen market. Kim *et al.* (1993) measured microscopically higher numbers of *Salmonella* on chicken carcasses after scalding at 60 °C than on those scalded at 52 °C or 56 °C. They failed, however, to confirm these results by bacteriological methods. Slavik *et al.* (1995), who scalded chicken carcasses at 52 °C, 56 °C or 60 °C, found that the incidence of *Salmonella* and *Campylobacter* can be influenced by the scald temperature (Table 5.4). Mulder and Dorresteyn (1977) have compared bacterial loads of scald water at different temperatures (Table 5.5).

It is evident from the above that, from a hygienic point of view, scalding is a hazardous operation – carcasses with a huge load of organic material and microbes are submerged thereby donating faecal material to the water

Table 5.4 Numbers of *Salmonella* and *Campylobacter* (log mpn/carcass) on chicken carcasses scalded at three different temperatures (Slavik *et al.*, 1995)

Bacteria	Scald temperature (°C)	Trial		
		1	2	3
Salmonella	52	3.00	3.17	3.09
	56	3.16	3.17	3.34
	60	3.50	3.48	3.36
Campylobacter	52	3.64	3.30	4.18
	56	3.39	2.94	3.39
	60	4.08	3.59	3.98

Table 5.5 Bacterial load of scald water at different temperatures and different slaughter capacities (Mulder and Dorresteyn, 1977)

Temperature (°C)	Slaughter capacity (birds/h)	Total counts (cfu/ml)	Enterobacteriaceae (cfu/ml)
52	6000	1.2×10^6	1.2×10^4
54	2700	1.0×10^5	1.5×10^3
55.5	5900	2.2×10^4	1.1×10^3
60	4800	1.7×10^4	3.2×10
60	3700	1.2×10^4	3.0×10

Table 5.6 Microbial counts of conventionally scalded and plucked broiler carcasses, compared with simultaneously scalded and plucked broilers (Veerkamp and Hofmans, 1973)

Method	Number	Water temperature (°C)	Total count (log cfu/g skin)	Entero-bacteriaceae
Simultaneous	90	65	3.82	2.54
Simultaneous	9	58	3.93	2.54
Conventional	20	58	4.93	4.30
Conventional	40	60	6.11	3.92

(Kotula and Pandya 1995). Consequently researchers have sought alternative methods or modifications of existing ones. Spray scalding with hot water (Veerkamp and Hofmans, 1973) or steam (Lillard *et al.*, 1973, Clouser *et al.*, 1995) has been investigated. Spray water should be used once only or pasteurized before reuse. Spraying and plucking in a single operation was developed by Veerkamp and Hofmans (1973). They noted lower bacterial contamination of the carcasses processed by this method than on

MULTI-STAGE SCALDING

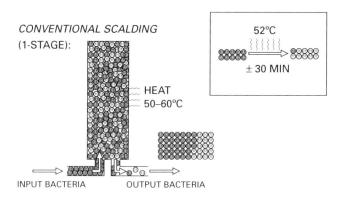

Figure 5.2 Comparison of traditional and multi-stage scalding system (Veerkamp, 1995. Credit: Meyn).

those that had been scalded and plucked conventionally (Table 5.6). The former was never implemented on a commercial scale, because line speeds increased and the switch from high to low temperature scalding enabled pluckers to remove feathers completely from the carcasses. Clouser *et al.* (1995) tested a spray scalding system for turkey carcasses. They concluded that less cross contamination with *Salmonella* occurred with this method *vis à vis* a traditional scalder.

Research has continued, therefore, on immersion scalding. Two main roads were followed: 1, modification of the existing technology, and 2, improvement of the efficiency and hygiene of the traditional scalding system. The traditional scalding tank was stirred in such a way that ideal mixing of water was achieved. A logical, and relatively simple modification, was the introduction of a counter-current flow of water and carcasses. In practice this option appeared to be improbable in a commercial context (Veerkamp, 1995) because the efficient mixing of water would spread the bacteria equally along the whole tank. Waldroup *et al.* (1992), however, compared traditional and counter-current scalding methods and found no difference in the bacterial quality of the end product.

Further developments led to division of the scald tank into several smaller ones (Veerkamp, 1991) with or without recirculation of water. The advantage of a cascade system using several tanks was that a counter-current flow of water and carcasses became feasible (Fig. 5.2). In a pilot system at the Spelderholt research centre in Beekbergen in The Netherlands, experiments showed clearly improvements in hygiene with increasing number of tanks (Fig. 5.3). The experiments showed moreover that recirculation and

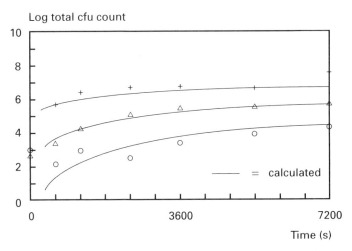

Figure 5.3 Results of measurements and theoretical calculation of total cfu counts (water supply is equal to adherent water). Marks are measurements: +, tank 1; △, tank 2; ○, tank 3.

pasteurization of scald water caused only a small additional reduction of the bacterial load (Veerkamp, 1991). Showering of carcasses post-scald did not lower the bacterial load significantly (Renwick *et al.*, 1993). High pressure treatment of scald water (>800 bar) has been considered as a possible treatment (Bolder, 1997).

Humphrey (1981) showed that pH of scald water and the presence of organic matter affected the Decimal Reduction value (D_{10}-value) of *Salmonella typhimurium*. During commercial processing the pH value decreases from pH 8.4 to pH 6.0, with a small peak during lunch break. The decrease in scald water pH from >8 to approximately 6 caused an increase in the D_{10} value of *Salmonella typhimurium* from 17 to 49.5 minutes. Bolder (1988, unpublished results) tested D_{10}-values of *Salmonella typhimurium*, *Staphylococcus aureus* and *Campylobacter jejuni* in scald water at pH 4–9 (Table 5.7) and found, in accordance with Humphrey *et al.* (1984), that the highest *D*-values for all micro-organisms was at pH 6. Deviation from pH 6, either higher or lower, resulted in a decrease of the D_{10}-value. Lillard *et al.* (1987) acidified scald water but found no effect on the bacterial load of carcasses. Lowering the pH with organic acids may affect the organoleptic quality of the product and may therefore be of questionable use in commerce. The addition of trisodium phosphate (TSP) at pH 12 to scald water could be considered although corrosion of equipment may be a problem.

5.4.3 Plucking

Defeathering was automated many years ago. The efficiency of feather removal is closely related with the scalding operation as pluckers are designed to treat many carcasses on line, the main problem being cross contamination (Mulder *et al.*, 1977). They found that the same bacterium could be isolated from up to 700 carcasses after the processing of a contaminated carcass. They also showed a beneficial effect of a water spray during plucking on the bacterial load of carcasses. Mead *et al.* (1993) demonstrated the colonization of plucking equipment with *Staphylococcus aureus*, by the increased levels of contamination of carcasses with this organism. In a

Table 5.7 D_{10}-values (min) of *Salmonella*, *Staphylococcus* and *Campylobacter* in scald water at 52°C (Bolder, 1988, unpublished results)

pH	Salmonella typhimurium	Staphylococcus aureus	Campylobacter jejuni
4	<2	2	<2
5	6	10	<2
6	>20	>20	3
7	>20	20	2
8	12	10	<2
9	5	5	<2

model system, Mead *et al.* (1995) investigated the attachment of staphylococci to rubber plucker fingers. The authors found attachment of the micro-organisms in spite of the rapid finger rotation. Mead and Scott (1994) isolated *E. coli* both from carcasses and machines, but they concluded that it was unlikely that *E. coli* colonized the equipment. Chlorination of spray water was effective in the control of cross contamination in pluckers (Mead *et al.*, 1994). Notermans *et al.* (1980) proved that attachment of bacteria both to equipment and product could be avoided while water was used during plucking. Purdy *et al.* (1988), who changed the lay-out of pluckers, increased line speeds and removed chlorine from spray water in pluckers. They concluded that bacterial loads on carcasses and the extent of colonization of plucker fingers with bacteria increased.

Clouser *et al.* (1995) found cross contamination of *Salmonella* during plucking of turkey carcasses. When steam-spray scalding and plucking was applied, the authors found that the high (> log 4 cfu before plucking) initial bacterial load on carcasses decreased throughout the slaughter process. Carcasses with lower counts (< log 4 cfu) were visually cleaner but did not show lower total viable counts at the end of plucking. They found also a relationship between total plate counts and incidence of *Salmonella* within flocks and an increase in this incidence during processing of *Salmonella*-positive flocks.

The introduction of a so-called closed-loop washing system during plucking was studied from an economic and hygiene point of view by Veerkamp and Pieterse (1993). This system is a further development of the simultaneous scalding and plucking system (Veerkamp and Hofmans, 1973). In the closed loop system water at 50 °C is sprayed on the carcasses during plucking and, after separation from the feathers and pasteurization, one part of the water is used in the scald tank and the other reused in the pluckers. The use of water in pluckers may cause the formation of highly contaminated aerosols. McDermid and Lever (1996) found aerosols were a possible way of transmission of *Salmonella* during poultry processing. Tinker *et al.* (1996) described model devices for plucking carcasses that would avoid aerosol formation and cross contamination to a great extent.

Stals (1992) indicated modifications in plucking operations by introducing additional tail feather removal. This affords less pressure of the pluckers, and avoids excretion of faecal material from the carcasses.

5.4.4 Evisceration

The evisceration process of poultry covers the process from plucking to chilling of the carcasses. There are several stages in the evisceration process, starting with head removal, opening of the body cavity, removal of intestines and finishing with cleaning of the carcass. With processing speeds of over 6000 birds per hour on a line, the inspectors can hardly check this number of carcasses properly.

The most important improvement in the hygiene of the evisceration process is the automated transfer of carcasses from the slaughter line to the evisceration line and, when appropriate, to the chilling line. This reduces product handling and thus cross contamination (Stals, 1992). Of course, cross contamination may occur due to the number of machines involved in the evisceration process. Opening the carcass is a two-step operation – cutting out the cloaca with the large intestine attached, followed by cutting open the body cavity. Then the eviscerator takes out the intestines and leaves them on the carcass for inspection or separates them from the carcass *via* a mechanical device for separate inspection. This latter system was introduced a few years ago and it has been claimed that it improves carcass quality, contributed to the prevention of cross contamination due to faecal material and facilitated the veterinarian inspection of the carcasses and intestines. No literature apart from that of equipment manufacturers is available (Anon, 1994, 1996) to support these claims. Less visible faecal contamination of carcasses was noted, but there was no relationship between bacterial counts and appearance of carcass (Veerkamp *et al.*, 1993, unpublished results).

5.4.5 Chilling

There are two methods for chilling poultry, immersion chilling and air chilling. Both are effective, but the application of each is closely related to the needs of the market. Frozen carcasses are normally water chilled; fresh, non-frozen poultry can either be water chilled and marketed 'wet', or air chilled and marketed 'dry'. Air chilled products have to be scalded at low temperatures so that the epidermis is left on the carcass and discoloration is avoided. Mulder and Veerkamp (1974) found no difference as measured by bacterial counts between spray chilling, or immersion chilling of broiler carcasses.

Mulder and Bolder (1987) found no difference in the shelf life of different types of poultry even though the initial bacterial load of the high scalded carcasses was higher. These authors used air chilling, combined with intermittent water spray on the carcasses so that evaporation of water facilitated chilling. Problems with air chilling can occur in large chilling tunnels having a number of cooling units. During operation, the water will freeze on the condensers, which have to be defrosted periodically. While defrosting, ventilators will blow contaminated water into the chiller and onto the product. Stephan and Fehlhaber (1994) investigated bacterial contamination in air chill-tunnels and found aerosols as a major cause of cross contamination when water was sprayed on carcasses.

Mead *et al.* (1993) screened three broiler processing plants with air chilling units and one broiler and one turkey plant with immersion chilling. Total viable counts on carcasses were lower after water chilling and equal or higher after air chilling. Levels of *Pseudomonas* spp. after chilling, however, were higher in all plants. In broiler processing plants during immersion chilling, both total viable counts and Enterobacteriaceae decreased significantly, but *Salmonella* incidence increased (Lillard, 1990).

Water treatment in immersion chillers can be used to control contamination. Chlorination was used successfully with chlorine concentrations >25 ppm (Mead and Thomas, 1973; James *et al.*, 1992). The use of 1.33 mg/l ClO_2 in chill water for poultry controlled the numbers of *Salmonella*, but did not reduce effectively bacterial counts (<0.5 log cfu) (Thiessen *et al.*, 1984). Eradication of *Salmonella* from carcasses occurred with a chlorine concentration of 300–400 ppm (Morrison and Fleet, 1985). Dickens and Whittemore (1995) immersed carcasses in chill water with or without acetic acid. They concluded that the level of Enterobacteriaceae could be reduced by 0.38 log cfu in the control group and by 1.4 log cfu when 0.6% (v/v) acid was used in the chill water. Blank and Powell (1995) found a substantial reduction in the levels of bacterial contamination of broiler carcasses after immersion chilling as such. Addition of acidic sodium pyrophosphate to chiller water (1.5%; pH 2.8) for poultry led to a significant reduction of coliform and *E. coli* counts in the water (Rathgeber and Waldroup, 1995).

Ozonation of chiller water did not reduce effectively bacterial contamination of carcasses (<1 log cfu) (Sheldon and Brown, 1986). The addition of 6600 ppm hydrogen peroxide reduced by 95% the total counts in water but 11000 ppm was needed to reduce carcass counts by 94%. Application in practice will be a problem, however, because catalase reactions in the water and on the carcasses leads to bleaching and bloating of carcasses and foaming of chiller water (Lillard and Thomson, 1983). Li *et al.* (1995) used pulsed electricity plus TSP or NaCl in poultry chiller water and found a correlation between D_{10}-values and concentration of chemicals and thus pH level.

5.5 Packing

Packing of poultry meat products varies with the type of slaughter process. Whole carcasses can be packed either individually in plastic bags, on trays wrapped in polyethylene foil for the fresh market, or in bulk without individual wrappings. The third method of packing is used for poultry carcasses processed with 'high' scalding and immersion chilling for fresh, nonfrozen products packed on ice in order to prevent discoloration. Low scalded and air chilled products are packed dry without ice. The maintenance of the cooling chain, especially for dry products is an essential aspect in packing and distribution of poultry products. During packaging and distribution the short residence times and constant low temperature are of great importance. Variations in temperature cause condensation on the product or on the inside of the wrapping material. This facilitates growth of spoilage micro-organisms.

Table 5.8 Effect of cleaning and disinfection of a slaughter line on bacterial load of carcasses (log N cfu/g skin) (Svendsen and Caspersen, 1981)

	After plucker	Entry evisceration room	After vent cutter	After eviscerator	After spray washer	After chill and pack
Total counts						
before	6.0	6.3	6.0	5.9	5.6	5.6
after	4.9	4.8	4.7	4.8	4.6	4.6
Pseudomonads						
before	3.3	3.1	3.0	3.3	2.8	3.3
after	2.5	2.6	2.4	2.8	2.8	2.5

The best way to avoid spoilage of products is production under hygienic conditions such that the product has a low initial bacterial load (Table 5.8). This can be achieved by application of HACCP in poultry processing operations; cleaning and disinfection may lead to a tenfold reduction in bacterial load (Svendson and Caspersen, 1981). Equipment construction may make hygienic production difficult and thus influence shelf life of the product. The implementation of a HACCP system will force poultry meat producers to study their production process and find, monitor and control the critical points (CPs) (Mulder, 1993b). Together with improvements in processing equipment, this should help in controlling product safety. Safe poultry products can be produced, either by slaughtering pathogen free animals or by giving the products a decontamination treatment during or after the process (Bolder, 1977; Corry *et al.*, 1995; Dickson and Anderson, 1992).

Hygienic processing must never be neglected and the social aspect of hygiene management must be emphasized; thus behaviour and information to workers, and training and feed-back of analytical findings, will improve awareness and engender a positive attitude (Gerats, 1987).

5.6 Further processing

Equipment for futher processing handles a high volume of products and thus can contribute to the cross contamination of products. Intermediate cleaning of this equipment is even more important because of small particles of meat or fat that may come in contact with other products. In this phase of the process, time and temperature control is important, with respect to condensation on the product.

References

Anon. (1994) Evisceration changes boost yield and quality. *Poultry Int.*, **33**, 16–18.

Anon. (1996) Meyn Giblet harvesting systems; High tech design for the future. *Poultry Int.*, **35**, 84–8.

Blank, G. and Powell, C. (1995) Microbiological and hydraulic evaluation of immersion chilling for poultry. *J. Food Protect.*, **58**, 1386–8.

Bolder, N.M. (1988) Reiniging en desinfectie van pluimveetransportkratten. *Vleesdistributie en Vleestechnologie*, **2**, 32–3.

Bolder, N.M. (1997) Decontamination of meat and poultry carcasses. *Trends Food Sci. Technol.* **8**, 221–7.

Bolder, N.M. and Mulder, R.W.A.W. (1983) Faecal materials in transport crates as source of *Salmonella* contamination of broiler carcasses. Proceedings of the 6th European symposium on quality of poultry meat (eds C. Lahellee, F.H. Ricard and P. Colin), Ploufragan. pp. 170–6.

Clouser, C.S., Doores, S., Mast, M.G. and Knabel, S.J. (1995) The role of defeathering in the contamination of turkey skin by Salmonella species and Listeria monocytogenes. *Poultry Sci.*, **74**, 723–31.

Corry, J.E.L., Adams, C., James, S.J. and Hinton, M.H. (1995) *Salmonella, Campylobacter* and *Escherichia coli* O157:H7 decontamination techniques for the future. *Int. J. Food Microbiol.*, **28**, 187–96.

Dickens, J.A. and Whittemore, A.D. (1995) The effects of extended chilling times with acetic acid on the temperature and microbiological quality of processed poultry carcasses. *Poultry Sci.*, **74**, 1044–8.

Dickson, J.S. and Anderson, M.E. (1992) Microbiological decontamination of food animal carcasses by washing and sanitizing systems: a review. *J. Food Protect.*, **55**, 133–40.

Firstenberg-Eden, R. (1981) Attachment of bacteria to meat surfaces: a review. *J. Food Protect.* **44**, 602–7.

Gerats, G.E. (1987) What hygiene can achieve – how to achieve hygiene, in *Elimination of Pathogenic Organisms from Meat and Poultry* (ed. F.J.M. Smulders), Elsevier, Amsterdam, pp. 269–79.

Goren, E., de Jong, W.A., Doorenbal, P., Bolder, N.M., Mulder, R.W.A.W. and Jansen, A. (1988) Reduction of salmonella infection of broilers by spray application of intestinal microflora: a longitudinal study. *Vet. Quart.*, **10**, 249–54.

Humphrey, T.J. (1981) The effects of pH and levels of organic matter on the death rates of Salmonellas in chicken scald-tank water. *J. Appl. Bacteriol.*, **51**, 27–39.

Humphrey, T.J., Lanning, D.G. and Leeper, D. (1984) The influence of scald water pH on the death rates of *Salmonella typhimurium* and other bacteria attached to chicken skin. *J. Appl. Bacteriol.*, **57**, 355–9.

Hupkes, H. (1996) Automation and hygiene in relation to poultry processing. Proceedings of a meeting in Concerted action CT94-1456: Microbial control in the meat industry, Perugia, Italy, (eds M.H. Hinton and C. Rowlings). University of Bristol Press, Bristol, pp. 95–8.

James, O.J., Brewer, R.L., Prucha, J.C., Williams, W.O. and Parham, D.R. (1992) Effects of chlorination of chill water on the bacteriologic profile of raw chicken carcasses and giblets. *J. Am. Vet. Med. Assoc.*, **200**, 60–3.

Kim, J.W., Slavik, M.F., Griffis, C.L. and Walker, J.T. (1993) Attachment of *Salmonella typhimurium* to skins of chicken scalded at various temperatures. *J. Food Protect.*, **56**, 661–71.

Kotula, K.L. and Pandya, Y. (1995) Bacterial contamination of broiler chickens before scalding. *J. Food Protect.*, **58**, 1326–9.

Li, J., Walker, J.T., Slavik, M.F. and Wang, H. (1995) Electrical treatment of poultry chiller water to destroy *Campylobacter jejuni*. *J. Food Protect.*, **58**, 1330–4.

Lillard, H.S. (1990) The impact of commercial processing procedures on the bacterial contamination and cross-contamination of broiler carcasses. *J. Food Protect.*, **53**, 202–4.

Lillard, H.S., Blankenship, L.C., Dickens, J.A., Craven, S.E. and Shackelford, A.D. (1987) Effect of acetic acid on the microbiological quality of scalded picked and unpicked broiler carcasses. *J. Food Protect.*, **50**, 112–14.

Lillard, H.S., Klose, A.A., Hegge, R.I. and Chew, V. (1973) Microbiological comparison of steam- (at sub atmospheric pressure) and immersion-scalded broilers. *J. Food Sci.*, **38**, 903–4.

Lillard, H.S. and Thomson, J.E. (1983) Efficacy of hydrogen peroxide as a bactericide in poultry chill water. *J. Food Sci.*, **48**, 125–6.

McDermid, A.S. and Lever, M.S. (1996) Survival of *Salmonella enteritidis* PT4 and *Salmonella typhimurium* Swindon in aerosols. *Letts Appl. Microbiol.*, **23**, 107–9.

McMeekin, T.A. and Thomas, C.J. (1979) Aspects of the microbial ecology of poultry processing and storage: a review. *Food Technol. Australia*, January, **31**, 35–41.

Mead, G.C., Hudson, W.R. and Hinton, M.H. (1993) Microbiological survey of five poultry processing plants in the UK. *Br. Poultry Sci.*, **34**, 497–503.

Mead, G.C., Hudson, W.R. and Hinton, M.H. (1994) Use of a marker organism in poultry processing to identify sites of cross contamination and evaluate possible control measures. *Br. Poultry Sci.*, **35**, 345–54.

Mead, G.C., Gibson, C. and Tinker, D.B. (1995) A model system for the study of microbial colonization in poultry defeathering machines. *Letts Appl. Microbiol.*, **20**, 134–6.

Mead, G.C. and Scott, M.J. (1994) Coagulase negative staphylococci and coliform bacteria associated with mechanical defeathering of poultry carcasses. *Letts Appl. Microbiol.*, **18**, 62–4.

Mead, G. and Thomas, N.L. (1973) Factors affecting the use of chlorine in the spin chilling of eviscerated poultry. *Br. Poultry Sci.*, **14**, 99–117.

Morrison, G.J. and Fleet, G.H. (1985) Reduction of *Salmonella* on chicken carcasses by immersion treatments. *J. Food Protect.*, **48**, 939–43.

Mulder, R.W.A.W. (1993a) Microbiology of poultry meat. *Poultry Int.* **32**, 26–30.

Mulder, R.W.A.W. (1993b) Hygiene control: an important part of integrated quality assurance in the poultry industry, in *Interruption of Bacterial Cycles in Animal Production*, (ed. B.A.P. Urlings), Veterinary University, Utrecht, The Netherlands, pp. 123–7.

Mulder, R.W.A.W. (1996) Impact of transport on the incidence of human pathogens in poultry. *World Poultry*, **12**, 18–19.

Mulder, R.W.A.W. and Bolder, N.M. (1987) Haltbarkeit kühlgelagerter Schlachthähnchen nach verschiedenen Brüh- und Kühlvorgängen. *Fleischwirtschaft*, **67**, 114–16.

Mulder, R.W.A.W. and Dorresteijn, L.W.J. (1977) Hygiene beim Brühen von Schlachtgeflügel. *Fleischwirtschaft*, **57**, 2220–2.

Mulder, R.W.A.W., Dorresteijn, L.W.J. and van der Broek, J. (1997) Kruisbesmetting tijdens het broeien en plukken van slachtkuikens. *Tijdschrift voor Diergeneeskunde*, **102**, 619–29.

Mulder, R.W.A.W. and Veerkamp, C.H. (1974) Microbiological aspects of water-cooling on poultry carcasses. *Lebensmittel Wissenschaft un Technologie*, **7**, 127–31.

Notermans, S., Terbijhe, R.J. and van Schothorst, M. (1980) Removing faecal contamination of broilers by spray cleaning during evisceration. *Br. Poultry Sci.*, **21**, 115–21.

Purdy, J., Dodd, C.E.R., Fowler, D.R. and Waites, W.M. (1988) Increase in microbial contamination of defeathering machinery in a poultry processing plant after changes in the method of processing. *Letts Appl. Microbiol.*, **6**, 35–8.

Rathgeber, B.M. and Waldroup, A.L. (1995) Antibacterial activity of a sodium acid pyrophosphate product in chiller water against selected bacteria on broiler carcasses. *J. Food Protect.*, **58**, 530–4.

Renwick, S.A., McNab, W.B. Ruffner Lowman, H. and Clarke, R.C. (1993) Variability and determinants of carcass bacterial load at a poultry abattoir. *J. Food Protect.*, **56**, 694–9.

Rigby, C.E., Pettit, J.R., Baker, M.F., Bently, A.H., Salomons, M.O. and Lior, H. (1980) Sources of Salmonellae in an uninfected commercially-processed broiler flock. *Can. J. Comp. Med.*, **44**, 267–74.

Sheldon, B.W. and Brown, A.C. (1986) Efficacy of ozone as a disinfectant for poultry carcasses and chill water. *J. Food Sci.*, **51**, 305–9.

Slavik, M.F., Kim, J.W. and Walker, J.T. (1995) Reduction of *Salmonella* and *Campylobacter* on chicken carcasses by changing scalding temperature. *J. Food Protect.*, **58**, 689–91.

Stals, P. (1992) From manual labour to automation. *Poultry Int.* **31**, 82–9.

Stephan, F. and Fehlhaber, K. (1995) Geflügelfleischgewinnig: Untersuchung zur Hygiene des Luft-Sprüh-Kühlverfahrens. *Fleischwirtschaft*, **74**, 870–3.

Svendsen, F.U. and Caspersen, C. (1981) Factors decreasing shelf life in production of fresh broilers, Proceedings of 5th European symposium on the quality of poultry meat, Apeldoorn, The Netherlands, May 17–23, 1981, (eds R.W.A.W. Mulder, C.W. Scheele and C.H. Veerkamp), pp. 459–67.

Thiessen, G.P., Usborne, U.R. and Orr, H.L. (1984) The efficacy of chlorine dioxide in controlling *Salmonella* contamination and its effect on product quality of chicken broiler carcasses. *Poultry Sci.*, **63**, 647–53.

Tinker, D.B., Gibson, C., Hinton, M.H., Allen, V.M. and Wathes, C.M. (1996) Reduction of cross contamination in defeathering machinery. *World Poultry*, **12**, 13–16.

Veerkamp, C.H. (1991) Multi stage scalding. *Poultry Int.* **30**, 62–7.

Veerkamp, C.H. (1995) Developments in primary processing technology. *Poultry Int.* **34**, 30–6.

Veerkamp, C.H., de Vries, A.W. and Bolder, N.M. (1993) Effect van een uithaalsysteem op de keuring en enkele kwaliteitskenmerken van slachtkuikens. *Spelderholt report 144* (unpublished).

Veerkamp, C.H. and Hofmans, G.J.P. (1973) New development in poultry processing: simultaneous scalding and plucking. *Poultry Int.*, **12**, 16–18.

Veerkamp, C.H. and Pieterse, C. (1993) Cleaning of broiler carcasses using a closed loop washing system during plucking. Proceedings of the 11th symposium on the quality of poultry meat, Tours, pp. 576–9.

Waldroup, A.L. (1996) Contamination of raw poultry with pathogens. *World Poultry Sci. J.*, **52**, 7–25.

Waldroup, A.L., Rathgeber, B.M. and Forsythe, R.H. (1992) Effects of six modifications on the incidence and levels of spoilage and pathogenic organisms on commercially processed postchill broilers. *J. Appl. Poultry Res.*, **1**, 226–34.

6 The microbiology of chill-stored meat

L.H. STANBRIDGE AND A.R. DAVIES

6.1 Introduction

Fresh red meat is an important part of the diet. In Britain alone the expenditure on fresh meat in 1996 was estimated at £1.5 billion on beef and veal, nearly £2 billion on pork, bacon and uncooked ham and £0.8 billion on mutton and lamb (Hilliam *et al.*, 1997). A general change in the lifestyle of the people of Western Europe has tended to decrease the patronage of local butchers shops, with supermarkets assuming a dominant role. Supermarkets also changed from the traditional packaging of fresh meat, either in grease-proof paper wrapping or plastic bags, to the display of meat on a rigid tray with a covering of a gas-permeable film. This protected the meat surfaces from casual contamination, retained moisture and, most importantly, allowed consumer selection. Initially, purchase decisions were based mainly on meat colour and price. Now a multitude of factors such as health (a desire for fresh/chilled, preservative-free products), convenience, ease of preparation, portion control and shelf-life (Rice, 1990; Farber, 1991) affect the decision.

New polymeric materials have been developed and energy costs have increased (Smith *et al.*, 1990a). These changes together with altered consumer habits have led to an increased interest in the use of modified atmosphere (MAP) and controlled atmosphere packaging (CAP). These systems have been defined by Koski (1988) as follows.

> *MAP*: 'Enclosure of food products in high gas barrier materials, in which the gas environment has been changed *once* to slow respiration rates, reduce microbiological growth, and retard enzyme spoilage – with the final effect of lengthening shelf life'. Vacuum packaging (VP) is included in MAP.
>
> *CAP*: 'Enclosure of food products in variable gas barrier materials, in which the gaseous environment has been changed and is selectively controlled to increase shelf life'.

It has been known for a long time that the shelf-life of meat can be extended by an increase in the concentration of carbon dioxide in the storage atmosphere. By the late 1930s most of the beef (60%) and some of the lamb (26%) shipped from Australia and New Zealand to the UK was stored in an atmosphere enriched with carbon dioxide (Lawrie, 1991). It was only during the 1970s, however, that there was renewed interest in the use of modified atmospheres particularly for individual primal joints. In 1979 Marks and Spencers launched a range of gas-flushed packs 'ATMOSPAK' of fresh meat products in the UK (Muller, 1986). High standards of hygiene and temperature control are essential prerequisites for the quality and safety of MAP packed meat.

6.2   Bacterial spoilage of chilled fresh meats stored in air

As noted elsewhere (Chapter 4), in a commercial context consideration needs only be given to surface contaminants. The moist, nutritious surface of meat is conducive to the growth of a range of bacteria. Various organisms spoil meat kept at >25 °C – see Chapter 1 also. A consortium of bacteria, commonly dominated by *Pseudomonas* spp., are usually responsible for spoilage at (–1 – 25 °C) providing the atmosphere is 'moist'. Odour and slime production cause spoilage in 10 d at 0 and 5 d at 5 °C (Hood and Mead, 1993). With a dry meat surface fungi (often *Thamnidium*, *Cladosporium*, *Sporotrichum*) are selected (Ingram and Dainty, 1971; Gill and Newton, 1980) – see Chapter 3 also. In the UK it would be very unusual to find meat stored at more than 25 °C, or under conditions conducive to fungal growth.

With storage, the odour of meat of normal pH changes gradually from a fresh 'meaty' smell ($\leq 10^7$ bacteria/g) to an inoffensive but definitely non-fresh one, to a dairy/buttery/fatty/cheesy (10^8), and eventually to a sweet/fruity and finally to putrid ($>10^9$) odour (Dainty *et al.*, 1985) – see Chapter 9 also. Slime becomes evident when the bacterial load is about 10^8/g; immediately before this the meat surface has a tacky feel (Ingram and Dainty, 1971). A deterioration of the colour of the meat due to a fall in the partial pressure of oxygen (Lambert *et al.*, 1991; Nychas *et al.*, 1988) under patches of micro-organisms (Gill and Molin, 1991) may cause customers to reject meat before spoilage is fully manifested. Pseudomonads tend to dominate the microbial consortium in aerobically stored meats (Harrison *et al.*, 1981) – see Chapter 1 also. The consortium is generally dominated by *Ps. fragi* (Table 6.1). *Pseudomonas fluorescens* and *Ps. lundensis* (Molin *et al.*, 1986) were present also in most studies of pseudomonad populations. As yet the reasons for the apparent uncompetitiveness of the last two is still unclear. Although pseudomonads tend to dominate the microbial population on air stored meats other bacterial groups may be important also.

Initially glucose is utilized for growth, but as the bacterial population approaches the carrying capacity ($>10^8$ cfu/g) the diffusion gradient from the underlying tissue to the surface of carcass meats fails to match the microbial demand. Other substrates are then used sequentially (Gill and Newton, 1977; Gill and Newton, 1978) until finally nitrogenous

Table 6.1 Prevalance of *Pseudomonas* spp. on meats stored aerobically

	Temperature of storage (°C)	Pseudomonas (%)		
		fragi	*fluorescens*	*lundensis*
Blickstad *et al.* (1981)	4	91	_[c]	–
	14	64	32	–
Erichsen and Molin (1981)	4	60	16	–
Molin and Ternström (1982)	N.S.[b]	56	13	10.5
Shaw and Latty (1982)	1	76	–	16
Banks and Board (1983)	4	65	15	–
Blickstad and Molin (1983)	0	93	–	–
	4	63	10	4
Shaw and Latty (1984)	0, 5, 10	78	4.8	11.1
Molin and Ternström (1986)	4	66	14	20
Gennari and Dragotto (1992)[a]	5	–	41	44
Prieto *et al.* (1992)	7	45	18	15
	30	65	4	21

[a] By design only fluorescent strains were isolated; [b] Not specified; [c] – Not isolated from the microbial association.

compounds lead to the formation of malodorous substances. This is of course a simplified overview, and this aspect is discussed in detail in Chapter 9. Antimicrobial compounds produced by some organisms (recently reviewed by Stiles and Hastings, 1991), the importance of micro-climates in different areas on the meat surface (Dainty *et al.*, 1979) as well as the metabolic attributes of particular organisms will contribute also to the rate and extent of spoilage.

In high pH Dark, Firm, Dry meat (DFD), the glucose level is low [0–33 μg/g wet weight of meat at pH 6 (Gill, 1982)], and the amino acids are used more quickly by *Shewanella putrefaciens, Aeromonas* spp., *Serratia liquefaciens, Yersinia enterocolitica* and lactic acid bacteria. Hydrogen sulphide production may lead to spoilage odours and the production of sulphmyo-globin (with associated greening of the meat). Both events lead to more rapid spoilage (Gill and Newton, 1979; Newton and Gill, 1980–81; Hood and Mead, 1993). With aerobic storage at 6 °C, DFD beef emitted off-odours after two days whereas that of normal pH did so after four days (Bem *et al.*, 1976). As would be expected the addition of glucose to DFD extended the shelf-life (Gill and Newton, 1979).

Brochothrix reached a maximum population of 10^9 after 9 d on beef of normal pH at 5 °C (Campbell *et al.*, 1979). The enumeration of Enterobac-teriaceae on selective media has shown that certain genera of this family are significant, but not dominant, members of the microbial associations on meats stored aerobically at chill temperatures (Blickstad *et al.*, 1981; Dainty *et al.*, 1985). *Serratia liquefaciens*, together with *Enterobacter aerogenes* and *Citrobacter* spp., were found on lamb chops (Newton *et al.*, 1977b).

Small numbers of the *Acinetobacter, Moraxella, Psychrobacter* complex occur on stored meat but they fail to compete effectively with pseudomonads. Although *Acinetobacter* would compete with pseudomonads for amino acids and lactic acid, their low oxygen affinity is such that the pseudomonads become dominant (Baumann, 1968). Contamination of meat with high numbers of *Acinetobacter/Moraxella* should be avoided as they may reduce the partial pressure of oxygen, allowing pseudomonads to utilize amino acids and cause spoilage odours (Gill, 1986). *Shewanella putrefaciens* (formerly *Alteromonas putrefaciens*) can be of high spoilage potential even if it is not the dominant species (Gill, 1986). It is more plentiful in fish and broilers than in meat – a possible link to the presence of water – a habitat in which these organisms are more numerous than on/in the abattoir environment. It is not found on normal pH meats but may be a problem in DFD meats. The organism is associated with odour production (often H_2S) and a greening of meat (sulphmyoglobin formation).

6.3 Modified atmosphere packaging

Modified atmosphere packaging extends the shelf-life of meat, minimizes spoilage losses, opens up new markets and provides a convenient packaging method for large scale distribution (Genigeorgis, 1985). It offers the supermarket a pack comparable to a grocery product allowing consumer selection with rigorous stock rotation. The method of packaging involves the use of gas-impermeable film and the introduction of a gaseous mixture differing from that of air (78.08% nitrogen, 20.95% oxygen, 0.93% argon, 0.03% carbon dioxide and small quantities of ozone and inert gases [Collins English Dictionary]). As stated previously vacuum packing is included in MAP; the trapped atmosphere is enriched with CO_2 derived from meat and microbial metabolism, the remaining atmosphere being nitrogen (Hood and Mead, 1993).

6.3.1 Methods

In MAP, meat is placed on a rigid pre-formed tray. A heat-sealable film is placed over the tray and the atmosphere modified by evacuation and/or flushing with the appropriate gas mixture before sealing. Close contact of meat and the web is an essential feature of vacuum packing if shelf-life is to be extended. Different methods are used to achieve this end (Taylor, 1985; Anon., 1987).

The CAPTECH system was developed to extend the shelflife of beef and lamb to more than 16 weeks. This allows shipment of fresh meat for example from New Zealand to the UK. The atmosphere in a tough metallized (e.g. aluminium) laminated barrier bag containing the meat (Gill,

1989) is modified with multiple gas flushing, using a snorkel system, which reduces the level of oxygen to <0.1% (Gill, 1989). Normal evacuation of a pack results in a residual 1% of oxygen. With the snorkel system less than 0.05% oxygen remains and a long shelf-life is achieved (Gill, 1990). The CAPTECH process can extend the shelf-life of beef and venison to five months and chicken and pork to three months (Bentley, 1991) providing the temperature of storage is –1 °C.

The importance of the plastic film properties, and particularly gas transmission rates, was demonstrated by Newton and Rigg (1979). They found that shelf-life was inversely proportional to oxygen permeability. Film permeability is expressed in theoretical gas transmission rates at specified temperature and humidity. Plastic-film laminates with low oxygen transmission rates (OTR) of <1 ml/m^2/24/h/atm allow the permeation of small amounts of oxygen (Stiles, 1991). A metallized layer must be included in the laminate for a film to be an effective barrier to oxygen. This is commonly achieved with aluminium.

6.3.2   Gases used and their effects

A combination of gases, commonly a mixture of oxygen, carbon dioxide and nitrogen, is normally used in modified atmosphere packaged meats.

Oxygen.   This gas restricts the growth of anaerobic micro-organisms, thereby enhancing the safety of the product with regard to *Clostridium* spp. High levels of oxygen promote lipid oxidation and rancidity (Ochi, 1987). At intermediate concentrations it may stimulate the growth of aerobic bacteria but, if high concentrations (e.g. 80%) are used, a decrease in the growth rate of these organisms was noted. At pressures above 240 mm there is an extension of the fresh appearance of meats. This is due to a decreased dissolution of oxygen from oxymyoglobin to myoglobin and a decreased reaction rate of myoglobin to metmyoglobin (Young *et al.*, 1988).

Nitrogen.   This inert, tasteless gas of low solubility is often used as a filler in MAP to reduce physical stress on the product and prevent pack collapse.

Table 6.2 Factors affecting the concentration and activity of carbon dioxide in meat

Meat attributes	Intracellular effects	General	Bacterial
Composition of meat	Protein binding	pH	Sensitivity
Proportion of fat	Metabolism	Temperature	Growth phase
Size and shape of cuts		pCO$_2$	
Composition of exposed surfaces		Solubility of CO$_2$ in system	
		Presence of other solutes	

Information from Gill (1988); Jones (1989); Stiles (1991).

Carbon dioxide. This water and lipid soluble gas inhibits product respiration. It is bacteriostatic also, extending the lag phase and/or increasing the generation time of susceptible bacteria. The effective concentration of carbon dioxide is dependent on the factors listed in Table 6.2. With an aqueous system the overall reaction is $CO_2 + H_2O \rightleftharpoons H_2CO_3 \rightleftharpoons HCO_3^- + H^+$ (Dixon and Kell, 1989), the proportions being dependent on pH (Fig. 6.1). Within the normal range of meat pH, carbon dioxide in solution is the predominant form. The solubility of carbon dioxide in meat increases by 360 ml/kg for each pH unit rise and approximately 19 ml/kg for each 1 °C rise between –1 °C and 10 °C (Gill, 1988).

The mechanism of bacteriostasis due to carbon dioxide is still unknown despite its use in food preservation for upwards of a century. The consensus opinion is of a synergistic action between some or all of the following in a product (Wolfe, 1980; Dixon and Kell, 1989; Jones, 1989; Farber, 1991; Lambert *et al.*, 1991):

(1) amendment of membrane function – nutrient uptake – (Gill and Tan, 1980; Tan and Gill, 1982);
(2) inhibition of enzymes or a decrease in enzyme reaction rates;
(3) an intracellular pH change – perturbation of enzyme equilibria;

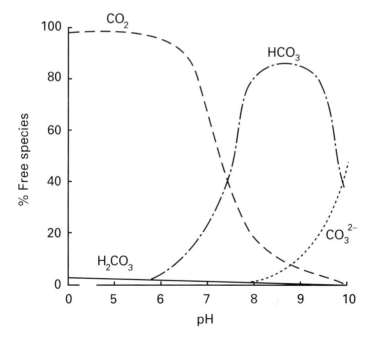

Figure 6.1 Proportions of dissolved carbon dioxide, carbonic acid, bicarbonate ions and carbonate ions as a function of pH. Data taken from Daniels *et al.* (1985).

(4) alteration of proton gradient (cf. the effect of decreased internal pH);
(5) a change in the physico-chemical properties of proteins due to a change in the internal electrostatic forces. The high reactivity of amines causes the formation of carbamic acids;
(6) feedback inhibition – internal pH affecting decarboxylating enzymes thereby producing more CO_2, and
(7) dissipation of energy – completion of a futile cycle. An increase in carboxylation and decarboxylation reactions results in a net loss of energy.

In aerobic systems, 20–30% carbon dioxide causes maximum inhibition; in anaerobic situations, the inhibition increases progressively with increased carbon dioxide concentration (Gill, 1988). Clark and Lentz (1972) found that carbon dioxide had to be present continuously in order to affect microbial growth. Silliker *et al.* (1977), who used this gas as a pre-treatment for pork, observed a residual effect; aerobic spoilage occurred at four days with controls whilst carbon dioxide-treated meat spoiled after seven days. The pressure of carbon dioxide applied to meat had a pronounced effect on shelf-life – 41 days with 1 atm. of carbon dioxide was extended to >121 days with 5 atm. at 4 °C. Pressure did not effect greatly the composition of the microflora (Blickstad *et al.*, 1981). High carbon dioxide alone can cause discoloration (browning at partial pressures greater than 0.2 atm.), a drop in the pH and a concomitant increase in the amount of exudate, a 'sharp' taste and collapse of the pack (Leeson, 1987). To ensure that the unsightly drip does not become obvious to the consumer, exudate in MAP packs containing high levels of carbon dioxide is absorbed in pads or trapped in the patterned bottom of the tray.

Carbon monoxide. Carbon monoxide has been used experimentally with MAP meats. As it causes the formation of the bright red carboxymyoglobin, the fresh appearance of meat is retained for periods longer than that stored in oxygen. Clark *et al.* (1976) found that if 1% (v/v) carbon monoxide was present in packaged meat throughout storage, then the shelf-life was extended as a consequence of a protracted lag phase and decreased growth rate of susceptible micro-organisms. The safety features of carbon monoxide have not been fully tested as yet, and potential problems of using this gas during preparation of packs needs to be evaluated.

Ozone. Ozone has been recommended for inhibition of mould growth but in practice is considered to inhibit Gram-positive rods > cocci > Gram-negative rods (Anon., 1980). It is relatively ineffective against bacteria in complex media (Genigeorgis, 1985). The main concern with use of this gas is the possibility of accelerated rancidity (Jones, 1988, 1989). It is involved in the destruction of amino compounds, the coagulation of proteins and inactivation of enzymes (Ingram and Barnes, 1954). The effect of ozone on the spoilage of beef was studied by Kaess and Weidemann (1968). Discoloration of the meat was observed with the application of >0.6 mg/m^3 of ozone and storage at 0.3 °C. At this level there was a small extension in the lag phase of the slow-growing strains of *Pseudomonas* spp.; the rapidly-growing ones were unaffected. It delayed the 'slime point' – visible manifestation of microbial growth – from day 12–13 in air to between days 20–24 with ozone.

6.3.3   The headspace in MAP

Gill and Penney (1988) determined the most efficient gas-to-meat ratio. This ratio was about 2 liters of gas per kilogramme of meat. The level of residual oxygen in the headspace of 'anoxic' packs is important. When various oxygen concentrations were included in packs having different headspace volumes, it was found that an initial level of >0.15% oxygen compromised the colour of beef and lamb but not pork (acceptable at 1%). Increasing headspace volumes tended to negate this problem but at a cost in terms of pack size (Penney and Bell, 1993). This problem may be overcome with oxygen scavengers such that the full potential of MAP is realized. 'Ageless' – a sachet containing loose, finely divided iron powder – reduces oxygen to less than 100 ppm by the formation of nontoxic iron oxide. Other proprietary brands of scavengers absorb oxygen with the concurrent production of equal volumes of carbon dioxide thereby preventing pack collapse (Smith *et al.*, 1990b). At present there is some consumer objection to the inclusion of sachets in fresh foods, probably due to suspicion of an intrusive addition to a pack and the fear of litigation arising from accidental ingestion of the scavenger. Oxygen-absorbing labels have been introduced into the UK market for smoked and processed delicatessen meat products (Anon., 1994). The product is presented with an outer paper label and an absorbing label on the lidstock under the paper one. Information on the outer label tells the consumer that the inside label helps retain freshness. It is seen by the user only after peeling back the lid. It is claimed that this application opens the door for future usage in the United States because the fear of ingestion is eliminated. The accurate sizing of the label for each application also allows the absolute minimum of reactive agent to be used (Anon., 1994).

6.3.4   Safety concerns of MAP

The microbial flora present on meat may include pathogenic bacteria because of the means by which it is obtained from the animal (Mead, 1994). Food-poisoning pathogens associated with chilled meat include *Aeromonas hydrophila*, *Bacillus cereus*, *Campylobacter* spp., *Clostridium botulinum*, *Clostridium perfringens*, *Escherichia coli* (notably verocytotoxigenic strains,

Table 6.3 Examples of the spoilage of MAP red meats

Meat		Storage characteristics							Reference
Species	Cut	Temperature (°C)	Atmosphere	OTR[a] of film	Gas to meat ratio	Shelf-life[b] (days)	Cause of spoilage	Dominant organisms[d]	
Beef	Roasts	1–3	VP*			34	O	Lb[si]	1, 2
			100% O_2	32	1:1	13	O/C	Ps/Lb	
			20% CO_2 + 80% N_2			34	O/C	Lb	
			50% CO_2 + 50% O_2			27	O/C	Lb	
			20% CO_2 + 80% O_2			20	O/C	Lb	
			25% CO_2 + 25% O_2 + 50% N_2			20	O/C	Lb/Ps	
			51% CO_2 + 30% O_2 + 18% N_2 + 1% CO			34	O	Lb	
Lamb	Chops	−1	Air	N.S.	N.S.	14	CO	B/Ps[si]	3
			80% Air + 20% CO_2			21	CF	B/Ps	
			80% O_2 + 20% N_2			21	CF	B/Ps	
			80% O_2 + 20% CO_2			21	CF	B	
			80% N_2 + 20% CO_2 Low O_2			42	C	B/Ps/Ent	
			80% N_2 + 20% CO_2 Oxygen free			56	F	B/Lb/Ent	
Pork	Roasts	1–3	VP	32	1:1	28	O	Lb[si]	4, 5
			100% O_2			14	O	Ps	
			20% CO_2 + 80% N_2			21	O	Leu	
			50% CO_2 + 50% O_2			14	O	Leu	
			20% CO_2 + 80% O_2			14	O	Leu/Ps	
			25% CO_2 + 25% O_2 + 50% N_2			14–21	O	Leu	
			51% CO_2 + 30% O_2 + 18% N_2 + 1% CO			14	C	Leu	

[a] OTR = oxygen transmission rate measured in ml/m²/24 h at 1 atmosphere. The temperature and relative humidity varies with film data; [b] Where detailed the time is taken from the time of slaughter. The methods and times of ageing differ; [c] O, off-odour; C, discoloration; F, off-flavour; [d] B, *Brochothrix thermosphacta*; Ps, *Pseudomonas* spp.; Leu, *Leuconostoc* spp.; Ent, Enterobacteriaceae; Lb, *Lactobacillus* spp.; [si] selective media were used but isolates taken from total counts were identified; * VP, Vacuum pack.

1, Christopher *et al.* (1979a); 2, Seideman *et al.* (1979a); 3, Newton *et al.* (1977a, b); 4, Seideman *et al.* (1979b); 5, Christopher *et al.* (1979b).

e.g. O157:H7), *Listeria monocytogenes*, *Salmonella* spp., *Staphylococcus aureus* and *Yersinia enterocolitica*.

In most cases the organisms are carried asymptomatically by the animals, and their presence is unknown unless specific tests are carried out. Fresh raw meats normally receive a bactericidal treatment before consumption and do not, therefore, present a direct food-poisoning hazard. They do, however, pose an indirect food-poisoning hazard through cross-contamination of cooked meats and other foods that are not cooked before consumption (Bell, 1996). This has resulted in meat and meat products having an unenviably high placing in the worldwide league tables of foods associated with food poisoning (Bell, 1996). A detailed description of the food-poisoning organisms and their significance is beyond the scope of this book; this chapter will consider their significance only in terms of the ecology of MAP fresh meats.

6.4   Spoilage of red meats stored in modified atmospheres

Beef in MAP packs has a shelf-life of more than two months at 1 °C and lamb and pork up to six weeks (Taylor, 1985). Shelf-life is determined by the choice of atmosphere, storage temperature and the meat type (Table 6.3 – for further information see the Appendix to this chapter). Spoilage of vacuum packed meat occurs between 10 and 12 weeks at 0 °C (Dainty, 1989) providing the meat is of normal pH and produced and processed under hygienic conditions, temperature control is adequate and a low oxygen permeability film is used (Egan, 1984).

With storage a sour/acid/cheesy odour develops, due to the production of organic acids from carbohydrates by lactic acid bacteria (LAB) (Egan *et al.*, 1991). Volatile compounds produced in MAP/VP are listed in Table 6.4. With vacuum packing (VP), spoilage occurs after bacteria have attained maximum numbers (Madden and Bolton, 1991). Greening and a sulphurous odour due to hydrogen sulphide production by some species of LAB (Dainty, 1989) may be evident also. Reactions within the meat may also occur and cause a bitter/liver-like odour and taint of the meat (Egan *et al.*, 1991).

6.4.1   Bacterial spoilage of modified atmosphere packaged meats

As the numbers of bacteria (particularly pseudomonads) are restricted by the relatively high concentration of carbon dioxide the spoilage of VP meat occurs later than that stored aerobically. Homo- and heterofermentative LAB (for example *Lactobacillus*, *Leuconostoc*, *Lactococcus* and

Table 6.4 The substrates used and volatiles produced by micro-organisms on MAP/VP meat

Organism	Substrate*	Volatiles	Odour
Lactic Acid Bacteria	Glucose[1] Amino acids especially arginine[2]	H_2S Methanethiol Dimethyl sulphide Ethanol	Sour
Brochothrix	Glucose[1] Ribose[a]	Acetoin/Diacetyl Branched chain alcohols Acetic acid	Malty Dairy Caramel
Enterobacteriaceae	Glucose[1a] Glucose-6-phosphate[2b] Amino acids[3**] Lactic acid[3] inosine-mono-phosphate[w]	Sulphides Amines Diamines H_2S	Boiled egg Putrid/rotten
Shewanella	Glucose[1a] Amino acids[1,2a] (serine and cysteine)	Sulphides	Putrid/faecal

* superscripts show order of utilization: 1–3 use on low pH meat (normal); a–b use on high pH meat; w, weak growth; ** except serine which may be used with glucose and glucose-6-phosphate, others used only after exhaustion of the other nutrients (Gill and Newton, 1979).
Based on information from Ingram and Dainty (1971), Gill (1976), Gill and Newton (1977), Gill and Newton (1978), Gill and Newton (1979), Dainty and Hibbard (1980), Dainty *et al.* (1985), Edwards *et al.* (1987), Grau (1988), Dainty (1989) and Dainty and Mackey (1992), Newton and Gill (1978).

Carnobacterium spp.) typically develop on meat under enriched carbon dioxide atmospheres. Huffman *et al.* (1975) did not recover lactic acid bacteria from meat stored in 100% carbon dioxide. The lactic acid produced by these bacteria inhibits Enterobacteriaceae, *Br. thermosphacta* and *Shew. patrefaciens* (Schillinger and Lücke, 1987b). With high pH meat (>pH 6.0) Enterobacteriaceae and other facultatively anaerobic bacteria (e.g. *Shewanella*, *Brochothrix*) may form an important part of the flora (Gill and Penney, 1986). Erichsen and Molin (1981) found that there was a mixed population including *Pseudomonas* spp., lactic acid bacteria and *Brochothrix* in MAP meat, whilst *Brochothrix* comprised 40% of the flora in VP and homofermentative lactobacilli were the predominant bacterial group in carbon dioxide packaged meat even in that having an initially high pH. As adipose tissue absorbs meat juices during processing, it manifests the spoilage characteristic of DFD meat when stored in MAP. With fat stored under VP, Patterson and Gibbs (1977) found a dominance of pseudomonads (55%) with 23% *Alcaligenes*, 9% LAB and 9% Enterobacteriaceae. On DFD meat, Enterobacteriaceae constituted 41%, pseudomonads 36%, *Aeromonas* 9% and *Alcaligenes* 14% of the microflora. *Clostridium* species (Collins *et al.*, 1992) have been isolated from spoiled vacuum packed beef on a number of occasions (Dainty, 1989, Kalchayanand *et al.*, 1989; Dainty and Mackey, 1992).

The dominance of a particular organism is determined by its relative affinity for a particular substrate (Newton and Gill, 1978). Growth, but not survival of pseudomonads is limited in very low oxygen conditions. As pseudomonads require oxygen, their growth in anaerobic packs is limited. There have been some reports of their growth on VP meat (e.g. Egan, 1984). The growth of this group is reduced by the addition of carbon dioxide (Enfors and Molin, 1981); in 0.5 atm. carbon dioxide the growth rate of *Ps. fragi* was 50% of that in air. In microculture studies at 2 °C and 6 °C no growth was detected in carbon dioxide, but small amounts were noted in VP and N_2 (Eklund and Jarmund, 1983). Erichsen and Molin (1981) isolated *Pseudomonas* spp. from pork of normal pH and stored in a modified atmosphere (78% N_2 + 20% CO_2 + 2% O_2) at 4 °C. *Pseudomonas fragi* comprised only 12% of the microflora; fluorescent strains were not detected. A similar study on MAP beef steaks showed that *Pseudomonas fragi* dominated the population (63.7% of pseudomonad strains) as would be expected from aerobically stored meats (Stanbridge, 1994). This species was inhibited by high levels of carbon dioxide, more so than the other meat pseudomonads, *Ps. fluorescens* (16.9% of the pseudomonad microflora) and *Ps. lundensis* (19.2%).

Pseudomonas spp. generally do not form a numerically significant proportion of microbial populations developing on MAP meats. They do, however, maintain a commercial importance due to their survival and potential growth in leaking packs and on meats stored aerobically after opening. If the impermeable film is removed, then the pseudomonads grow rapidly (Roth and Clark, 1972). Madden and Bolton (1991) found that in vacuum-packed beef, pseudomonads colonized the inner surface of the vacuum pouch, possibly due to the transferred oxygen being trapped at the meat-film interface. Growth was supported by carbohydrates in the meat exudate as well as catabolites of LAB metabolism.

Lactic acid bacteria.   The adoption of modified atmospheres for the packaging of fresh meats has shown that lactic acid bacteria generally become numerically dominant in such environments. In MAP meats lactobacilli, leuconostocs and carnobacteria are of primary importance, but lactococci and pediococci occur also (Table 6.5). The lactic acid bacteria developing on MAP beef steaks were identified during storage in a variety of atmospheres at 5 °C (Table 6.6).

Lactic acid bacteria in general do not produce malodourous substances (Dainty *et al.*, 1975), but tyramine is produced by *Carnobacterium* (Edwards *et al.*, 1987). Small concentrations of dimethylsulphide and methanethiol have been associated with the sour odour typical of vacuum packed/CO_2 stored meat stored for long periods (Dainty and Mackey, 1992). The growth and activity of a homofermentative *Lactobacillus* and a strain of *Leuconostoc* on meat stored at 4 °C in 5% CO_2 and 95% N_2 were studied (Borch

Table 6.5 Lactic acid bacteria isolated from meat. For additional details see Chapter 2.

Meat	Packing method[a]	Isolation medium[b]	Test methods[c]	No. of strains	Identification (%)	Authors[d]
Beef	VP	TSA	B/M/API	177	10 *Leuconostoc mesenteroides* / 65 Heterofermentative rods / 25 Homofermentative rods	1
Meat	VP	PCA/AA MRS5.5/BQ	B/M/API	100	31 Non-aciduric rods / 57 *Lactobacillus sake/curvatus* / 7 possibly *Leuconostoc paramesenteroides*	2
Meats/products	–	Briggs MRS LBS AA	B/M/Ferm	690	2 *Leuconostoc* spp. / 64 Streptobacteria. possibly *Lact. sake/curvatus* / 18 Thermobacteria. some *Lact. acidophilus* / 17 Betabacteria. *Lact. fermentum, viridescens, brevis*	3
Meat	VP radurized	mod MRS	B/M/Ferm	113	88 *Lact. sake* with 4 sub-groups / 7 *Lact. sake/curvatus*, 3 *Lact. curvatus*, 2 *Lact. farciminis*	4, 5[e]
Meat/products	–	AA MRS	B/M/Ferm – miniaturized	229	*Lactobacillus* spp. only: / 57 *sake*, 22 *curvatus*, 7 *divergens*, 3 *brevis, plantarum, viridescens*, 2 *hilgardii, carnis, casei*, 1 *halotolerans. farciminis*, 0.5 *alimentarius, coryneformis*	6
Pork Beef	VP	AA MRS	B/M/Ferm	246	30 *Leuconostoc*, 13 *Lactococcus* / 10 Heterofermentative lactobacilli (*Lact. divergens*) / 47 *Lact. sake/curvatus*	7
Meats/products	–	TGE	B/M/Ferm/ Ass	94	5 *Carnobacterium piscicola* / 9 *Carnobacterium divergens* / 10 *Lact.* sp. biovar 1, 18 *Leuconostoc* / 4 *Lact.* sp. biovar 2, 12 *Lact.* sp. biovar 3 / 30 *Lact.* sp. biovar 4, 4 *Lact.* sp. biovar 5	8
Pork Chicken	MAP irradiated	BHIYE	API + 4 tests	94	84 *Lact.* possibly *sake*, 3 *Carnobacterium* sp. / 2 *Lact.* possibly *curvatus*, 1 *Leu.* possibly *dextranicum* / 4 Organisms similar to *Carnobacterium* / 2 *Leuconostoc* sp.	9
Carcass	–	mod MRS	API + 1 test	35	Ten groups of ropy, slime-producing bacteria described / No species identification	10

Table 6.5 Continued

Meat	Packing method[a]	Isolation medium[b]	Test methods[c]	No. of strains	Identification (%)	Authors[d]
Beef	MAP	APT	B/M/Ferm	Total 469	56 *Carnobacterium divergens* 2.6 *Carnobacterium mobile* 23.2 *Lactobacillus sake/curvatus** 4.4 Other lactobacilli 5.9 *Leuconostoc gelidum* 5.5 *Leuconostoc mesenteroides*	11

[a] VP, Vacuum pack; MAP Modified atmosphere packs in a variety of atmospheres.

[b] TSA, Tryptone soy agar; MRS, de Man, Rogosa, Sharp; PCA, Plate count agar; AA, Acetate agar (Rogosa); MRS 5.5, MRS at pH 5.5; BQ, see Fournaud *et al.* (1973); Briggs, Tomato juice agar; LBS, Lactobacilli selective agar from BBL, mod MRS, modified MRS; TGE, Tryptone glucose extract agar; BHIYE, Brain heart infusion yeast extract.

[c] B, biochemical; M, morphological; API, API 50CH kit; Ferm, Carbohydrate fermentation tests; Ass, assimilation tests.

[d] 1 Hitchener *et al.* (1982); 2 Shaw and Harding (1984); 3 Morishita and Shiromizu (1986); 4 Hastings and Holzapfel (1987a); 5 Hastings and Holzapfel (1987b); 6 Schillinger and Lücke (1987a); 7 Schillinger and Lücke (1987b); 8 Borch and Molin (1988); 9 Grant and Patterson (1991a, b); 10 Mäkelä *et al.* (1992); 11 Stanbridge (1994).

[e] Results from two studies were very similar (both 88% *Lact. sake*) so only one study has been included.

* There is a problem distinguishing *Lact. sake* from *Lact. curvatus* (Hastings and Holzapfel, 1987a, b) so in this study no attempt was made to do so.

Table 6.6 Lactic acid bacteria developing on modified atmosphere packaged beef steaks during storage at 5°C

Species/group[1]	Percentage of each bacterial group				
	Time zero	End of storage[2]			
		Vacuum pack	50% N_2 + 50% CO_2	80% O_2 + 20% CO_2	100% CO_2
Unidentified LAB	5	5	10		5
Others[3]	60		10		
Brochothrix thermosphacta	10			100	
Leuconostoc mesenteroides group 2		10	10		10
Leuconostoc mesenteroides group 1			5		5
Leuconostoc gelidum		5	5		40
Carnobacterium mobile		30	5		
Carnobacterium divergens groups 2	15	15	40		
Carnobacterium divergens groups 1	10	5	10		5
Lactobacillus sake/curvatus group 2		30	5		20
Lactobacillus sake/curvatus group 1					15

[1] Identification relates to clusters determined by numerical taxonomy of biochemical and fermentation tests.
[2] The end of storage was taken as the time at which the beef steak was quite obviously 'off'; this varied between atmospheres and even duplicate packages within an atmosphere.
[3] Others (includes Gram-negative rods, Gram-positive, catalase-positive cocci and yeast).
Information taken from Stanbridge (1994).

and Agerhem, 1992). A maximum population of 10^7 cfu/g was present after two weeks, at which time flavour changes were noted with the *Lactobacillus* sp. which produced acetic acid and D-lactate from L-lactate and glucose. Formate or ethanol was not detected, although some hydrogen sulphide was evident towards the end of storage. An analogous flavour change was detected before the *Leuconostoc* sp. had attained maximum numbers. This organism produced D-lactate and ethanol from glucose. Some hydrogen sulphide was also formed but no malodour was detected in either case (Borch and Agerhem, 1992). A homofermentative rod attained a maximum population of 10^8 on VP beef. Off-flavours developed 7–13 days after this population had been achieved (Egan and Shay, 1982). Gill and Newton (1979) found the generation time of a *Lactobacillus* sp. was unaffected by high or low pH meat in VP.

McMullen and Stiles (1993) identified the LAB species prevalent in MAP pork. *Lactobacillus alimentarius*, *L. farciminis* and *L. sake* were the three most common isolates. In VP or MAP (75% O_2 + 25% CO_2) beef stored for 14 or 28 days, leuconostocs were the dominant organisms (Hanna *et al.*, 1981).

Brochothrix thermosphacta.   Variable growth of *Brochothrix* occurs under conditions simulating MAP or VP. Its growth is affected by the tempera-ture, pH and gaseous atmosphere obtaining in the storage conditions (Gardner, 1981). In an anaerobic atmosphere, numbers decreased on low pH meat (Campbell *et al.*, 1979). Transfer after two days from aerobic to anaerobic conditions resulted in growth stasis (Campbell *et al.*, 1979). On high pH meat, however, 10^8 cfu/g were present after 10 days under anaer-obic conditions at 5 °C (Campbell *et al.*, 1979). The maximum yield depended on the substrate (Newton and Gill, 1978). The lag phase of *Br. thermosphacta* was longer than that of a *Lactobacillus* sp. or *Enterobacter* sp. under nitrogen (Newton and Gill, 1978).

Enterobacteriaceae.   The enumeration of Enterobacteriaceae on selective media has shown that certain genera of this family are significant, but not dominant, members of the microbial associations on meats stored in modi-fied atmospheres at chill temperatures (Dainty *et al.*, 1979; Blickstad *et al.*, 1981; Lee *et al.*, 1985; Nortjé and Shaw, 1989). Blickstad *et al.* (1981) com-pared the microflora on pork stored in atmospheres enriched with carbon dioxide (Table 6.7). Only one species of Enterobacteriaceae, *Providencia rettgeri*, was found initially and then on only one of the six pork loins tested; it comprised 5% of the microbial flora on that loin. By the end of storage the numbers of Enterobacteriaceae (determined with Violet Red Bile Dex-trose medium) had increased by a factor of 10^4–10^5 on pork stored in air or in 1 atmosphere of carbon dioxide at 4 or 14 °C. Much less multiplication

Table 6.7 Prevalence of Enterobacteriaceae on pork under various storage conditions

Storage conditions		Numbers of micro-organisms at the end of storage (log n/cm^2)		Main species of Enterobacteriaceae (% total population)
Atmosphere	Temperature (°C)	Total count (TGE[a])	Entero-bacteriaceae (VRBD[b])	
Initial[c]	–	3.2	0.5	*Providencia rettgeri* 1%
Air	4	7.0	4.4	*Enterobacter cloacae* 2%
	14	7.8	5.7	*Serratia liquefaciens* 4%
1 atm CO_2	4	7.1	5.5	–
	14	6.3	5.4	*Serratia liquefaciens* 16%
5 atm CO_2	4	7.0	1.1	–
	14	7.0	<4.0	–

Data from Blickstad *et al.* (1981).
[a] Tryptone glucose extract agar; [b] Violet red bile dextrose agar; [c] Initial values taken as an average of six sampled pork loins; – Enterobacteriaceae were not identified.

Table 6.8 Enterobacteriaceae isolated from meats

Meat	Conditions of storage			Enterobacteriaceae[a]		Authors[b]
	Temperature (°C)	Atmosphere	Time (d)	%[c]	Species	
Lamb	−1	Low oxygen	42	*	Serratia liquefaciens	1
		No oxgen	56	*	Citrobacter spp.	
				*	Enterobacter aerogenes	
N.S.	Ab[d]	Ab	0	51	Ent. cloacae	2
				22	Klebsiella pneumoniae	
				15	Ent. aerogenes, also Ent. liquefaciens, Serratia spp.	
Beef (High pH)	0–2	VP	56	93	Ser. liquefaciens	3
				7	Hafnia spp.	
Pork	14	Air	3	4[c]	Serratia spp.	4
		CO₂	6	16	Serratia spp.	
	4	Air	8	2	Ent. cloacae	
N.S.	Ab	Ab	0	19	Escherichia coli	5
				10	Citrobacter spp.	
				3	Ent. aerogenes	
				9	Ent. agglomerans	
				9	Ent. cloacae	
				4	Ent. hafniae	
				22	Klebsiella spp.	
				14	Ser. liquefaciens	
Sausage	f	f	f	40	Ent. agglomerans	6
				16	Citrobacter freundii	
				29	Haf. alvei	
				15	Ser. liquefaciens	
Pork	0–3	VP	14	25[e]	Ser. liquefaciens	7
				3	Ent. agglomerans	
			70	5	Ser. liquefaciens	
				2	Ent. agglomerans	
		VP + N₂ flush	14	24	Ser. liquefaciens	
			70	6	Ser. liquefaciens	
				3	Ent. agglomerans	
				11	Hafnia alvei	
Beef	4	Barrier bag –	3	13[c]	Ser. liquefaciens	8
		no evacuation	9	60	Ser. liquefaciens	
Beef	6	25% CO₂ +	11[g]	20[e]	Enterobacteriaceae	9
		75% O₂	11[h]	5	Enterobacteriaceae	
Beef	2	50% CO₂ +	0	28	Enterobacter spp.	10
		15% O₂ + 35%		48	Serratia spp.	
		N₂		20	Hafnia spp.	
				16	Yersinia spp.	
				8	Citrobacter spp.	
			28	7	Enterobacter spp.	
				52	Serratia spp.	
				19	Hafnia spp.	
				21	Yersinia spp.	

Table 6.8 Continued

Meat	Conditions of storage			Enterobacteriaceae		Author s[b]
	Temperature (°C)	Atmosphere	Time (d)	%[c]	Species	
Beef[i]	5	Variety	End of shelf-life	21	*Pantoea agglomerans*	11
				28	*Serratia liquefaciens*	
				6	*Enterobacter aerogenes*	
				<1	*Enterobacter cloacae*	
				2	*Escherichia/Shigella*	
				<1	*Escherichia vulneris*	
				37	*Hafnia alvei*	
				2	*Providencia alcalifaciens*	
				<1	*Klebsiella pneumoniae*	
				<1	*Citrobacter freundii*	
				<1	*Buttiauxella agrestis*	

[a] *Ent. liquefaciens* is synonymous with *Ser. liquefaciens* (see Jones, 1988), *Ent. hafniae* with *Haf. alvei* (see Greipsson and Priest, 1983), and *Ent. agglomerans* with *Pant. agglomerans* (Garini *et al.*, 1989).

[b] 1 Newton *et al.* (1977a); 2 Newton *et al.* (1977b); 3 Patterson and Gibbs (1977); 4 Blickstad *et al.* (1981); 5 Stiles and Ng (1981); 6 Banks and Board (1982); 7 Lee *et al.* (1985); 8 Ahmad and Marchello (1989); 9 Nortjé and Shaw (1989); 10 Manu-Tawiah *et al.* (1991); 11 Stanbridge (1994).

[c] Percentage of species within the Enterobacteriaceae population.

[d] Abattoir study.

[e] Proportion of species within the total microbial population.

[f] Study included meat from fresh and stored (4, 10, 15 or 22 °C for up to 8 days) unsulphited sausages.

[g] Meat aged by hanging for 7 days.

[h] Meat aged in vacuum pack for 7 days.

[i] For more information see Table 6.9.

N.S. Not specified. * Proportion not specified.

had occurred with 5 atm. of carbon dioxide. Two species only were detected at the end of storage – *Serratia liquefaciens* and *Enterobacter cloacae* (Table 6.7). The analysis of species was done with isolates taken from plates used for the total aerobic count rather than from the selective medium. In all instances the viable counts on the former were larger than those on the latter.

The incidence of Enterobacteriaceae from the time of slaughter and processing in an abattoir through to meats prepared for retail sale was studied by Stiles and Ng (1981). *Escherichia coli* and *Serratia liquefaciens* were present at all stages. The latter together with *Pantoea agglomerans* were predominant in ground beef supplied to retail outlets. *Serratia liquefaciens* has been found by many investigators to be the most common member of this family on meat taken from abattoirs or stored in atmospheres of different composition (Table 6.8). Thus, with vacuum packaged pork stored at 0–3 °C, the proportion of *Ser. liquefaciens* decreased during storage (Lee *et al.*, 1985). *Hafnia alvei*, another common contaminant of meats (see

Table 6.9), became the dominant member of the Enterobacteriaceae in vacuum packed pork flushed initially with nitrogen (Lee *et al.*, 1985). It was also dominant on beef steaks stored in modified atmospheres at 5 °C (Table 6.9). *Hafnia alvei* did not compete well in the high oxygen atmosphere, where pseudomonads tended to be prevalent, even in Violet Red Bile Glucose agar. It also was inhibited by the modified atmospheres more at 0 °C than 5 °C. The reasons for this have not yet been established. Relatively high proportions of *Enterobacter cloacae* and *Ent. aerogenes* were found occasionally (Newton *et al.*, 1977a; Newton *et al.*, 1977b; Blickstad *et al.*, 1981; Stiles and Ng, 1981; Manu-Tawiah *et al.*, 1991). *Citrobacter* (Newton *et al.*, 1977a; Stiles and Ng, 1981; Banks and Board, 1983; Manu-Tawiah *et al.*, 1991), *Yersinia* (Manu-Tawiah *et al.*, 1991) and *Klebsiella* (Newton *et al.*, 1977b; Stiles and Ng, 1981) were minor contaminants only of a range of meats and meat products.

Table 6.9 Enterobacteriaceae developing on modified atmosphere packaged beef steaks stored at 0° or 5°C

(a) 0°C

Species isolated from violet red bile glucose agar	Percentage of bacterial groups isolated from storage environments				
	Time zero	End of storage			
		Vacuum pack	50% N_2 + 50% CO_2	80% O_2 + 20% CO_2	100% CO_2
Pseudomonas spp.	10.5		2.5	47.5	77.5
Escherichia coli	2.6				
Citrobacter freundii	2.6				3.0
Providencia alcalifaciens	21.1				19.5
Aeromonas spp.	23.7				
Enterobacter aerogenes			2.5	7.5	
Serratia liquefaciens	2.6	7.5	35.0	32.5	
Hafnia alvei		92.5	42.5	7.5	
Pantoea agglomerans	36.8		17.5	5.0	

(b) 5°C

Pseudomonas spp.	10.5	3		42.1	2.5
Escherichia coli	2.6				
Citrobacter freundii	2.6				
Providencia alcalifaciens	21.1				
Aeromonas spp.	23.7			12.5	
Enterobacter aerogenes			7.5	10.0	2.5
Serratia liquefaciens	2.6		17.5	22.7	2.5
Hafnia alvei		97	80.0	12.8	92.5
Pantoea agglomerans	36.8				

Data taken from Stanbridge (1994).

Although members of the family Enterobacteriaceae do not become a numerically dominant part of the microbial association on meats, they may contribute to spoilage. Thus, *Haf. alvei* and *Ser. liquefaciens* produced malodourous diamines – putrescine and cadaverine – in vacuum packed beef. Putrescine levels may be enhanced through ornithine production by arginine-utilizing strains of lactic acid bacteria (Dainty *et al.*, 1986). These two species together with *Pantoea agglomerans* produced hydrogen sulphide during the aerobic storage of beef (Dainty *et al.*, 1989). *Serratia liquefaciens* did so in vacuum packed beef of high pH – 6.6 (Patterson and Gibbs, 1977). Dainty *et al.* (1989) associated an 'eggy' odour with the growth of Enterobacteriaceae. This odour (ranging from boiled to rotten egg) was considered to be due to the formation of sulphur compounds, including hydrogen sulphide. Methanethiol and its derivatives were also found in the headspace of packs of meat inoculated with *Haf. alvei*, *Ser. liquefaciens* and *Pant. agglomerans*. A green discoloration of the meat was associated with the growth of the two first-named organisms. This was probably due to the formation of sulphmyoglobin by combination of hydrogen sulphide and myoglobin (Nicol *et al.*, 1970). The presence of members of the Enterobacteriaceae, especially *Haf. alvei* and *Ser. liquefaciens*, in large numbers in meats is, therefore, of commercial importance. The effect of modified atmospheres on the growth/survival of members of this group in the packs of beef steaks was included in the recent study by Stanbridge (1994).

Serratia liquefaciens and *Hafnia alvei* grew and produced cadaverine and putrescine in vacuum packed meat (Edwards *et al.*, 1985; Dainty *et al.*, 1986); *Pantoea* (formerly *Enterobacter*) *agglomerans* and *Klebsiella* grew, but did not produce the diamines. *Citrobacter freundii*, *Proteus* and *Yersinia* did not grow on similar meat in vacuum packs at 1 °C (Dainty *et al.*, 1986). Synergism was found between *Serratia* and *Hafnia* and those lactic acid bacteria utilizing arginine to produce ornithine. Ornithine decarboxylase of the first two was involved in the conversion of ornithine to putrescine (Edwards *et al.*, 1985). Although arginine could be utilized, the Enterobacteriaceae generally do not contain the required dihydrolase (Holmes and Costas, 1992). In VP, Enterobacteriaceae produced branched chain esters, methanethiol and derivatives, dimethyldisulphide, dimethyltrisulphide, methylthioacetate and bis(methylthio)methane, the last being produced from high pH meat only (Dainty and Mackey, 1992). *Hafnia* also produced hydrogen sulphide in mince (Hanna *et al.*, 1983).

Others. *Shewanella putrefaciens* grown on high pH VP meat produced dimethyldisulphide, bis(methylthio)methane and methylthiopropionoate. A putrid/faecal odour was produced (Dainty and Mackey, 1992). This organism may cause greening (sulphmyoglobin) on VP DFD meat through hydrogen sulphide production from cysteine. Its growth may be inhibited by addition of an acid to high pH meat – it is unable to grow on low pH meat (Gill and Newton, 1979). *Aeromonas* also is able to grow in high pH VP meat and produce putrid odours, probably due to methylthiopropionoate production (Dainty *et al.*, 1989). In VP meat *Yersinia enterocolitica* was recovered after four weeks and *Shew. putrefaciens* after two weeks storage at 0–2 °C. The latter reached 10^6 cfu/g by week 10 at which time greening was evident (Seelye and Yearbury, 1979).

6.4.2 Safety of meats stored in modified atmospheres

Whilst the potential benefits of MAP in extending shelf-life are apparent, several workers have expressed concerns about the microbiological safety of MAP foods, particularly with respect to the psychrotrophic pathogens (Hintlian and Hotchkiss, 1986; Gill and Reichel, 1989; Farber, 1991; Lambert *et al.*, 1991; Church and Parsons, 1995). Research indicates that MAP storage does not increase the hazards from *Salmonella*, *Staphylococcus aureus*, *Campylobacter* or *Vibrio parahaemolyticus* (Lambert *et al.*, 1991). Until recently, however, few studies had addressed the psychrotrophic pathogens: *Aeromonas hydrophila*, nonproteolytic *Clostridium botulinum*, *Listeria monocytogenes* and *Yersinia enterocolitica*. The relative growth rates of spoilage and pathogenic organisms in MAP is a critical factor in determining the safety of MAP. If the normal aerobic spoilage flora is suppressed, while pathogen growth continues, foods could become unsafe before being rejected owing to the development of overt spoilage (Sheridan *et al.*, 1995).

Aeromonas hydrophila. Whilst *A. hydrophila* is most commonly associated with water supplies, it has been hypothesized that foodborne dissemination may play a major role in the transmission of this suspected pathogen (Buchanan and Palumbo, 1985). Surveys of raw meats have found them to be contaminated frequently with this organism (Fricker and Tompsett, 1989; Hudson and DeLacy, 1991; Hudson *et al.*, 1992; Gobat and Jemmi, 1993; Walker and Brooks, 1993).

Several mathematical models predict the growth of *A. hydrophila* at pH 5.5 (Gill *et al.*, 1997), although studies on muscle foods have found inhibition or death on normal pH (<6.0) muscle. Palumbo (1988) observed that *A. hydrophila* was sensitive to pH values less than 6.0 in ground pork, and this has been substantiated by Doherty *et al.* (1996) for lamb, Davies (1995) for beef and Gill *et al.* (1997) for pork muscle. Palumbo (1988) observed a nonspecific inhibition of *A. hydrophila* by the background microflora and it has been suggested that this may account for the inhibition of growth on normal pH muscle tissue. Gill *et al.* (1997) conclude that this suggestion can be discounted in view of studies with *A. hydrophila* inoculated onto sterile muscle. Doherty *et al.* (1996) also found no evidence that the background flora was antagonistic to the growth of *A. hydrophila* on lamb. In contrast to the situation for normal-pH tissue, the ability of *A. hydrophila* to grow on high-pH tissue stored both aerobically and in modified atmospheres has been clearly demonstrated. Palumbo (1988) observed significant growth (>4 $\log_{10}$ cfu/g) on ground pork (pH 6.40) stored aerobically at 5 °C whilst in a vacuum pack there was only a small increase (about 1.5 $\log_{10}$ cfu/g). Gill and Reichel (1989) observed growth of *A. hydrophila* on vacuum-packaged high-pH (>6.0) beef stored at –2, 0, 2, 5 or 10 °C. When the beef was packaged in a carbon dioxide atmosphere, growth only occurred with storage at 10 °C. These authors concluded that carbon dioxide packaging is likely to be safe with respect to *A. hydrophila*, *L. monocytogenes* and *Y. enterocolitica* for storage at all temperatures below 5 °C. Doherty *et al.* (1996) studied the growth of *A. hydrophila* on high pH (>6.0) lamb packaged in air, vacuum, 80% O_2/20% CO_2, 50% CO_2/50% N_2 or 100% CO_2 and stored at 0 °C or 5 °C. Storage at 5 °C allowed significant increases in *A. hydrophila* numbers under all the atmospheres, except 100% CO_2. After storage at 0 °C, significant increases in *A. hydrophila* numbers were observed only when the lamb was stored in air or vacuum.

Overall, as concluded by Palumbo (1988) and Doherty *et al.* (1996), the factors affecting growth of *A. hydrophila* on meat appear to be complex and interrelated. Storage in modified atmospheres does not, however, appear to present any greater hazard than from aerobic storage.

Clostridium botulinum *(non-proteolytic, psychrotrophic).* Whilst the hazards from non-proteolytic, psychrotrophic *Cl. botulinum* in MAP fish and fish products has received considerable attention, raw meats have received only scant attention. Lücke and Roberts (1993) have reviewed the control of *Cl. botulinum* (including nonproteolytic strains) in meat and meat products. They comment that, although psychrotrophic, nonproteolytic strains have a higher growth potential on fresh meats than mesophilic strains, they have not been responsible for any outbreaks of meat-borne botulism worldwide other than among the Inuit population in northern Canada and Alaska. Whilst consumption of raw unprocessed meats is common in some other countries (e.g. Germany, Belgium and the Netherlands), only the Inuits appear to eat raw, putrid meat regularly (Lücke and Roberts, 1993). In the United Kingdom, the Advisory Committee on the Microbiological Safety of Food (ACMSF, 1992) in its report on vacuum packaging and associated processes, whilst including raw animal products, categorize them as having a low priority for attention, with chill temperatures, shelf-life and spoilage by aerobic bacteria as usual controlling factors. Products categorized as low-priority for attention include those where psychrotrophic strains of *Cl. botulinum* are either unlikely to occur, or those foods where the product composition results in the presence of a number of controlling factors, some of which may, on their own, be at or above the level required to inhibit growth and toxin production by *Cl. botulinum*. The foods in this category are regarded as presenting a low risk to the public of the botulinum hazard occurring. Whilst the risks from psychrotrophic *Cl. botulinum* in raw meats appears to be low, the severe nature of the disease means that this risk must not be taken lightly.

Listeria monocytogenes.   Of the psychrotrophic pathogens, *L. monocytogenes* has probably received the widest attention. Johnson *et al.* (1990) reviewed the incidence of *Listeria* on meat and meat products and found that *L. monocytogenes* occurred frequently on raw meats. Johnson *et al.* (1990) and Ryser and Marth (1991) have reviewed the growth and survival of *L. monocytogenes* on raw meat. From these reviews and other publications (Gill and Reichel, 1989; Manu-Tawiah *et al.*, 1993; Sheridan *et al.*, 1995), it is apparent that the ability of *L. monocytogenes* to grow on both aerobically and MAP raw meat is unpredictable. Whilst several studies have observed growth of *L. monocytogenes* in aerobic (Khan *et al.*, 1973; Johnson *et al.*, 1988; Shelef, 1989; Sheridan *et al.*, 1995), and MAP (including vacuum-pack) stored meats (Gill and Reichel, 1989; Grau and Vanderlinde, 1990; Avery *et al.*, 1994; Sheridan *et al.*, 1995), others have observed that numbers have remained static or declined (Buchanan and Klawitter, 1992; Kaya and Schmidt, 1991; Sheridan *et al.*, 1995). It has been suggested that temperature of storage, type of tissue, i.e. lean or fat, pH, associated microflora, bacteriocin production, and condition of the inoculum used in the experiments may affect the growth of *L. monocytogenes* on meats.

Few of the above studies examined the physicochemical changes that occurred during storage. This was one aspect addressed by Drosinos and Board (1994), who examined the growth of *L. monocytogenes* in meat (lamb) juice stored under an aerobic or a modified gas atmosphere. They identified a key role of glucose in limiting the growth of *L. monocytogenes* and concluded that their results supported the concept of the emergent property of an ecosystem. Namely, the inhibition of *L. monocytogenes* is caused by synergistic effects of: (i) an atmosphere enriched with carbon dioxide; (ii) incubation at chill temperatures; (iii) interspecific antagonism via competition for carbohydrates; and (iv) the accumulation of acetate in the ecosystem, with the inhibition and decline of the pathogen's population. These, or similar, interacting effects may account for the discrepancies observed in the studies with *L. monocytogenes* on meats and also in studies using the other psychrotrophic pathogens.

Yersinia enterocolitica.   As with *A. hydrophila* and *L. monocytogenes, Y. enterocolitica* are frequent contaminants of fresh meats (Manu-Tawiah *et al.*, 1993). There is, however, conflicting information as to the ability of *Y. enterocolitica* to grow on aerobically and modified-atmosphere-packaged meats. Kleinlein and Untermann (1990) studied aerobically stored minced meat with a high background flora and showed a marked inhibition of *Y. enterocolitica* growth by the flora. This effect was so marked that only a slight additional inhibition was observed when the mince was stored in a 20% CO_2/80% O_2 atmosphere. This was in contrast to the situation where meat with a low total bacterial count was used, when the growth of *Y. enterocolitica* was adversely affected by the gaseous atmosphere. Gill and Reichel (1989) observed significant growth of *Y. enterocolitica* in vacuum-packaged high-pH (>6.0) beef stored at –2, 0, 2, 5 or 10 °C. In carbon dioxide packs, *Yersinia* also grew at 10 °C and 5 °C but not at the lower temperatures. Doherty *et al.* (1995) observed growth of *Y. enterocolitica* at 5 °C on minced lamb in air, in vacuum pack and in 50% CO_2/50% N_2, but not in 100% CO_2, whilst at 0 °C, growth was observed in air only. These studies contrast with those of Manu-Tawiah *et al.* (1993) and Fukushima and Gomyoda (1986), who observed inhibition of *Y. enterocolitica* on pork stored aerobically at 4 °C and 6 °C, respectively. Whilst Manu-Tawiah *et al.* (1993) did not observe growth on aerobically stored product, there was significant growth when product was stored in vacuum, 20% CO_2/80% N_2, 40% CO_2/60% N_2 and 40% CO_2/10% O_2/50% N_2. In this study and in that of Doherty *et al.* (1995), the inclusion of oxygen in the atmosphere resulted in an inhibition of the growth of *Yersinia*.

Predictive microbiology.   From the above it is obvious that pathogen growth on raw meats, as with many other foods, is controlled by several inter-related factors. There is considerable interest in the development of predictive mathematical models to describe the effects of different parameters on the growth of potential pathogens (Grau and Vanderlinde, 1993). Predictive microbiology uses mathematical equations to estimate the growth, survival or death of micro-organisms as affected by extrinsic (processing and storage conditions) and intrinsic parameters of the food (e.g. pH, salt concentration, or a_w). Two personal computer based programs are widely available: Food MicroModel (Food MicroModel Ltd, Leatherhead, Surrey, UK) and MFS Pathogen Modeling Program (USDA, Philadelphia, USA). These programs currently have only limited scope for the inclusion of modified atmospheres. The MFS Pathogen Modeling Program has the choice of aerobic or anaerobic growth for some models, whilst Food MicroModel currently has the effect of carbon dioxide concentration for three organisms (verocytotoxigenic *Escherichia coli*, *Listeria monocytogenes* and mesophilic *Bacillus cereus*). One of the reasons for this is the difficulty in modelling the very dynamic nature of the atmosphere, particularly for products such as raw meats. It is necessary to have a good understanding of the changes in the gaseous environment with time (which itself is dependent on numerous other variables), and a sufficiently sophisticated model to be able to predict the effect of this on microbial growth/survival. Food validation of models is assessed by comparing their predictions with the reported behaviours of pathogens in foods and/or with the observed growth of pathogens in inoculated foods (Gill *et al.*, 1997). McClure *et al.* (1994) reported that, in general,

Table 6.10 Future developments of MAP/VP

Development	Hurdle	Notes	Reference
Natural preservatives	√	As an additive or by inoculation/ selection of appropriate organisms	1
Organic acids	√	As a decontaminating spray*/dip or by selection of LAB**	2, 3, 4
Bacteriocins	√	Added in a pure form or by inoculation/ selection of LAB**	5, 6
Irradiation	√	Decreases microbial load and meat reactions	7, 8
and acid addition	√	Adds another selective pressure	9
High pressure	√	For reduction of microbial load – colour rejection causes problems	10
Bacteriophage	√	Reduction of particular bacteria, e.g. pseudomonads	11
Glucose addition		To extend the shelf-life of aerobic or MAP DFD meats	12, 13
Anti-oxidants		For further colour stability, e.g. sodium erythorbate, ascorbic acid	14, 15
Edible films with or without additives		Reduction of exudate	16, 1
Time-temperature Indicators		Records temperature history	17

* see Chapter 4.
** Localized effects in meat from production by LAB.
1, Ooraikul (1993); 2, Baird-Parker (1980); 3, Brown and Booth (1991); 4, Labadie *et al.* (1975); 5, Stiles and Hastings (1991); 6, Kim (1993); 7, Lambert *et al.* (1992); 8, Grant and Patterson (1991b); 9, Farkas and Andrássy (1993); 10, Carlez *et al.* (1993); 11, Greer (1986); 12, Shelef (1977); 13, Newton and Gill (1980–81); 14, Gill and Molin (1991); 15, Manu-Tawiah *et al.* (1991); 16, Farouk *et al.* (1990); 17, Taoukis *et al.* (1991).

the models (developed in broth culture) generated predictions relevant to most food groups, showing excellent agreement, and that deviations from the model were usually explicable by other preservative factors. Gill *et al.* (1997), however, examined the growth of *A. hydrophila* and *L. monocytogenes* on pork and in broth and concluded that current models were likely to be highly unreliable guides to the behaviour of these organisms on pork and other raw meats. It must be noted that growth models are generally designed to be 'fail-safe', as it is accepted that they cannot include all possible factors.

6.5 Potential future developments in MAP

Many adaptations and developments of MAP technology are under study (Table 6.10). Some of these attempt to combine different preservation methods – the hurdle effect (Leistner, 1985; Earnshaw, 1990; Ooraikul, 1993; Gould, 1992). As yet none is exploring the use of additives or colour enhancers. With current legislation, approval of such amendments would be restricted to meat products (Gill, 1989).

Table A6.1 Appendix: The modified atmosphere packaging of fresh red meats

Meat			Storage characteristics			Spoilage characteristics			Comments	Reference
Species	Muscle type	Temperature (°C)	Atmosphere	OTR[a] of film	Gas to meat ratio	Shelf-life[b] (days)	Cause of spoilage	Dominant organisms[d]		
Beef	Round	1	60% CO_2 + 20% O_2 + 20% N_2 + aerobic	Drum	N.S.	>14	–	G-ves	Found residual effect	8
			20% CO_2 + 25% O_2 + 55% N_2 + aerobic			>14				
		0	Air				O			
			VP*			>14				
				0		>105	O	LAB/B[si]	Colour changes sooner but no rejection	13
				190		>105				
				290		77–105				
				532		42–63				
				818		28–42				
				920		14–28				
	DFD	10	VP	300		5	OC	N.D.	Thought *Alteromonas* – greening *S liq* odour	12
			VP + glucose			5	C			
			VP + citrate			8	C			
			VP + citrate + lactate			>14	–			
	Rump	0	100% N_2	0.4	N.S.	0	C	N.D.	Meat inoculated with 10^4 of *Moraxella*/ *Pseudomonas* mixture	1
			0.5% CO + N_2			24	O			
			1% CO + N_2			>30	–			
			10% CO + N_2			>30	C			
		5	100% N_2			0	C			
			0.5% CO + N_2			20	O			
			1% CO + N_2			24	O			
			10% CO + N_2			>30	–			
		10	100% N_2			0	C			
			0.5% CO + N_2			8	O			
			1% CO + N_2			10	O			
			10% CO + N_2			12	O			

Table A6.1 Continued

Meat			Storage characteristics			Spoilage characteristics			Comments	Reference
Species	Muscle type	Temperature (°C)	Atmosphere	OTR[a] of film	Gas to meat ratio	Shelf-life[b] (days)	Cause of spoilage	Dominant organisms[d]		
	Strip loin	−1.5	CO_2 $CO_2 + 0.1\% \ O_2$ $CO_2 + 0.2\% \ O_2$ CO_2 $CO_2 + 0.1\% \ O_2$ $CO_2 + 0.2\% \ O_2$	Foil lam.	1:1/2:1	>7/>7 1/>7 <1/<7 >7/>7 >7/>7 <7/<7	– C/– C/C –/– –/– C/C	N.D.	Residual oxygen level very important	38
	High pH	0–2	VP then retail at 4 °C	30		35 + 72h 42 + 72h 49 + 72h 63 + 60h 70 + 36h	O C + O	LAB/Ps[s,i] LAB[i]		2
	Low pH	1	VP VP 100% CO_2	30–40 0	0.4:1 1:1 2:1	70–84 70–84 105–126 126–147 126–147	F/O F F F	LAB[i]		22
	High pH		VP VP 100% CO_2	30–40 0	0.4:1 1:1 2:1	<49 <49 70–84 84–105 84–105	C/O F O O F		Ents caused spoilage in some cases	
	High pH	3	VP 100% CO_2 40% CO_2 + 60% N_2 20% CO_2 + 80% O_2	<20	N.S.	22–30 22–30 22 7–14	O O O O	Lb[i] Leu Lb Lb/Leu/Ps	Ps. putida unusually common	36

Table A6.1 Continued

Meat			Storage characteristics			Spoilage characteristics			Comments	Reference
Species	Muscle type	Temperature (°C)	Atmosphere	OTR[a] of film	Gas to meat ratio	Shelf-life[b] (days)	Cause of spoilage	Dominant organisms[d]		
	Ribeye rolls	−1	100% air	N.S.	N.S.	<23	C	Aerobic, not LAB[s]		3
			100% CO_2			<23				
			100% N_2			<23				
			100% O_2			<23				
			70% N_2 + 25% CO_2 + 5% O_2			<23				
	Knuckles	1–3	High VP (25.8 mmHg)	0.08–0.15		>28	−	Lb[si]		6
			Medium VP (16.6 mmHg)			14–21	O		Ents 12%	
			Low VP (11.2 mmHg) Partial evacuation only			>28	−			
			Aerobic			>4	−	N.D.		
	Roasts	1–3	VP	32	1:1	34	O	Lb[si]		9, 10
			100% O_2			13	O/C	Ps/Lb		
			20% CO_2 + 80% N_2			34	O/C	Lb		
			50% CO_2 + 50% O_2			27	O/C			
			20% CO_2 + 80% O_2			20	O/C			
			25% CO_2 + 25% O_2 + 50% N_2			20	O/C	Lb/Ps		
			51% CO_2 + 30% O_2 + 18% N_2 + 1% CO			34	O	Lb		
	Steaks	1	75% O_2 + 25% CO_2	8.1	N.S.	13d	App	Leu[i]		15, 16
			VP 14d then 75% O_2 + 25% CO_2			14 + 13	App			
			VP 28d then 75% O_2 + 25% CO_2			14 + 13	App			
		1	VP	40		70	N.D.	LAB[s]	Ps. growth on film	33
		4				56				

Table A6.1 Continued

Meat		Storage characteristics				Spoilage characteristics			Comments	Reference
Species	Muscle type	Temperature (°C)	Atmosphere	OTR[a] of film	Gas to meat ratio	Shelf-life[b] (days)	Cause of spoilage	Dominant organisms[d]		
	Hung 7d	1	75% O_2 + 25% CO_2	<2–4	2:1	21–25	O/C	Leu[si]		23
	VP 7d					21–28	O			
	VP 21d					35	O/C			
	Hung 7d	6				11–13	O	Lb/Ps/B/Ent		26
	VP 7d					11–13	O	Leu		
	VP 21d					25	O			
		2	50% CO_2 + 15% O_2 + 35% N_2	<15.5	N.S.	0+8 retail	N.S.	Ps/Lb/Ach[si]	Master pack	26
						7+8				
						14+4				
						21+2				
	High pH	0–2	VP	N.S.	N.S.	42	O	LAB/G-ve		32
	Low pH	3	VP	2–4	N.S.	>45	–	LAB[s]	Colour better with storage in CO_2	29
			100% CO_2			>45	–	LAB/Ent		
	High pH	VP	100% CO_2			34	O	LAB		
						>45	–			
	Pieces	4	Air	10	10:1	14	O	B[si]		17
	Low pH		VP			>21	–	LAB		
			100% CO_2			>51	–			
			78% N_2 + 20% CO_2 + 2% O_2			21	O	B		
	Pieces	4	Air			14	O	LAB		
	High pH		VP			>21	–			
			100% CO_2			>51	–			
			78% N_2 + 20% CO_2 + 2% O_2			21	O	B		
	Sirloin	1	75% O_2 + 25% CO_2	<10	2:1	18–22	O	LAB[s]		24
			VSP	<5		38	O			

Table A6.1 Continued

	Meat			Storage characteristics			Spoilage characteristics				
Species	Muscle type	Temperature (°C)		Atmosphere	OTR[a] of film	Gas to meat ratio	Shelf-life[b] (days)	Cause of spoilage	Dominant organisms[d]	Comments	Reference
	Chucks										
	Low pH	0–1		VP	N.D.		>66	–	LAB[si]		34
	High pH						<66	O			
	Loin Steaks	2		VP then retail in Air	N.D.		24 + 6	N.D.	LAB[i]		35
				Medium O_2 barrier	N.D.		24 + 30				
				High O_2 barrier			24 + 30		Lb		
	Rounds	0		Air (retail)	Foil lam.	>10:1	N.S.	N.D.	LAB[i]	Used anaerobic conditions throughout for anaerobic counts. Unusual numbers of staphs	31
				100% CO_2			15				
				VP			9				
		2		Air (retail)			N.S.				
				100% CO_2			15				
				VP			9				
		4		Air (retail)			N.S.				
				100% CO_2			15				
				VP			9				
	Cuts	1		VP	30–40	2:1	108	O/F	LAB	E.S.	20
	Low pH				2		108	F	LAB		
				CO_2	0		>108	–	LAB		
				VP	0		>108	–	Lb		
	High pH				30–40		66	O	Lb	High pH higher Ents	
					2		87	O	Lb		
				CO_2	0		87	O/F	Lb		
					0		>108	–	Lb		
	Trim/ Mince	1–2		Trim VP with solid CO_2, then mince and retail display	40		30 + 7	–	N.D.	2g CO_2 absorbed in 24 h; colour better	21

Table A6.1 Continued

Meat		Storage characteristics				Spoilage characteristics			Comments	Reference
Species	Muscle type	Temperature (°C)	Atmosphere	OTR[a] of film	Gas to meat ratio	Shelf-life[b] (days)	Cause of spoilage	Dominant organisms[d]		
Ground Beef		2	50% CO$_2$ + 15% O$_2$ + 35% N$_2$	<15.5	N.S.	0 + 8 retail; 7 + 6; 14 + 2	N.S.	N.S.	Master Pack	26
Lamb	Chops	−1	Air	N.S.	N.S.	14	CO	B/Ps^{si}		7
			80% Air + 20% CO$_2$			21	CF	B/Ps		
			80% O$_2$ + 20% N$_2$			21	CF	B/Ps		
			80% O$_2$ + 20% CO$_2$			21	CF	B		
			80% N$_2$ + 20% CO$_2$			42	C	B/Ps/Ent		
			80% H$_2$ + 20% CO$_2$ Low O$_2$			42	C	B/Ps/Ent		
			80% N$_2$ + 20% CO$_2$			56	F	B/Lb/Ent		
			80% H$_2$ + 20% CO$_2$ Oxygen free			56	F	B/Lb/Ent		
	Long loins	−1.5	CO$_2$	Foil lam.	1:1/2:1	>7/>7	-/-	N.D.		38
			CO$_2$ + 0.1% O$_2$			<7/>7	C/-			
			CO$_2$ + 0.2% O$_2$			<1/1	C/C			
Pork	Loins	4	100% Air	N.D.	2:1	3–7	O	Ps^{si}		11
			100% CO$_2$			>35		LAB		
			100% N$_2$			approx 7		Ps		
	Roasts	1–3	VP	32	1:1	28	O	Lb^{si}		4, 5
			100% O$_2$			14		Ps		
			20% CO$_2$ + 80% N$_2$			21		Leu		
			50% CO$_2$ + 50% O$_2$			14		Leu		
			20% CO$_2$ + 80% O$_2$			14		Leu/Ps		
			25% CO$_2$ + 25% O$_2$ + 50% N$_2$			14–21		Leu		
			51% CO$_2$ + 30% O$_2$ + 18% N$_2$ + 1% CO			14	C	Leu		
		1	50% CO$_2$ + 25% O$_2$ + 25% N$_2$	Drum	>2:1	>14	C	N.D.	Residual effect	8
			50% CO$_2$ + 25% O$_2$ + 25% N$_2$ + aerobic retail			14 + 3–7	C			

Table A6.1 Continued

Meat		Storage characteristics				Spoilage characteristics			Comments	Reference
Species	Muscle type	Temperature (°C)	Atmosphere	OTRᵃ of film	Gas to meat ratio	Shelf-lifeᵇ (days)	Cause of spoilage	Dominant organismsᵈ		
Chops		0	Aerobic retail display at 0 °C			>7	–	Psˢ	Difficult to assess data	37
			Aerobic retail display at 5 °C			5	App	LAB		
			VP + retail display at 0 °C + retail display at 5 °C	67		14 + 7	App	Ps		
			100% CO₂ + retail display at 0 °C + retail display at 5 °C		N.S.	14 + 2–5 / 14 + >7	App / –	Ps / Ps		
		2	50% CO₂ + 15% O₂ + 35% N₂	<15.5	N.S.	14 + 2–5 / 0 + 8 retail / 7 + 8 / 14 + 6 / 21 + 2	App / N.S.	LAB/Ps / Ps/Entˢⁱ	Master Pack	26
		4	Air	Imperm.	N.S.	2–3		Psˢⁱ	Irradiation treatment 12d	25
			CO₂			8		LAB + B		
			N₂			8				
			50% CO₂ + 50% N₂			8			Yeasts dominate flora in air	
			25% CO₂ + 75% N₂			8				
			10% O₂ + 70% CO₂ + 20% N₂			8				
		5	Aerobic retail display at 0 °C	67		5		LABˢ		37
			Aerobic retail display at 5 °C			5		LAB		
			VP + retail display at 0 °C + retail display at 5 °C		N.S.	14 + 2–5 / 14 + 2–5 / 14 + >7 / 14 + 0–2		LAB/Ps / Ps/LAB/Ent / Ps/LAB / LAB/Ent		
			100% CO₂ + retail display at 0 °C + retail display at 5 °C							
		10	Air	N.S.	N.S.	N.S.		LAB + Bˢⁱ		25
			CO₂			N.S.				
			N₂			N.S.				
			50% CO₂ + 50% N₂			N.S.				
			25% CO₂ 75% N₂			N.S.				
			10% O₂ + 70% CO₂ + 20% N₂			N.S.				

Table A6.1 Continued

Meat		Storage characteristics				Spoilage characteristics			Comments	Reference
Species	Muscle type	Temperature (°C)	Atmosphere	OTR[a] of film	Gas to meat ratio	Shelf-life[b] (days)	Cause of spoilage	Dominant organisms[d]		
	Steaks	−1	40% CO_2 + 60% N_2		5:1	56	C/O	B[s]	Aseptic vs commercial meat	27
		4.4				35	C/O	LAB		
		10				14–21	O	Ent		28
		2	20% CO_2 + 80% air	25	>2:1	16–21	O	B[si]		
			20% CO_2 + 80% O_2			16–21	O	B		
	Long loins	−1.5	CO_2	Foil lam.	1:1	>7	–	N.D.		38
			CO_2 + 1.0% O_2			<7	C			
			CO_2		2:1	>7	–			
			CO_2 + 1.0% O_2			>7	–			
	Loins	0	VP	8	N.S.	15	App	LAB[si]		19
			100% Air	10	2:1	15–20	O	Ps/B[si]		18
			100% CO_2			79–119	O	LAB		18
		1	75% O_2 + 25% CO_2	<10	2:1	10–14	O	LAB/B[s]		24
			VSP	<5		20	O	LAB		24
		3	VP	8	N.S.	8	App	LAB[si]		19
		4	100% Air	10	2:1	8–12	O	Ps[si]		18
			100% CO_2			27–40	O	LAB		18
		4	100% Air	Foil lam.	>2:1	<11	O	Ps[si]		14
			1 atm CO_2			41	O	LAB		14
			5 atm CO_2			>121	–	LAB		14
		7	VP	8	N.S.	8	App	LAB[si]		19
		14	100% Air	Foil lam.	>2:1	3	O	Ps[si]		14
			1 atm CO_2			7	O	LAB		14
			5 atm CO_2			>15	–	LAB		14

Table A6.1 Continued

Meat		Storage characteristics				Spoilage characteristics			Comments	Reference
Species	Muscle type	Temperature (°C)	Atmosphere	OTR[a] of film	Gas to meat ratio	Shelf-life[b] (days)	Cause of spoilage	Dominant organisms[d]		
	Loins then retail chops	-1.5	CO_2 (CAPTECH) then retail overwrap	Foil lam.	2.5:1	168	–	LAB[s]		30
						84 + 4	C/App	–		
						168 + 3	C/App	–		

[a] OTR = Oxygen transmission rate. Measured in $ml/m^2/24$ h at 1 atmosphere. The temperature and relative humidity at the time of measurement varies with film data. Drum – Atmosphere created in a drum. Imperm: impermeable film. Foil lam. = foil laminate – theoretically impermeable.

[b] Where detailed the time is taken from the time of slaughter. The methods and times of aging differ. Where retail shelf-life was tested, time is given as time in modified atmosphere + time in retail display.

[c] – No spoilage noted, O Off-odour, C Discoloration, F Off-flavour, App. General appearance.

[d] G-ve Gram-negative organisms, LAB Lactic acid bacteria, B *Brochothrix thermosphacta*, Ps. *Pseudomonas* spp., Leu *Leuconostoc* spp., Ent Enterobacteriaceae. Lb *Lactobacillus* spp., Ach *Achromobacter* spp., S liq *Serratia liquefaciens. Staphylococcus* tested with [s] selective medium only, [i] identification of isolates from total counts or [si] selective media were used but isolates were also taken from total counts and identified.

* VP Vacuum pack.

N.S. Not specified; N.D. Not determined; E.S. Electrical stimulation.

1, Clark *et al.* (1976); 2, Sutherland *et al.* (1975); 3, Huffman *et al.* (1975); 4, Seideman *et al.* (1979b); 5, Christopher *et al.* (1979b); 6, Seideman *et al.* (1976); 7, Newton *et al.* (1977a); 8, Silliker *et al.* (1977); 9, Christopher *et al.* (1979a); 10, Seideman *et al.* (1979a); 11, Enfors *et al.* (1979); 12, Gill and Newton (1979); 13, Newton and Rigg (1979); 14, Blickstad *et al.* (1981); 15, Hanna *et al.* (1981); 16, Savell *et al.* (1981); 17, Erichsen and Molin (1981); 18, Blickstad and Molin (1983); 19, Lee *et al.* (1985); 20, Gill and Penney (1986); 21, Madden and Moss (1987); 22, Gill and Penney (1988); 23, Nortjé and Shaw (1989); 24, Taylor *et al.* (1990); 25, Grant and Patterson (1991b); 26, Manu-Tawiah *et al.* (1991); 27, McMullen and Stiles (1991); 28, Ordóñez *et al.* (1991); 29 Rousset and Rennerre (1991); 30, Greer *et al.* (1993); 31, Venugopal *et al.* (1993); 32, Patterson and Gibbs (1977); 33 Madden and Bolton (1991); 34, Dainty *et al.* (1979); 35, Vanderzant *et al.* (1982); 36, Jackson *et al.* (1992); 37, Buys *et al.* (1993); 38, Penney and Bell (1993).

Table A6.2

Meat			Storage characteristics			Spoilage characteristics			Comments	Reference
Species	Muscle type	Temperature (°C)	Atmosphere	OTR^a of film	Gas to meat ratio	Shelf-life^b (days)	Cause of spoilage	Dominant organisms^d		
Beef	Round	1	60% CO_2 + 20% O_2 + 20% N_2 + aerobic	Drum	N.S.	>14	–	G-ves	Found residual effect	8
			20% CO_2 + 25% O_2 + 55% N_2 + aerobic			>14	–			
			Air			<14	O			
		0	VP*	0		>105	O	LAB/B^{si}	Colour changes sooner but no rejection	13
				190		>105				
				290		77–105				
				532		42–63				
				818		28–42				
				920		14–28				
	Strip loin	–1.5	CO_2	Foil lam.	1:1/2:1	>7/>7	–	N.D.	Residual oxygen level very important	38
			CO_2 + 0.1% O_2			1/>7	C/–			
			CO_2 + 0.2% O_2			<1/<7	C/C			
	Low pH	1	VP	30–40		70–84	F/O	LAB^i		22
			VP	0		70–84	F			
			100% CO_2		0.4:1	105–126	F			
					1:1	126–147	F			
					2:1	126–147	F			
	High pH		VP	30–40		<49	C/O		Ents caused spoilage in some cases	
			VP	0		<49	F			
			100% CO_2		0.4:1	70–84	O			
					1:1	84–105	O			
					2:1	84–105	F			

Table A6.2 Continued

Meat		Storage characteristics				Spoilage characteristics			Comments	Reference
Species	Muscle type	Temperature (°C)	Atmosphere	OTR[a] of film	Gas to meat ratio	Shelf-life[b] (days)	Cause of spoilage	Dominant organisms[d]		
	DFD	10	VP			5	OC	N.D.	Thought *Alteromonas* – greening S liq odour	12
			VP + glucose	300		5	C	N.D.		
			VP + citrate			8	C			
			VP + citrate + lactate			>14	–			
	Roasts	1–3	VP	32		34	O	Lb[si]		9, 10
			100% O₂		1:1	13	O/C	Ps/Lb		
			20% CO₂ + 80% N₂			34	O/C	Lb		
			50% CO₂ + 50% O₂			27	O/C			
			20% CO₂ + 80% O₂			20	O/C			
			25% CO₂ + 25% O₂ + 50% N₂			20	O/C	Lb/Ps		
			51% CO₂ + 30% O₂ +18% N₂ +1% CO			34	O	Lb		
	Knuckles	1–3	High VP (25.8 mmHg)	0.08		>28	–	Lb[si]	Ents 12%	6
			Medium VP (16.6 mmHg)	0.15		14–21	O			
			Low VP (11.2 mmHg)			>28	–			
			Partial evacuation only			>4	–	N.D.		
			Aerobic							
	Trim/ Mince	1–2	Trim VP with solid CO₂, then mince and retail display	40		30 + 7	–	N.D.	2g CO₂ absorbed in 24 h; colour better	21
	Cuts Low pH	1	VP	30–40	2:1	108	O/F	LAB	E.S.	20
				2		108	F	LAB		
			CO₂	0		>108	–	LAB		
			VP	0		>108	–	Lb		
	High pH			30–40		66	O	Lb	High pH higher Ents	
				2		87	O	Lb		
				0		87	O/F	Lb		
			CO₂	0		>108	–	Lb		

Table A6.2 Continued

| Meat | | Storage characteristics | | | | Spoilage characteristics | | | Comments | Reference |
Species	Muscle type	Temperature (°C)	Atmosphere	OTR[a] of film	Gas to meat ratio	Shelf-life[b] (days)	Cause of spoilage[c]	Dominant organisms[d]		
Lamb	Chops	-1	Air	N.S.	N.S.	14	CO	B/Ps^{si}		7
			80% Air + 20% CO_2			21	CF	B/Ps		
			80% O_2 + 20% N_2			21	CF	B/Ps		
			80% O_2 + 20% CO_2			21	CF	B		
			80% N_2 + 20% CO_2 } Low O_2			42	C	B/Ps/Ent		
			80% H_2 + 20% CO_2 }			42	C	B/Ps/Ent		
			80% N_2 + 20% CO_2 } Oxygen free			56	F	B/Lb/Ent		
			80% H_2 + 20% CO_2 }			56	F	B/Lb/Ent		
Pork	Steaks	-1	40% CO_2 + 60% N_2		5:1	56	C/O	B^s	Aseptic vs commercial meat	27
		4.4				35	C/O	LAB		
		10				14–21	O	Ent		
	Loins	0	100% Air	10	2:1	15–20	O	Ps/B^{si}		18
		0	100% CO_2			79–119	O	LAB		18
			75% O_2 + 25% CO_2	<10	2:1	10–14	O	LAB/B^s		24
		1	VSP	<5		20	O	LAB		24
		1	100% Air	10	2:1	8–12	O	Ps^{si}		18
		3	100% CO_2			27–40	O	LAB		18
		4	100% Air	Foil lam.	>2:1	<11	O	Ps^{si}		14
		4	1 atm CO_2			41	O	LAB		14
			5 atm CO_2			>121	–	LAB		14
		4	CO_2 (CAPTECH) then retail overwrap		2.5:1	168	–	LAB^s		30
	Loins then retail chops	-1.5		Foil lam		84 + 4	C/App			
						168 + 3	C/App			

[a] OTR = oxygen transmission rate measured in ml/m^{-2}/24 h at 1 atmosphere. The temperature and relative humidity at the time of measurement varies with film data. Drum = atmosphere created in a drum. Imperm = impermeable film. Foil lam. = foil laminate, theoretically impermeable.

[b] Where detailed the time is taken from the time of slaughter. The methods and times of aging differ. Where retail shelf-life was tested, time is given as time in modified atmosphere + time in retail display.

[c] – no spoilage noted; O, off-odour; C, discoloration; F, off-flavour; App, general appearance.

[d] G-ve, Gram-negative organisms; LAB, Lactic acid bacteria; B, *Brochothrix thermosphacta*; Ps, *Pseudomonas* spp.; Leu, *Leuconostoc* spp.; Ent, Enterobacteriaceae; Lb, *Lactobacillus* spp.; Ach, *Achromobacter* spp; Tested with ^{s} selective medium only; ^{i} identification of isolates from total counts or ^{si} selective media were used but isolates were also taken from total counts and identified.

* VP, Vacuum pack.

6. Seideman *et al.* (1976); 7. Newton *et al.* (1977a, b); 8. Silliker *et al.* (1977); 9. Christopher *et al.* (1979a, b); 10. Seideman *et al.* (1979a, b); 10. Seideman *et al.* (1979a, b); 12. Gill and Newton (1979); 13. Newton and Rigg (1979); 14. Blickstad *et al.* (1981); 18. Blickstad and Molin (1983); 20. Gill and Penney (1986); 21. Madden and

Table A6.3

Meat			Storage characteristics			Spoilage characteristics			Comments	Reference
Species	Muscle type	Temperature (°C)	Atmosphere	OTR[a] of film	Gas to meat ratio	Shelf-life[b] (days)	Cause of spoilage[c]	Dominant organisms[d]		
Beef	Roasts	1–3	VP	32	1:1	34	O	Lb[si]		9, 10
			100% O_2			13	O/C	Ps/Lb		
			20% CO_2 + 80% N_2			34	O/C	Lb		
			50% CO_2 + 50% O_2			27	O/C			
			20% CO_2 + 80% O_2			20	O/C			
			25% CO_2 + 25% O_2 + 50% N_2			20	O/C	Lb/Ps		
			51% CO_2 + 30% O_2 + 18% N_2 + 1% CO			34	O	Lb		
Lamb	Chops	–1	Air	N.S.	N.S.	14	CO	B/Ps[si]		7
			80% Air + 20% CO_2			21	CF	B/Ps		
			80% O_2 + 20% N_2			21	CF	B/Ps		
			80% O_2 + 20% CO_2			21	CF	B		
			80% N_2 + 20% CO_2 (Low O_2)			42	C	B/Ps/Ent		
			80% H_2 + 20% CO_2 (Low O_2)			42	C	B/Ps/Ent		
			80% N_2 + 20% CO_2 (Oxygen free)			56	F	B/Lb/Ent		
			80% H_2 + 20% CO_2 (Oxygen free)			56	F	B/Lb/Ent		
Pork	Roasts	1–3	VP	32	1:1	28	O	Lb[si]		4, 5
			100% O_2			14		Ps		
			20% CO_2 + 80% N_2			21		Leu		
			50% CO_2 + 50% O_2			14		Leu		
			20% CO_2 + 80% O_2			14		Leu/Ps		
			25% CO_2 + 25% O_2 + 50% N_2			14–21		Leu		
			51% CO_2 + 30% O_2 + 18% N_2 + 1% CO			14	C	Leu		

[a] OTR = oxygen transmission rate measured in ml/m^{-2}/24 h at 1 atmosphere. The temperature and relative humidity at the time of measurement varies with film data. Drum = atmosphere created in a drum. Imperm = impermeable film. Foil lam. = foil laminate, theoretically impermeable.

[b] Where detailed the time is taken from the time of slaughter. The methods and times of aging differ. Where retail shelf-life was tested, time is given as time in modified atmosphere + time in retail display.

[c] – no spoilage noted; O, off-odour; C, discoloration; F, off-flavour; App, general appearance.

[d] G-ve, Gram-negative organisms; LAB, Lactic acid bacteria; B, *Brochothrix thermosphacta*; Ps, *Pseudomonas* spp.; Leu, *Leuconostoc* spp.; Ent, Enterobacteriaceae; Lb, *Lactobacillus* spp.; Ach, *Achromobacter* spp; Tested with [s] selective medium only; [i] identification of isolates from total counts or [si] selective media were used but isolates were also taken from total counts and identified.

* VP, Vacuum pack.

4. Seideman et al. (1979a, b); 5. Christopher et al. (1979a, b); 7. Newton et al. (1977a, b); 9. Christopher et al. (1979a, b); 10. Seideman et al. (1979a, b).

References

ACMSF. (1992) Report on vacuum packaging and associated processes. Advisory Committee on the Microbiological Safety of Food, HMSO, London.

Ahmad, H.A. and Marchello, J.A. (1989) Microbial growth and succession on steaks as influenced by packaging procedures. *J. Food Protect.*, **52**, 236–9, 243.

Anonymous (1980) In *Microbial ecology of foods*, Vol. 1. *Factors affecting life and death of micro-organisms*, International Commission on Microbiological Specifications for Foods (eds J.H. Silliker, R.P. Elliot, A.C. Baird-Parker, F.L. Bryan, J.H.B. Christian, D.S. Clark, J.C. Olson and T.A. Roberts), Academic Press, New York, pp. 189–92.

Anonymous (1987) New packaging technologies. *The National Provisioner*, **197**, 75–87.

Anonymous (1994) Britons report successful use of an oxygen-removing label. *Food Eng.*, (March) 68, 70.

Avery, S.M., Hudson, J.A. and Penney, N. (1994) A research note. Inhibition of *Listeria monocytogenes* on normal ultimate pH beef (pH 5.3–5.5) at abusive storage temperatures by saturated carbon dioxide controlled packaging. *J. Food Protect.*, **57**, 331–3, 336.

Baird-Parker, A.C. (1980) In *Microbial Ecology of Foods* Vol. 1, International Commission on Microbiological Specifications for Foods (eds J.H. Silliker, R.P. Elliott, A.C. Baird-Parker, F.L. Bryan, J.H.B. Christian, D.S. Clark, J.C. Olson Jr., and T.A. Roberts), Academic Press, New York, pp. 126–35.

Banks, J.G. and Board, R.G. (1982) Sulfite-inhibition of Enterobacteriaceae including *Salmonella* in British fresh sausage and in culture systems. *J. Food Protect.*, **45**, 1292–7.

Banks, J.G. and Board, R.G. (1983) The classification of pseudomonads and other obligately aerobic Gram-negative bacteria from British pork sausage and ingredients. *System. Appl. Microbiol.*, **4**, 424–38.

Baumann, P. (1968) Isolation of *Acinetobacter* form soil and water. *J. Bacteriol.*, **96**, 39–42.

Bell, R.G. (1996) Chilled and frozen raw meat, poultry and their products. In *Leatherhead Food Researh Association Microbiology Handbook*, Vol. 2, *Meat Products* (ed. J. Milner), Leatherhead Food Research Association, Leatherhead, UK.

Bem, Z., Hechelmann, H. and Leistner, L. (1976) Mikrobiologie des DFD-fleisches. *Fleischwirtschaft*, **56**, 985–7.

Bentley, R. (1991) Dressing upside down, down under. *Food Sci. Technol. Today*, **5**, 111–13.

Blickstad, E., Enfors, S-O. and Molin, G. (1981) Effect of hyperbaric carbon dioxide pressure on the microbial flora of pork stored at 4 or 14 °C. *J. Appl. Bacteriol.*, **50**, 493–504.

Blickstad, E. and Molin, G. (1983) Carbon dioxide as a controller of the spoilage flora of pork, with special reference to temperature and sodium chloride. *J. Food Protect.*, **46**, 756–63.

Borch, E. and Agerhem, H. (1992) Chemical, microbial and sensory changes during the anaerobic cold storage of beef inoculated with a homofermentative *Lactobacillus* sp. or a *Leuconostoc* sp. *Int. J. Food Microbiol.*, **15**, 99–108.

Borch, E. and Molin, G. (1988) Numerical taxonomy of psychrotrophic lactic acid bacteria from prepacked meat and meat products. *Antonie van Leeuwenhoek*, **54**, 301–23.

Brown, M.H. and Booth, I.R. (1991) Acidulants and low pH, in *Food Preservatives* (eds N.J. Russell and G.W. Gould), Blackie, London, pp. 22–43.

Buchanan, R.L. and Klawitter, L.A. (1992) Effect of temperature history on the growth of *Listeria monocytogenes* Scott A at refrigeration temperatures. *Int. J. Food Microbiol.*, **12**, 235–45.

Buchanan, R.L. and Palumbo, S.A. (1985) *Aeromonas hydrophila* and *Aeromonas sobria* as potential food poisoning species: A review. *J. Food Safety*, **7**, 15–29.

Buys, E.M., Nortjé, G.L. and Steyn, P.L. (1993) The effect of wholesale vacuum and 100% CO_2 storage on the subsequent microbiological, colour and acceptability attributes of PVC-overwrapped pork loin chops. *Food Res. Int.*, **26**, 421–9.

Campbell, R.J., Egan, A.F., Grau, F.H. and Shay, B.J. (1979) The growth of *Microbacterium thermosphactum* on beef. *J. Appl. Bacteriol.*, **47**, 505–9.

Carlez, A., Rosec, J-P., Richard, N. and Cheftel, J-C. (1993) High pressure inactivation of *Citrobacter freundii*, *Pseudomonas fluorescens* and *Listeria innocua* in inoculated minced beef muscle. *Lebensmittel-Wissenschaft-Technologie*; *Food Science and Technology*, **26**, 357–63.

Christopher, F.M., Seideman, S.C., Carpenter, Z.L., Smith, G.C. and Vanderzant, C. (1979a) Microbiology of beef packaged in various gas atmospheres. *J. Food Protect.*, **42**, 240–4.

Christopher, F.M., Vanderzant, C., Carpenter, Z.L. and Smith, G.C. (1979b) Microbiology of pork packaged in various gas atmospheres. *J. Food Protect.*, **42**, 323–7.

Church, I.J. and Parsons, A.L. (1995) Modified atmosphere packaging technology: A review. *J. Sci. Food Agric.*, **67**, 143–52.

Clark, D.S. and Lentz, C.P. (1972) Use of carbon dioxide for extending shelf-life of pre-packaged beef. *Can. Inst. Food Sci. Technol. J.*, **5** 175–8.

Clark, D.S., Lentz, C.P. and Roth, L.A. (1976) Use of carbon monoxide for extending shelf-life of pre-packaged fresh beef. *Can. Inst. Food Sci. Technol. J.*, **9**, 114–17.

Collins, M.D., Rodrigues, U.M., Dainty, R.H., Edwards, R.A. and Roberts, T.A. (1992) Taxonomic studies on a psychrotrophic *Clostridium* from vacuum-packed beef: description of *Clostridium esterheticum* sp. nov. *FEMS Microbiol. Letts*, **96**, 235–40.

Dainty, R.H. (1989) Spoilage microbes on meat and poultry. *Food Sci. Technol. Today*, **3**, 250–1.

Dainty, R.H., Edwards, R.A. and Hibbard, C.M. (1985) Time course of volatile compound formation during refrigerated storage of naturally contaminated beef in air. *J. Appl. Bacteriol.*, **59**, 303–9.

Dainty, R.H., Edwards, R.A., Hibbard, C.M. and Marnewick, J.J. (1989) Volatile compounds associated with microbial growth on normal and high pH beef stored at chill temperatures. *J. Appl. Bacteriol.*, **61**, 117–23.

Dainty, R.H., Edwards, R.A., Hibbard, C.M. and Ramantanis, S.V. (1986) Bacterial sources of putrescine and cadaverine in chill stored vacuum-packaged beef. *J. Appl. Bacteriol.*, **61**, 117–23.

Dainty, R.H. and Hibbard, C.M. (1980) Aerobic metabolism of *Brochothrix thermosphacta* growing on meat surfaces and in laboratory media. *J. Appl. Bacteriol.*, **48**, 387–96.

Dainty, R.H. and MacKey, B.M. (1992) The relationship between the phenotypic properties of bacteria form chill-stored meat and spoilage processes, in *Ecosystems: Microbes: Food* (eds R.G. Board, D. Jones, R.G. Kroll and G.L. Pettipher), Society for Applied Bacteriology Symposium Series, Number 21, supplement to Volume **73**, pp. 103S–114S.

Dainty, R.H., Shaw, B.G., de Boer, K.A. and Scheps, E.S.J. (1975) Protein changes caused by bacterial growth on beef. *J. Appl. Bacteriol.*, **39**, 73–81.

Dainty, R.H., Shaw, B.G., Harding, C.D. and Michanie, S. (1979) The spoilage of vacuum-packed beef by cold tolerant bacteria, in *Cold Tolerant Microbes in Spoilage and the Environment* (eds A.D. Russell and R. Fuller), The Society for Applied Bacteriology Technical Series No. 13, Academic Press, London. pp. 83–100.

Daniels, J.A., Krishnamurthi, R. and Rizvi, S.H. (1985) A review of effects of carbon dioxide on microbial growth and food quality. *J. Food Protect.*, **48**, 532–7.

Davies, A.R. (1995) Fate of food-borne pathogens on modified-atmosphere packaged meat and fish. *Int. Biodeter. Biodegrad.*, **36**, 407–10.

Dixon, N.M. and Kell, D.B. (1989) The inhibition by CO_2 of the growth and metabolism of micro-organisms. *J. Appl. Bacteriol.*, **67**, 109–36.

Doherty, A., Sheridan, J.J., Allen, P., McDowell, D.A., Blair, I.S. and Harrington, D. (1995) Growth of *Yersinia enterocolitica* O:3 on modified atmosphere packaged lamb. *Food Microbiol.*, **12**, 251–7.

Doherty, A., Sheridan, J.J., Allen, P., McDowell, D.A., Blair, I.S. and Harrington, D. (1996) Survival and growth of *Aeromonas hydrophila* on modified atmosphere packaged normal and high pH lamb. *Int. J. Food Microbiol.*, **28**, 379–92.

Drosinos, E.H. and Board, R.G. (1994) Growth of *Listeria monocytogenes* in meat juice under a modified atmosphere at 4 °C with or without members of a microbial association from chilled lamb under a modified atmosphere. *Letts Appl. Microbiol.*, **19**, 134–7.

Earnshaw, R. (1990) Use of combination processes – their benefits, in *Irradiation and Combination Treatments*, Proceedings of the conference 1–2 March, 1990, London, UK Technical Services, London.

Edwards, R.A., Dainty, R.H. and Hibbard, C.M. (1985) Putrescine and cadaverine formation in vacuum packed beef. *J. Appl. Bacteriol.*, **58**, 13–19.

Edwards, R.A., Dainty, R.H., Hibbard, C.M. and Ramantanis, S.V. (1987) Amines in fresh beef of normal pH and the role of bacteria in changes in concentration observed during storage in vacuum packs at chill temperatures. *J. Appl. Bacteriol.*, **63**, 427–34.

Egan, A.F. (1984) Microbiology and storage life of chilled fresh meats, in *Proceedings of the European Meeting of Meat Research Workers, 1984*, **30**, 211–14.

Egan, A.F., Eustace, I.J. and Shay, B.J. (1991) Meat packaging: maintaining the quality and prolonging the shelf-life of chilled beef, pork and lamb. *Meat Focus Int.*, 25–33.

Egan, A.F. and Shay, B.J. (1982) Significance of lactobacilli and film permeability in the spoilage of vacuum-packed beef. *J. Food Sci.*, **47**, 1119–22, 1126.

Eklund, T. and Jarmund, T. (1983) Microculture model studies on the effect of various gas atmospheres on microbial growth at different temperatures. *J. Appl. Bacteriol.*, **55**, 119–25.

Enfors, S-O. and Molin, G. (1981) The influence of temperature on the growth inhibitory effect of carbon dioxide on *Pseudomonas fragi* and *Bacillus cereus. Can. J. Microbiol.*, **27**, 15–19.

Enfors, S-O., Molin, G. and Ternström, A. (1979) Effect of packaging under carbon dioxide, nitrogen or air on the microbial flora of pork stored at 4 °C. *J. Appl. Bacteriol.*, **47**, 197–208.

Erichsen, I. and Molin, G. (1981) Microbial flora of normal and high pH beef stored at 4 °C in different gas environments. *J. Food Protect.*, **44**, 866–9.

Farber, J.M. (1991) Microbiological aspects of modified atmosphere packaging technology – a review. *J. Food Protect.*, **54**, 58–70.

Farkas, J. and Andrássy, É. (1993) Interaction of ionising radiation and acidulants on the growth of the microflora of a vacuum-packaged chilled meat product. *Int. J. Food Microbiol.*, **19**, 145–52.

Farouk, M.M., Price, J.F. and Salih, A.M. (1990) Effect of an edible collagen film overwrap on exudation and lipid oxidation in beef and round steak. *J. Food Sci.*, **55**, 1510–12.

Fournaud, J., Salé, P. and Valin, C. (1973) *Conservation de la viande borne sous emballage plastique sous vide ou en atmospheres controlées. Aspects biologiques et microbiologiques*, in Proceedings of the 19th Meeting of European Meat Research Worker, Paris, pp. 287–315.

Fricker, C.R. and Tompsett, S. (1989) *Aeromonas* spp. in foods: A significant cause of food poisoning? *Int. J. Food Microbiol.*, **9**, 17–23.

Fukushima, H. and Gomyoda, M. (1986) Inhibition of *Yersinia enterocolitica* serotype O3 by natural microflora of pork. *Appl. Environ. Microbiol.*, **51**, 990–4.

Gardner, G.A. (1981) *Brochothrix thermosphacta (Microbacterium thermosphactum)* in the spoilage of meats: A review, in *Psychrotrophic Micro-organisms in Spoilage and Pathogenicity*, (eds T.A. Roberts, G. Hobbs, J.H.B. Christian and N. Skovgaard). Academic Press, London, pp. 139–73.

Gavini, F., Mergaert, J., Beji, A., Mielcarek, C., Izard, D., Kersters, K. and De Ley, J. (1989) Transfer of *Enterobacter agglomerans* (Beijerinck 1888) Ewing and Fife 1972 to *Pantoea* gen. nov. as *Pantoea agglomerans* comb. nov. and description of *Pantoea dispersa* sp. nov. *Int. J. System. Bacteriol.*, **39**, 337–45.

Genigeorgis, C.A. (1985) Microbial and safety implications of the use of modified atmospheres to extend the storage life of fresh meat and fish. *Int. J. Food Microbiol.*, **1**, 237–51.

Gennari, M. and Dragotto, F. (1992) A study of the incidence of different fluorescent *Pseudomonas* species and biovars in the microflora of fresh and spoiled meat and fish, raw milk, cheese, soil and water. *J.Appl. Bacteriol.*, **72**, 281–8.

Gill, C.O. (1976) Substrate limitation of bacterial growth at meat surfaces. *J. Appl. Bacteriol.*, **41**, 401–10.

Gill, C.O. (1982) Microbial interaction with meats, in *Meat Microbiology*, (ed. M.H. Brown). Applied Science, London, pp. 225–64.

Gill, C.O. (1986) Control of microbial spoilage in fresh meats, in *Advances in Meat Research*, Vol. 2. *Meat and Poultry Microbiology* (eds A.M. Pearson and T.R. Dutson). AVI Publishing, Westport, Connecticut, pp. 49–88.

Gill, C.O. (1988) The solubility of carbon dioxide in meat. *Meat Sci.*, **22**, 65–71.

Gill, C.O. (1989) Packaging meat for prolonged chilled storage: the CAPTECH process. *Br. Food J.*, **91**, 11–15.

Gill, C.O. (1990) Controlled atmosphere packaging of chilled meat. *Food Control*, **1**, 74–8.

Gill, C.O., Greer, G.G. and Bilts, B.D. (1997) The aerobic growth of *Aeromonas hydrophila* and *Listeria monocytogenes* in broths and on pork. *Int. J. Food Microbiol.*, **35**, 67–74.

Gill, C.O. and Molin, G. (1991) Modified atmospheres and vacuum packaging, in *Food Preservatives*, (eds N.J. Russell and G.W. Gould). Blackie, Glasgow, pp. 172–99.

Gill, C.O. and Newton, K.G. (1977) The development of aerobic spoilage on meat stored at chill temperatures. *J. Appl. Bacteriol.*, **43**, 189–95.

Gill, C.O. and Newton, K.G. (1978) The ecology of bacterial spoilage of fresh meat at chill temperatures. *Meat Sci.*, **2**, 207–17.

Gill, C.O. and Newton, K.G. (1979) Spoilage of vacuum-packaged dark, firm, dry meat at chill temperatures. *Appl. Environ. Microbiol.*, **37**, 362–4.

Gill, C.O. and Newton, K.G. (1980) Growth of bacteria on meat at room temperature. *J. Appl. Bacteriol.*, **49**, 315–23.

Gill, C.O. and Penney, N. (1986) Packaging conditions for extended storage of chilled dark, firm, dry beef. *Meat Sci.*, **18**, 41–53.

Gill, C.O. and Penney, N. (1988) The effect of the initial gas volume to meat weight ratio on the storage life of chilled beef packaged under carbon dioxide. *Meat Sci.*, **22**, 53–63.

Gill, C.O. and Reichel, M.P. (1989) Growth of the cold-tolerant pathogens *Yersinia enterocolitica*, *Aeromonas hydrophila* and *Listeria monocytogenes* on high-pH beef packaged under vacuum or carbon dioxide. *Food Microbiol.*, **6**, 223–30.

Gill, C.O. and Tan, K.H. (1980) Effect of carbon dioxide on growth of meat spoilage bacteria. *Appl. Environ. Microbiol.*, **39**, 317–19.

Gobat, P. and Jemmi, T. (1993) Distribution of mesophilic *Aeromonas* species in raw and ready-to-eat fish and meat products in Switzerland. *Int. J. Food Microbiol.*, **20**, 117–20.

Gould, G.W. (1992) Ecosystem approaches to food preservation. *J. Appl. Bacteriol.*, Symp. Suppl., **73**, 585S–685S.

Grant, I.R. and Patterson, M.F. (1991a) A numerical taxonomic study of lactic acid bacteria isolated from irradiated pork and chicken packaged under various gas atmospheres. *J. Appl. Bacteriol.*, **70**, 302–7.

Grant, I.R. and Patterson, M.F. (1991b) Effect of irradiation and modified atmosphere packaging on the microbiological and sensory quality of pork stored at refrigeration temperatures. *Int. J. Food Sci. Technol.*, **26**, 507–19.

Grau, F.H. (1988) Substrates used by *Brochothrix thermosphacta* when growing on meat. *J. Food Protect.*, **51**, 639–42.

Grau, F.H. and Vanderlinde, P.B. (1990) Growth of *Listeria monocytogenes* on vacuum-packaged beef. *J. Food Protect.*, **53**, 739–41, 746.

Grau, F.H. and Vanderlinde, P.B. (1993) Aerobic growth of *Listeria monocytogenes* on beef lean and fatty tissue: equations describing the effects of temperature and pH. *J. Food Protect.*, **56**, 96–101.

Greer, G.G. (1986) Homologous bacteriophage control of *Pseudomonas* growth and beef spoilage. *J. Food Protect.*, **49**, 104–9.

Greer, G.G., Dilts, B.D. and Jeremiah, L.E. (1993) Bacteriology and retail case life of pork after storage in carbon dioxide. *J. Food Protect.*, **56**, 689–93.

Greipsson, S. and Priest, F.G. (1983) Numerical taxonomy of *Hafnia alvei*. *Int. J. System. Bacteriol.*, **33**, 470–5.

Hanna, M.O., Savell, J.W., Smith, G.C., Purser, D.E., Gardner, F.A. and Vanderzant, C. (1983) Effect of growth of individual meat bacteeria on pH, colour and odour of aseptically prepared vacuum packaged round steaks. *J. Food Protect.*, **46**, 216–21.

Hanna, M.O., Vanderzant, C., Smith, G.C. and Savell, J.W. (1981) Packaging of beef loin steaks in 75% O_2 plus 25% CO_2. II. Microbiological properties. *J. Food Protect.*, **44**, 928–33.

Harrison, M.A., Melton, C.C. and Draughon, F.A. (1981) Bacterial flora of ground beef and soy extended ground beef during storage. *J. Food Sci.*, **46**, 1088–90.

Hastings, J.W. and Holzapfel, W.H. (1987a) Conventional taxonomy of lactobacilli surviving radurization of meat. *J. Appl. Bacteriol.*, **62**, 209–16.

Hastings, J.W. and Holzapfel, W.H. (1987b) Numerical taxonomy of lactobacilli surviving radurization of meat. *Int. J. Food Microbiol.*, **4**, 33–49.

Hilliam, M.A., James, S.J., Worman, L.A. and Boyle, C. (1997) The UK Food and Drinks Report, Market, Industry and New Product Trends, Volume I: the UK Food Market. Leatherhead Food Research Association, UK. March, 1997.

Hintlian, C.B. and Hotchkiss, J.H. (1986) The safety of modified atmosphere packaging: a review. *Food Technol.*, **40**, 70–6.

Hitchener, B.J., Egan, A.F. and Rogers, P.J. (1982) Characteristics of lactic acid bacteria isolated from vacuum-packaged beef. *J. Appl. Bacteriol.*, **52**, 31–7.

Holmes, B. and Costas, M. (1992) Identification and typing of Enterobacteriaceae using computerised methods, in *Identification Methods for Microbiologists*, 3rd edn, (eds R.G. Board and D. Jones). Technical Series of the Society of Applied Bacteriology, Academic Press, London, pp. 127–49.

Hood, D.E. and Mead, G.C. (1993) Modified atmosphere storage of fresh meat and poultry, in *Principles and Applications of Modified Atmosphere Packaging of Food*, (ed. R.T. Parry). Blackie Academic and Professional, London, pp. 269–98.

Hudson, J.A. and DeLacy, K.M. (1991) Incidence of motile aeromonads in New Zealand retail foods. *J. Food Protect.*, **54**, 696–9.

Hudson, J.A., Mott, S.J., DeLacy, K.M. and Eldridge, A.L. (1992) Incidence and coincidence of *Listeria* spp., motile aeromonads and *Yersinia enterocolitica* on ready-to-eat fleshfoods. *Int. J. Food Microbiol.*, **16**, 99–108.

Huffman, D.L., Davis, K.A., Marple, D.N. and McGuire, J.A. (1975) Effect of gas atmospheres on microbial growth, colour and pH of beef. *J. Food Sci.*, **40**, 1229–31.

Ingram. M. and Barnes, E.M. (1954) Sterilization by means of ozone. *J. Appl. Bacteriol.*, **17**, 246–71.

Ingram. M. and Dainty, R.H. (1971) Changes caused by microbes in spoilage of meats. *J. Appl. Bacteriol.*, **34**, 21–39.

Jackson, T.C., Acuff, G.R., Vanderzant, C., Sharp, T.R. and Savell, J.W. (1992) Identification and evaluation of volatile compounds of vacuum and modified atmosphere packaged beef strip loins. *Meat Sci.*, **31**, 175–90.

Johnson, J.L., Doyle, M.P. and Cassens, R.G. (1988) Survival of *Listeria monocytogenes* in ground beef. *Int. J. Food Microbiol.*, **6**, 243–7.

Johnson, J.L., Doyle, M.P. and Cassens, R.G. (1990) *Listeria monocytogenes* and other *Listeria* spp. in meat and meat products. A review. *J. Food Protect.*, **53**, 81–91.

Jones, D. (1988) Composition and properties of the family Enterobacteriaceae. *J. Appl. Bacteriol.*, Symp. Suppl., 1S–19S.

Jones, M.V. (1989) Modified atmosphere, in *Mechanisms of Action of Food Preservation Procedures*, (ed. G.W. Gould). Elsevier Applied Science, London, pp. 247–84.

Kaess, G. and Weidemann, J.F. (1968) Ozone treatment of chilled beef I. Effect of low concentrations of ozone on microbial spoilage and surface colour of beef. *J. Food Technol.*, **3**, 325–34.

Kalchayanand, N., Ray, B., Rield, R.A. and Johnson, M.C. (1989) Spoilage of vacuum-packaged refrigerated beef by *Clostridium*. *J. Food Protect.*, **52**, 424–6.

Kaya, M. and Schmidt, U. (1991) Behaviour of *Listeria monocytogenes* on vacuum-packed beef. *Fleischwirtsch*, **71**, 424–6.

Khan, M.A., Palmas, C.V., Seaman, A. and Woodbine, M. (1973) Survival versus growth of a facultative psychrotroph. Meat and products of meat. *Zbl. Bakt. Hug., I Abt. Orig. B*, **157**, 277–82.

Kim, W.J. (1993) Bacteriocins of lactic acid bacteria: their potentials as food biopreservatives. *Food Rev. Int.*, **9**, 299–313.

Kleinlein, N. and Untermann, F. (1990) Growth of pathogenic *Yersinia enterocolitica* strains in minced meat with and without protective gas with consideration of the competitive background flora. *Int. J. Food Microbiol.*, **10**, 65–72.

Koski, D.V. (1988) Is current modified/controlled atmosphere packaging technology applicable to the U.S. food market? *Food Technol.*, **42**, 54.

Labadie, J., Fournaud, J. and Dumont, B.L. (1975) Relations entre le pH et la microflore des viandes hachées de bovins. *Annales de Technologie Agricole*, **24**, 193–203.

Lambert, A.D., Smith, J.P. and Dodds, K.L. (1991) Shelf-life extension and microbiological safety of fresh meat – a review. *Food Microbiol.*, **8**, 267–97.

Lambert, A.D., Smith, J.P., Dodds, K.L. and Charbonneau, R. (1992) Microbiological changes and shelf-life of MAP irradiated fresh pork. *Food Microbiol.*, **9**, 231–44.

Lawrie, R.A. (1991) *Meat Science*. 5th edn. Pergamon Press, Oxford, England.

Lee, B.H., Simard, R.E., Laleye, L.C. and Holley, R.A. (1985) Effects of temperature and storage duration on the microflora, physicochemical and sensory changes of vacuum- or nitrogen-packed pork. *Meat Sci.*, **13**, 99–112.

Leeson, R. (1987) The use of gaseous mixtures in controlled and modified atmosphere packaging. *Food Technol. New Zealand*, **22**, 24–5.

Leistner, L. (1985) Hurdle technology applied to meat products of the shelf stable product and intermediate moisture food types, in *Properties of Water in Foods*, (eds D. Simatos and Multon, J.L.). Martinus Nijhoff, Dordrecht, pp. 309–29.

Lücke, F-K. and Roberts, T.A. (1993) Control in meat and meat products, in *Clostridium botulinum. Ecology and Control in Foods*, (eds A.H.W. Hauschild and K.L. Dodds), Marcel Dekker, New York, pp. 177–207.

Madden, R.H. and Bolton, G. (1991) Influence of aerobes on volatiles accumulating in vacuum packaged beef. *SAB Conference Poster*, 1991.

Madden, R.H. and Moss, B. (1987) Extension of shelf-life of minced beef by storage in vacuum packages with carbon dioxide. *J. Food Protect.*, **50**, 229–33.

Mäkelä, P.M., Korkeala, H.J. and Laine, J.J. (1992) Ropy slime-producing lactic acid bacteria contamination at meat processing plants. *Int. J. Food Microbiol.*, **17**, 27–35.

Manu-Tawiah, W., Ammann, L.L., Sebranek, J.G. and Molins, R.A. (1991) Extending the colour stability and shelf life of fresh meat. *Food Technol.*, **45**, 94–102.

Manu-Tawiah, W., Myers, D.J., Olson, D.G. and Molins, R.A. (1993) Survival and growth of *Listeria monocytogenes* and *Yersinia enterocolitica* in pork chops packaged under modified gas atmospheres. *J. Food Sci.*, **58**, 475–9.

McClure, P.J., Blackburn, C. de W., Cole, M.B., Curtis, P.S., Jones, J.E., Legan, J.D., Ogden, I.D., Peck, M.W., Roberts, T.A., Sutherland, J.P. and Walker, S.J. (1994) Modelling the growth, survival and death of microorganisms in foods: the UK Food Micromodel approach. *Int. J. Food Microbiol.*, **23**, 265–75.

McMullen, L.M. and Stiles, M.E. (1991) Changes in microbial parameters and gas composition during modified atmosphere storage of fresh pork loin cuts. *J. Food Protect.*, **54**, 778–83.

McMullen, L.M. and Stiles, M.E. (1993) Microbial ecology of fresh pork stored under modified atmosphere at –1, 4.4 and 10 °C. *Int. J. Food Microbiol.*, **18**, 1–14.

Mead, G.C. (1994) Microbiological hazards from red meat and their control. *Br. Food J.*, **96**, 33–6.

Molin, G. and Ternström, A. (1982) Numerical taxonomy of psychrotrophic pseudomonads. *J. Gen. Microbiol.*, **128**, 1249–64.

Molin, G., Ternström, A. and Ursing, J. (1986) *Pseudomonas lundensis*, a new bacterial species isolated from meat. *Int. J. System. Bacteriol.*, **36**, 339–42.

Morishita, Y. and Shiromizu, K. (1986) Characterisation of lactobacilli from meats and meat products. *Int. J. Food Microbiol.*, **3**, 19–29.

Muller, N.J. (1986) Longer product shelf life using modified atmosphere packaging. *The National Provisioner*, **194**, 19–23.

Newton, K.G. and Gill, C.O. (1978) The development of the anaerobic spoilage flora of meat stored at chill temperatures. *J. Appl. Bacteriol.*, **44**, 91–5.

Newton, K.G. and Gill, C.O. (1980–81) The microbiology of DFD fresh meats: a review. *Meat Sci.*, **5**, 223–32.

Newton, K.G., Harrison, J.C.L. and Smith, K.M. (1977a) The effect of storage in various gaseous atmospheres on the microflora of lamb chops held at –1 °C. *J. Appl. Bacteriol.*, **43**, 53–9.

Newton, K.G., Harrison, J.C.L. and Smith, K.M. (1977b) Coliforms from hides and meat. *Appl. Environ. Microbiol.*, **333**, 199–200.

Newton, K.G. and Rigg, W.J. (1979) The effect of film permeability on the storage life and microbiology of vacuum-packed meat. *J. Appl. Bacteriol.*, **47**, 433–41.

Nicol, D.J., Shaw, M.K. and Ledward, D.A. (1970) Hydrogen sulphide production by bacteria and sulfmyoglobin formation in prepacked chilled beef. *Appl. Microbiol.*, **19**, 937–9.

Nortjé, G.L. and Shaw, B.G. (1989) The effect of ageing treatment on the microbiology and storage characteristics in modified atmosphere packs containing 25% CO_2 plus 75% O_2. *Meat Sci.*, **25**, 43–58.

Nychas, G.J., Dillon, V.M. and Board, R.G. (1988) Glucose, the key substitute in the microbiological changes occurring in meat and certain meat products. *Biotechnol. Appl. Biochem.*, **10**, 203–31.

Ochi, H. (1987) Development of MA storage packaging system in pursuit of health and natural taste. *Packaging Japan*, **8**, 29–34.

Ooraikul, B. (1993) Further research in modified atmosphere packaging, in *Modified Atmosphere Packaging of Food*, (eds B. Ooraikul and M.E. Stiles). Ellis Horwood, New York, pp. 261–81.

Ordóñez, J.A., de Pablo, B., de Castro, B.P., Asensio, M.A. and Sanz, B. (1991) Selected chemical and microbiological changes in refrigerated pork stored in carbon dioxide and oxygen enriched atmospheres. *J. Agric. Food Chem.*, **39**, 668–72.

Palumbo, S.A. (1988) The growth of *Aeromonas hydrophila* K144 in ground pork at 5 °C. *Int. J. Food Microbiol.*, **7**, 41–8.

Patterson, J.T. and Gibbs, P.A. (1977) Incidence and spoilage potential of isolates from vacuum packaged meat of high pH value. *J. Appl. Bacteriol.*, **43**, 25–38.

Penney, N. and Bell, R.G. (1993) Effect of residual oxygen on the colour, odour and taste of carbon dioxide-packaged beef, lamb and pork during short term storage at chill temperatures. *Meat Sci.*, **33**, 245–52.

Prieto, M., García-Armesto, M.R., García-López, M.L., Alonso, C. and Otero, A. (1992) Species of *Pseudomonas* obtained at 7 °C and 30 °C during aerobic storage of lamb carcasses. *J. Appl. Bacteriol.*, **73**, 317–23.

Rice, J. (1990) Meat and poultry packaging trends. *Food Processing (Chicago)*, **51**, 74–83.

Roth, L.A. and Clarke, D.S. (1972) Studies on the bacterial flora of vacuum packaged fresh beef. *Can. J. Microbiol.*, **18**, 1761–66.

Rousset, S. and Renerre, M. (1991) Effect of CO_2 or vacuum packaging on normal and high pH meat shelf-life. *Int. J. Food Sci. Technol.*, **26**, 641–52.

Ryser, E.T. and Marth, E.M. (1991) Incidence and behaviour of *Listeria monocytogenes* in meat products, in *Listeria, Listeriosis, and Food Safety*, (eds E.G. Ryser and E.H. Marth). Marcel Dekker, New York. pp. 405–62.

Savell, J.W., Smith, G.C., Hanna, M.O. and Vanderzant, C. (1981) Packaging of beef loin steaks in 75% O_2 plus 25% CO_2. I. Physical and sensory properties. *J. Food Protect.*, **44**, 923–7.

Schillinger, U. and Lücke, F-K. (1987a) Identification of lactobacilli from meat and meat products. *Food Microbiol.*, **4**, 199–208.

Schillinger, U. and Lücke, F-K. (1987b) Lactic acid bacteria on vacuum-packaged meat and their influence on shelf-life. *Fleischwirtschaft*, **67**, 1244–8.

Seelye, R.J. and Yearbury, B.J. (1979) Isolation of *Yersinia enterocolitica*-resembling organisms and *Alteromonas putrefaciens* from vacuum-packed chilled beef cuts. *J. Appl. Bacteriol.*, **46**, 493–9.

Seideman, S.C., Carpenter, Z.L., Smith, G.C., Dill, C.W. and Vanderzant, C. (1979a) Physical and sensory characteristics of beef packaged in modified gas atmospheres. *J. Food Protect.*, **42**, 233–9.

Seideman, S.C., Carpenter, Z.L., Smith, G.C., Dill, C.W. and Vanderzant, C. (1979b) Physical and sensory characteristics of pork packaged in various gas atmospheres. *J. Food Protect.*, **42**, 317–22.

Seideman, S.C., Vanderzant, C., Smith, G.C., Hanna, M.O. and Carpenter, Z.L. (1976) Effect of degree of vacuum and length of storage on the microflora of vacuum packaged beef wholesale cuts. *J. Food Sci.*, **41**, 738–42.

Shaw, B.G. and Harding, C.D. (1984) A numerical taxonomic study of lactic acid bacteria from vacuum-packed beef, pork, lamb and bacon. *J. Appl. Bacteriol.*, **56**, 25–40.

Shaw, B.G. and Latty, J.B. (1982) A numerical taxonomic study of *Pseudomonas* strains from spoiled meat. *J. Appl. Bacteriol.*, **52**, 219–28.

Shaw, B.G. and Latty, J.B. (1984) A study of the relative incidence of *Pseudomonas* groups on meat using a computer-assisted identification technique employing only carbon source tests. *J. Appl. Bacteriol.*, **57**, 59–67.

Shelef, L.A. (1977) Effect of glucose on the bacterial spoilage of beef. *J. Food Sci.*, **42**, 1172–5.

Shelef, L.A. (1989) Survival of *Listeria monocytogenes* in ground beef or liver during storage at 4 and 25 °C. *J. Food Protect.*, **52**, 379–83.

Sheridan, J.J., Doherty, A., Allen, P., McDowell, D.A., Blair, I.S. and Harrington, D. (1995) Investigations on the growth of *Listeria monocytogenes* on lamb packaged under modified atmospheres. *Food Microbiol.*, **12**, 259–66.

Silliker, J.H., Woodruff, R.E., Lugg, J.R., Wolfe, S.K. and Brown, W.D. (1977) Preservation of refrigerated meats with controlled atmospheres: treatment and post-treatment effects of carbon dioxide on pork and beef. *Meat Sci.*, **1**, 195–204.

Smith, J.P., Ramaswamy, H.S. and Simpson, B.K. (1990a) Developments in food packaging technology. Part I: Processing/cooking considerations. *Trends Food Sci. Technol.*, **1**, 107–10.

Smith, J.P., Ramaswamy, H.S. and Simpson, B.K. (1990b) Developments in food packaging technology. Part II: Storage aspects. *Trends Food Sci. Technol.*, **1**, 111–18.

Stanbridge, L.H. (1994) Microbial associations developing on modified atmosphere packaged beef steaks. PhD Thesis. University of Bath, England.

Stiles, M.E. (1991) Modified atmosphere packaging of meat, poultry and their products, in *Modified Atmosphere Packaging of Food* (eds B. Ooraikul and M.E. Stiles). Ellis Horwood, New York, pp. 118–47.

Stiles, M.E. and Hasting, J.W. (1991) Bacteriocin production by lactic acid bacteria: potential for use in meat preservation. *Trends Food Sci. Technol.*, **2**, 247–51.

Stiles, M.E. and Ng, L-K. (1981) Enterobacteriaceae associated with meats and meat handling. *Appl. Environ. Microbiol.*, **41**, 867–72.

Sutherland, J.P., Patterson, J.T. and Murray, J.G. (1975) Changes in the microbiology of vacuum-packaged beef. *J. Appl. Bacteriol.*, **39**, 227–37.

Tan, K.H. and Gill, C.O. (1982) Physiological basis of CO_2 inhibition of a meat spoilage bacterium, *Pseudomonas fluorescens*. *Meat Sci.*, **7**, 9–17.

Taoukis, P.S., Fu, B. and Labuza, T.P. (1991) Time-temperature indicators. *Food Technol.*, **45**, 70–82.

Taylor, A.A. (1985) Packaging fresh meat, in *Developments in Meat Science*, Volume 3, (ed. R. Lawrie). Elsevier, Applied Science, London, pp. 89–113.

Taylor, A.A., Down, N.F. and Shaw, B.G. (1990) A comparison of modified atmosphere and vacuum skin packing for the storage of red meats. *Int. J. Food Sci. Technol.*, **25**, 98–109.

Vanderzant, C., Hanna, M.O., Ehlers, J.G., Savell, J.W., Smith, G.C., Griffin, D.B., Terrell, R.N., Lind, K.D. and Galloway, D.E. (1982) Centralised packaging of beef loin steaks with different oxygen-barrier films: microbiological characteristics. *J. Food Sci.*, **47**, 1070–9.

Venugopal, R.J., Ingham, S.C., McCurdy, A.R. and Jones, G.A. (1993) Anaerobic microbiology of fresh beef packaged under modified atmosphere or vacuum. *J. Food Sci.*, **58**, 935–8.

Walker, S.J. and Brooks, J. (1993) Survey of the incidence of *Aeromonas* and *Yersinia* species in retail foods. *Food Control*, **4**, 34–40.

Wolfe, S.K. (1980) Use of CO- and CO_2-enriched atmospheres for meats, fish and products. *Food Technol.*, **34**, 55–8, 63.

Young, L.L., Reviere, R.D. and Cole, A.B. (1988) Fresh red meats: A place to apply modified atmospheres. *Food Technol.*, **42**, 65–9.

7 Meat microbiology and spoilage in tropical countries

D. NARASIMHA RAO, K.K.S. NAIR AND P.Z. SAKHARE

7.1 Introduction

Meat from sheep, goat, chicken and pigs is mainly consumed in the domestic market in India while meat from buffalo and cattle is exported. Meat is produced in 3000 traditional abattoirs and 20 modern meat complexes. In recent years, the Government of India has taken steps to modernize these traditional abattoirs to improve the quality of meat and provide training programmes to upgrade the skills of operators in meat hygiene. Indian meat consumers do not know how the meat is produced, processed and marketed in retail and institutional outlets. In recent times, however, consumers' awareness of hygienic meat has increased due to education and exposure to public media. This has emphasized the need to produce meat under carefully controlled conditions.

The safety and hygienic quality of meat are largely determined by the presence of micro-organisms, which are ubiquitous in nature Fung (1987). Thus meat can not be produced in a germ-free environment, and hence it cannot escape contamination from micro-organisms, such as viruses, bacteria, yeasts, moulds, protozoa and algae. Micro-organisms can be broadly classified into three major groups: (i) those beneficial to humans as in fermented products such as fermented sausages; (ii) harmful ones that cause health hazards in humans (especially food poisoning) and animals; (iii) spoilage ones that bring about deterioration of meats.

Meats provide most of the basic necessities for the growth of micro-organisms. Temperature is a unique factor that influences the growth of micro-organisms. Based on temperature requirements, micro-organisms are classified as psychrotrophs, mesophiles and thermophiles (ICMSF, 1980). The distribution of micro-organisms in different zones depends on environmental temperature. Mesophiles and thermophiles occur predominantly in tropical climates while psychrotrophs are more common in temperate zones. Thus microbial contamination of meat is likely to be different in warm and cold climates. This will have a different impact on spoilage and health problems in warm climates, such as in India. Thus microbial ecology will have a unique role in meat handled in warm climates. The region of tropical climates has been designated as the belt around the earth lying between latitudes 30° north and 30° south (N–S 30°). This region has a characteristic climate that causes problems directly or indirectly on meat handling. Meat handling and meat consumption patterns are different in India as compared to western parts of the world. Animals are slaughtered either early (05.00–08.00 hrs) or late in the day (15.00–23.00 hrs) in traditional abattoirs. Meat is marketed and sold for long periods in the day at ambient temperatures. Consumption of hot meat (unchilled meat) is a common practice throughout the country. Meat from buffalo and cattle for export is chilled. The intention of this chapter is to present the available information on meat microbiology and spoilage in tropical countries.

7.2   Characteristics of meat

As discussed in Chapter 9, meat is the flesh of the slaughtered animals used for food. Meat provides the nutrients required for the growth of a wide range of micro-organisms. In general micro-organisms do not attack protein, fat and connective tissue but they use low molecular substances (Gill, 1986). The concentrations of these components (mg/g) in *post rigor* meat are creatin 6.5; inosine monophosphate 3.0; glycogen 1.0; glucose 0.1; glucose-6-phosphate 0.2; lactic acid 9.0; amino acids 3.5 and dipeptides (carnosine, anserine) 3.0.

The unique phenomenon that takes place following slaughter of the animal is the conversion of muscle to meat, a topic discussed in Chapter 9. The process is biochemical in nature. Following the death of the animal, blood supply to muscles is stopped. This results in oxygen depletion of tissues. Thus there is a shift in muscle metabolism from an aerobic to an anaerobic environment. A major change that occurs in muscle *post-mortem* is glycolysis under anaerobic conditions. This results in conversion of glycogen to lactic acid and depletion of adenosine triphosphate (ATP). During *rigor mortis* the accumulation of lactic acid results in a fall in muscle pH. The ultimate pH of meat from a normal healthy animal is 5.8 to 5.5. The following ranges were reported for the ultimate pH for meats of various animals: beef 5.1–6.2; lamb 5.4–6.7; and pork 5.3–6.9 (Callow, 1949). These data show the differences in ultimate pH of meats of different species. Both the rate and extent of *post-mortem* pH fall are influenced by intrinsic factors such as species, the type of muscle, environmental temperature and variability between animals (Lawrie, 1985). The ultimate pH in pork and veal was higher than that in cattle (Lawrie, 1985). It has been observed that buffalo meat had a lower pH (5.6–5.9) than meat from cattle (Valin *et al.*, 1984). The pH is an important determinant of microbial growth.

The other physical characteristics of meat are colour, water holding capacity, texture and tenderness. These are likely to be perceived as desirable by meat consumers. Colour and water-holding capacity provide the consumer with a more prolonged sensation. Texture and tenderness are the important attributes of eating quality.

7.3 Micro-organisms associated with meat

As already noted, micro-organisms can be broadly classified into three major groups: (i) pathogenic micro-organisms; (ii) spoilage micro-organisms; and (iii) beneficial micro-organisms.

7.3.1 Pathogenic micro-organisms

These organisms cause infections or intoxications in man. The major groups of pathogenic micro-organisms associated with meat and meat products in India as elsewhere are *Salmonella, Staphylococcus, Escherischia coli, Listeria monocytogenes, Bacillus cereus, Clostridium botulinum, Clostridium perfringens, Yersinia Enterocolitica, Campylobacter* spp. and *Aspergillus flavus*.

7.3.2 Spoilage micro-organisms

The micro-organisms that contribute to the spoilage of meat and meat products are considered in Chapters 1, 2 and 3. Of those discussed in those chapters, the following are important in India: *Pseudomonas, Micrococcus*, Lactic acid bacteria, *Brochothrix thermosphacta, Acinetobacter/Moraxella*, Enterobacteriaceae, *Shewanella putrefaciens, Aspergillus, Penicillium, Thamnidium, Rhizopus, Cladosporium, Sporotrichum, Debaryomyces, Candida, Torulopsis* and *Rhodotorula*.

7.3.3 Beneficial micro-organisms

These micro-organisms are mainly responsible for fermentation of meat and meat products such as sausages. The major groups of micro-organisms in this category are discussed in Chapter 2.

7.4 Sources of microbial contamination on meat

Muscle of healthy animals is free from micro-organisms because of defensive mechanisms associated with: (a) skin and mucous membranes, hair and cilia, gastric juices, the intestines and urine; (b) inflammatory processes. Humeral antibodies play a part also (Ayres, 1955). All these defence mechanisms present barriers to the entry of micro-organisms into the muscle of live animals. Micro-organisms inevitably gain access to the meat at slaughter, when the defences break down and during processing. The minimization of microbial contamination is essential in meat handling systems in order to retard meat spoilage as well as to prevent health hazards that may arise from meat consumption. Therefore there is a need to know how micro-organisms enter meat and to determine the critical points of contamination. These points would provide basis for development of HACCP for meat production. Sources of microbial contamination in fresh meat processing have been well documented (Ayres, 1955; Grau, 1986; Newton *et al.*, 1978; Nottingham, 1982; Narasimha Rao, 1982; Narasimha Rao and Ramesh, 1992; Tarwate *et al.*, 1993). Hides and skins, hooves, fleece and hair of live animals, gut microflora, the stick-knife, scald tank, equipment, instruments and tools (overhead rail, gambrels, stainless steel platforms, S-hooks, trays, tables, knives, axe, saw-blade), chopping blocks (wooden), floors, walls, air, water, cloths, hands and gumboots have been identified as sources of microbial contamination of carcasses and meat cuts, as discussed in Chapter 4.

Bacteria enter meat during the following operations: sticking, skinning, scalding, dehairing, evisceration, and splitting and quartering.

In a study on the sources of microbial contamination prior to slaughter in a local slaughter house, a retail shop and a CFTRI modern abattoir, it was found that knives, the floor, the butchers' hands and water showed high microbial numbers (Table 7.1) (Narasimha Rao, 1982). The microbial counts of water ranged from 4.0 to 5.0 log cfu/ml. Chopping blocks (wooden) from the local retail shop gave the highest microbial numbers (5.5–7.5 log cfu/cm^2). Knives, water, floor, butchers' hands and cutting table in the modern abattoir showed the lowest microbial counts. Floor washings, which consist of carcass wash water, blood, hair, tissues and debris would obviously be a major source of microbial contamination (Narasimha Rao and Ramesh, 1992). A detailed investigation was conducted on the environmental sources of microbial contamination in the buffalo slaughter line in Deonar abattoir, the largest municipal slaughter house in India (Tarwate *et al.*, 1993). Nine different points (knife, axe, saw-blade, hooks, floor, wall,

Table 7.1 Sources of microbial contamination (Narasimha Rao, 1982)

Source	Total counts (log cfu/cm^2)
Local slaughterhouse (Mysore) prior to slaughter (sheep)	
Knives	3.6–4.0
Floor	3.0–4.5
Hands	2.5–3.0
Water (log cfu/ml)	4.0–5.0
Local retail shop	
Chopping knife	3.8–4.3
Chopping block (wooden)	5.5–7.5
Hands	2.4–3.2
CFTRI modern abattoir (prior to slaughter)	
Knives	1.0–1.5
Water (log cfu/ml)	1.0–2.0
Floor	1.5–2.0
Hands	1.0–1.5
Cutting table	1.0–2.0

Table 7.2 Bacterial counts enumerated at different points in Deonar abattoir (Tarwate *et al.*, 1993)

Source	Bacterial counts (log cfu/cm^2)					
	Total viable	Entero-bacteriaceae	Faecal coliforms	*Bacillus* spp.	*Staphylo-coccus* spp.	*Clostridium* spp.
Knives	3.2	5.9	5.7	5.5	4.9	2.1
Axe	3.8	5.6	5.9	5.4	4.9	2.4
Saw-blade	3.1	4.7	5.5	5.8	4.8	2.8
Hooks	4.2	5.6	5.6	5.7	4.6	2.3
Floor	6.7	6.9	7.3	6.9	5.9	4.1
Wall	5.3	6.2	6.6	6.4	5.2	3.5
Platform	5.6	6.5	6.8	6.7	5.4	3.9
Handswabs	2.9	5.4	5.5	4.6	4.2	1.3
Water*	2.1	4.4	4.5	3.9	3.4	0.0

* log cfu/ml.

platform, handsaw, and water) in the buffalo slaughterline were analysed for bacterial contamination (Table 7.2). Floors, platforms, walls, knives, axes, saw-blades, hooks and handsaws were identified as the critical points of microbial contamination. Bacteria from these sources were isolated and identified (Table 7.3). A total of 651 isolates were identified in this study. Pathogenic organisms (*S. aureus, B. cereus, Clostridium* spp., *E. coli, Shigella* spp.) as well as spoilage organisms were among the isolates.

7.5 Microbiology of market meat in India

It is interesting to note that the microbiology of meat differs from one region to another, it being largely influenced by the feeding habits of animals, means of transport of animals, environment, temperature, abattoir conditions, slaughter and handling practices. Few studies are available on the microbial contamination of meat in India. A study conducted on microbial quality of sheep meat from Mysore city gave counts (log cfu/g) as follows: total plate counts 4.9–6.0; coliforms 2.6–5.1; lactobacilli 2.7–5.4; staphylococci 4.6–5.3; and yeasts and moulds 1.9–3.6 (Krishnaswamy and Lahiry, 1964). Another detailed study was made on microbial contamination of sheep carcasses from the slaughterhouse and from Mysore retail shops by Narasimha Rao (1982). The results are presented in Tables 7.4 and 7.5. Total plate counts, and coliforms, staphylococci and enterococci counts were high. This could be attributed to the floor-slaughter practice, contamination from skin, floor washings, water and mishandling during evisceration. Five million aerobic counts/g and 50 *E. coli*/g were suggested as microbiological limits for fresh meat (Carl, 1975). Season did not affect the microbial quality of the carcasses (Table 7.4). No differences were recorded

Table 7.3 Bacterial cultures (samples positive) isolated from different points in a slaughter house (Tarwate *et al.*, 1993)

Organisms	Source							
	Knives	Axe	Saw-blades	Hooks	Floor	Walls	Plat-forms	Hand-swabs
Bacillus								
*cereus**	3	3	4	5	6	6	5	0
*subtilis***	4	2	5	4	7	6	6	1
*megaterium***	4	3	3	4	4	5	4	0
*circulans***	3	3	4	4	6	5	6	2
*coagulans***	5	4	6	5	5	7	6	1
Bacillus spp.**	5	4	6	6	7	7	7	3
Staphylococcus								
*aureus**	1	1	3	2	4	5	4	1
*epidermidis***	2	3	4	4	7	5	6	3
Micrococcus								
*roseus***	4	2	5	4	5	6	7	2
*luteus***	3	5	4	5	3	5	5	1
Micrococcus spp.**	6	2	5	3	7	6	7	3
*Streptococcus faecalis**	0	1	0	0	3	2	2	0
*mitis***	1	0	0	1	2	1	1	0
*faecium***	2	1	1	2	2	1	2	0
*bovis***	1	3	0	1	3	2	3	0
Streptococcus spp.**	2	1	2	1	5	5	6	2
Clostridium spp.*	4	3	3	4	7	7	7	0
*Escherichia coli**	5	6	5	3	7	7	2	2
Klebsiella spp.**	2	2	4	4	5	6	5	3
*Citrobacter freundii***	1	2	1	1	3	1	2	0
*Enterobacter aerogenes***	2	1	1	1	2	2	2	0
*E. cloacae***	1	0	0	0	2	1	1	0
*Pseudomonas aeruginosa**	2	0	0	1	3	2	3	1
Alcaligenes spp.**	0	0	0	0	1	1	1	0
Shigella spp.**	0	1	2	1	4	3	4	0
*Serratia marcescens***	2	1	0	0	2	2	3	0
Proteus								
*mirabilis**	0	0	0	1	5	5	6	0
*vulgaris***	2	1	1	1	6	4	6	1

* Pathogenic, ** Non-pathogenic spoilage micro-organism. One *Micrococcus* spp. was isolated from a water sample.

Table 7.4 Microbial counts on sheep carcasses processed in Mysore city slaughter house (Narasimha Rao, 1982)

Microbial counts	Mean counts (log cfu/cm^2)*	
	Summer	Winter
Total plate counts	6.2	6.1
Coliforms	4.4	4.3
Staphylococci	5.2	5.0
Enterococci	4.4	4.4
Psychrotrophs	3.5	3.6

*One sample from each of the 30 carcasses in winter and one from each of 30 carcasses in summer.

Table 7.5 Microbial counts on carcasses from a local retail shop, Mysore city (Narasimha Rao, 1982)

Microbial counts	Mean counts (log cfu/cm^2)*	
	Summer	Winter
Total plate counts	6.1	6.2
Coliforms	5.3	4.3
Staphylococci	5.0	5.2
Enterococci	4.3	4.4
Psychrotrophs	3.7	3.6

*Average of 30 samples from 30 carcasses in winter and 30 samples from 30 carcasses in summer. Carcasses obtained from one retail shop.

between slaughterhouse carcasses and those in retail shops. In another study, it was found that the larger the slaughter capacity, the higher the microbial contamination of the sheep and goat carcasses (Nair *et al.*, 1991).

The types of micro-organisms associated with sheep carcasses in the market were *Staphylococcus* (42.5%), *Pseudomonas* (12.0%), *Micrococcus* (8.5%), *Escherichia* (10.5%), *Enterococcus* (13.5%), *Flavobacterium* (1.5%), *Aeromonas* (1.5%), *Serratia* (1.0%), *Bacillus* (1.0%), *Proteus* (2.0%) and *Aerobacter* (3.0%) (Narasimha Rao, 1982).

Buffalo carcasses at an abattoir in Mysore city were examined for microbiological quality (Syed Ziauddin *et al.*, 1994). Bacterial counts are presented in Table 7.6. The differences in bacterial counts on the different regions of the carcasses as well as between two slaughter units were marginal. Further it was observed that the counts of psychrotrophs were generally lower than for other types of bacteria. High levels of coliform and staphylococci contamination were observed. The total bacterial counts of fresh beef were found to be 4.70 log cfu/cm^2 (Stringer *et al.*, 1969). Cattle

Table 7.6 Bacterial number at four sites on buffalo carcasses from local slaughter units in Mysore city (Syed Ziauddin *et al.*, 1994)

	Log cfu/cm^2*									
	Total plate counts		Coliforms		Staphylococci		Enterococci		Psychrotrophs	
	I	II	I	II	I	II	I	II	I	II
Leg	4.82	4.92	3.44	2.99	4.08	3.89	3.44	2.84	2.85	2.70
Loin	5.13	4.71	3.18	2.91	3.92	3.75	3.76	2.95	2.92	2.62
Shoulder	5.49	5.41	3.78	3.65	4.30	3.87	3.68	3.75	2.88	2.99
Neck	4.80	4.52	3.35	3.30	4.25	4.35	2.85	3.49	2.83	3.22
Mean	5.00	4.89	3.43	3.21	4.13	3.97	3.43	3.50	2.87	2.88
	±0.42	±0.41	±0.28	±0.28	±0.24	±0.26	±0.25	±0.15	±0.19	±0.19

I = Slaughter unit I, II = Slaughter unit II. *Average counts from 12 samples from 12 carcasses in each unit.

and sheep carcasses will have on their surfaces about 10^3–10^5 aerobic mesophiles/cm^2 (Grau, 1986). Ingram and Roberts (1976) reported counts of 10^3–10^5 aerobic mesophiles/cm^2 on cattle and sheep carcasses. The following micro-organisms were isolated from buffalo carcasses: *Staphylococcus* (37.0%), *Micrococcus* (10.0%), *Enterococcus* (12.0%), *Escherichia* (15.0%), *Enterobacter* (10.0%), *Pseudomonas* (8.0%), *Acinetobacter* (4.0%) and *Bacillus* (4.0%) (Syed Ziauddin, 1988).

Three striking features in the Indian studies were: (i) the high number of microbes on carcasses; (ii) the predominance of staphylococci; and (iii) the low number of psychrotrophs. The latter two indicate the shift in microbial ecology of meat in tropical countries as compared to that of the temperate (western) parts of the world where psychrotrophic micro-organisms, especially *Pseudomonas* predominate on meat (see Chapter 1). These differences have a significant effect on meat spoilage.

7.5.1   Pathogenic micro-organisms isolated from meat

It has been well documented that meat and meat products are sources of pathogenic micro-organisms that constitute health hazards in humans. Such organisms gain access to meat surfaces by poor hygienic practices during slaughter, evisceration and further handling. Indeed the practices of meat handling determine ultimately the incidence of pathogenic micro-organisms on meat. Thus, studies on the occurrence of pathogenic micro-organisms would help improve hygienic practices and thereby protect the consumer from diseases acquired from meat and meat products.

Salmonella.   There are a few reports on the incidence of *Salmonella* in meat in India. It was reported that 8% of goat carcasses were contaminated with *Salmonella* (Randhawa and Kalra, 1970). The serotypes isolated were *S. anatum, S. dublin, S. weltevreden, S. virginia* and nonmotile serotype belonging to group E1. Three strains of *S. newport, S. anatum* and *S. weltevreden* were isolated from the chopping blocks. *Salmonella* were not isolated from instruments, floor surface, water samples and from meat handlers. It was found that 3% of dressed chickens were contaminated with *Salmonella* (Panda, 1971). An incidence of 4.4% of *Salmonella* was reported on sheep and goat carcasses (Mandokhot *et al.*, 1972). The presence of *Salmonella* was detected in intestines, gall bladder, lymph nodes, liver and muscle of cattle and goat (Das Guptha, 1974, 1975). *Salmonella* was reported in mutton (4.9–5.5%), beef (5.0–5.9%), chicken carcass (2.8%) and pork (0%) (Manickam and Victor, 1975). The serotypes isolated from these meats were *S. bareilly, S. senftenberg, S. typhimurium* and *S. bredeney*. In a review, it was emphasized that there was a need for an in-depth study on the incidence of *Salmonella* in meats in India (Rao and Panda, 1976).

A detailed study was made on the presence of *Salmonella* in 200 mutton carcasses (sheep meat) from Mysore city slaughter house, 100 mutton carcasses from retail shops and 100 mutton carcasses processed in the CFTRI modern abattoir (Narasimha Rao, 1983). The results indicated that 9% of the carcasses from the city slaughter house and 8% of the meat samples from the retail shops harboured *Salmonella*. The serotypes *S. gaminara, S. adelaide, S. virchow, S. newport* and *S. paratyphi* B were isolated. Chopping blocks and knives from the retail shops and rectal swabs from the animals prior to slaughter revealed the presence of *Salmonella*. This incidence of *Salmonella* may be attributed to improper handling during slaughtering, dressing and evisceration of sheep. Interestingly, *Salmonella* was not detected on the carcasses processed under hygienic conditions in the modern abattoir. The hygienic conditions in the latter included: (a) resting of animals after transportation; (b) starving the animals prior to slaughter; (c) dressing and evisceration of carcasses over the rail pipe system; (d) careful removal of viscera; (e) use of sterile knives during operations; (f) frequent washing of butchers' hands and use of clean water.

In one study, two out of eleven quarters of buffalo carcasses examined for the presence of *Salmonella* (Nair *et al.*, 1983) were positive. *Salmonella newport* was isolated. In another study the incidence of *Salmonella* in buffalo meat and meat products was assessed (Bachhil and Jaiswal, 1988a). Five percent of samples of fresh and frozen buffalo meat, 6.6% minced meat, 10% kabab (traditional meat product) and 5% each of prescapular and poplilteal lymph nodes were positive for *Salmonella*. The following were isolated: *S. anatum, S. weltevreden, S. typhimurium, S. poona* and *S. newport*.

The largest Indian abattoir, Deonar, is located in Bombay (presently known as Mumbai), India. Seventy meat samples from this abattoir were examined for the presence of *Salmonella* (Paturkar *et al.*, 1992). Seven isolates (one from beef, three from mutton, one from pork and two from buffalo beef) belonged to the one serotype *S. saintpaul*. These authors examined also 96 meat samples from large municipal meat markets; eight samples (beef, 4; mutton, 1; pork, 2; and chicken, 1) were positive for *Salmonella*. *Salmonella mbandaka* and *S. adelaide* was isolated from beef, *S. liverpool* from mutton, *S anatum* and *S. derby* from pork and *S. butantan* from chicken. From this study, it was found that fifteen of the 166 (9%) meat samples listed were positive for *Salmonella*.

The presence of *Salmonella* in meat is a world-wide phenomenon (Hobbs, 1974). Thus in summary it appears *Salmonella* is present in 3–9% of Indian meats. *Salmonella* contamination rates were reported elsewhere as follows: 12–35.6% (Ingram and Simonson, 1980); 1.9% (Nazer and Osborne, 1976); and less than 1% (Childers *et al.*, 1973; Roberts, 1976).

Staphylococcus aureus.   *Staphylococcus aureus* is an organism that causes concern in human beings (Minor and Marth, 1971). Certain strains of

S. aureus produce thermostable enterotoxins (Bergdoll, 1967) that affect the gastro-intestinal system of humans. Meats are often contaminated with *S. aureus* from the major depots of this organism, skins, hides and nose. *Staphylococcus aureus* must multiply in a food product to a population level in excess of 10^6 per gram before sufficient toxin is produced to cause illness (gastro-enteritis) in humans (Evans, 1986).

It was reported that 26.6% of sheep and goat carcasses in India were contaminated with coagulase-positive staphylococci (Mandokhkot *et al.*, 1972) while 27.27% of samples of buffalo meat at the retail level contained coagulase-positive *S. aureus* (Panduranga Rao, 1977). All 11 of the buffalo carcasses examined showed the presence of coagulase-positive *S. aureus* (Nair *et al.*, 1983). It was observed in another detailed study that 60% of the mutton carcasses from the local slaughter house showed coagulase-positive *S. aureus* whereas 5% of the mutton carcasses processed in the modern abattoir did so (Narasimha Rao, 1982); 36% of the strains of *S. aureus* produced enterotoxins. The incidence of producers of enterotoxin A was highest followed by that for enterotoxin D producer (Narasimha Rao, 1982). Staphylococci predominated among the types of micro-organisms. There is no information available on enterotoxin production in meats under market conditions. Chances of enterotoxin production may be limited in raw meats due to the competitive growth of other mesophilic organisms. However, high incidence of *S. aureus* in raw meats may pose problems of cross contamination in cooked meat products in the kitchen. In other parts of the world, it was reported that 39% of 173 meat samples showed coagulase-positive staphylococci (Jay, 1962). The counts of *S. aureus* varied <100 to 4500 per gram in ground beef (Stiles and Ng, 1981).

Bacillus cereus.   Meat and meat products are also contaminated with *B. cereus*. Since the spores of the organism are resistant to heat, *B. cereus* has a significant contaminant role in cooked meat products. Gastro-enteritis in humans caused by *B. cereus* has been reported in many countries around the world (Gilbert, 1979; Johnson, 1984; and Foegeding, 1986). Almost all the *B. cereus* strains isolated from vomiting and diarrhoeal material produce enterotoxin.

A variety of ready-to-eat meat products in India contained large numbers of *B. cereus* (Sherikar *et al.*, 1979). A study of the incidence, prevalence and enterotoxigenicity of *B. cereus* in Indian meats (Bachhil and Jaiswal, 1988b) showed that 35% of fresh buffalo meats, 100% of kabab and 30% of the curry samples harboured 9.65×10^4, 1.07×10^3 and 6.4×10^2 *B. cereus*/g respectively, the counts in cooked kabab and curry ranging from 1.10×10^2 to 1.10×10^4/g. Ninety percent of the strains were enterotoxigenic; consequently 30% of the buffalo meat and 60% of the cooked meat products harboured enterotoxigenic *B. cereus*. The prevalence of these strains of *B. cereus* reflects the insanitary conditions during the preparation and storage of meat and meat products in open markets and emphasizes also the importance of this organism in food poisoning through meat and meat products.

In a survey of microbiology of ready-to-cook pork products (Tables 7.7–7.9), high microbial counts were noted. The presence of potential pathogens highlights the potential public health hazard in the ready-to-cook pork products. A strong need is felt for microbiological control of meat foods. A study on pork products (bacon, ham and sausage) sold in Indian markets also indicated the need for improving the microbiological status of the products (Prasad et al., 1983). Interestingly, pork products (fresh pork, sausage, ham and bacon) processed under hygienic conditions in a bacon factory contained 2–4 log organisms/g (Varadarajulu and Narasimha Rao, 1975).

Table 7.7 Total viable counts (TVC) of various ready-to-cook pork products samples (Sherikar et al., 1979)

Product	Number of samples	Mean TVC of surface washings per ml × 10⁶	Range of TVC of blended sample/g	Mean TVC of refrigerated sample plates × 10⁵
Cocktail sausages	29	10	1×10^5 to 4.5×10^6	–
Oxford sausages	16	10	8×10^5 to 5×10^6	5.5
Porkies (pork and breakfast sausages)	16	20	4×10^5 to 9×10^6	–
Ham	8	20	3×10^5 to 1×10^6	4.0
Kababs (mutton and pork)	16	40	5×10^5 to 30×10^6	–
Bacon	10	10	3×10^5 to 3×10^6	–

Table 7.8 Mean of bacterial (differential count) counts per gram of blended products (Sherikar et al., 1979)

Bacteria	Cocktail sausages (× 10⁵)	Oxford sausages (× 10⁵)	Porkies (× 10⁵)	Ham (× 10⁵)	Kababs (× 10⁵)	Bacon (× 10⁵)
Staphylococcus aureus	2	3	6	6	3	5
Micrococcus spp.	2	2	2	3	2	1
Bacillus spp.	3	5	3	3	5	3
Escherichia coli	2	4	4	4	1	2
Klebsiella spp.	1	–	4	1	2	2
Serratia spp.	2	–	–	–	7	–
Lactobacillus spp.	–	3	–	2	–	2
Enterococcus faecalis	1	–	2	2	4	1
Clostridium (sulphite reducing)	–	–	–	–	3	4
Pseudomonas spp.	–	–	–	–	–	1
Salmonella	–	–	2	–	3	–

Table 7.9 Micro-organisms isolated from different types of pork products (Sherikar *et al.*, 1979; discussed in Chapter 2)

Type and name of organisms	Number of isolates						
	Cocktail sausages (29)*	Oxford sausages (16)	Porkies (16)	Hams (8)	Bacons (10)	Kababs (16)	Total isolates
Spoilage:							
Staphylococcus epidermidis	9	7	–	5	5	9	35
Micrococcus luteus	4	5	3	–	–	3	15
Micrococcus spp.	4	–	4	4	4	6	22
Bacillus subtilis	10	–	5	–	–	–	15
Bacillus megaterium	–	–	–	–	5	4	9
Bacillus spp.	3	3	4	3	4	6	23
Proteus vulgarus	10	4	6	1	3	–	24
Proteus rettgeri	–	–	–	–	–	6	6
Klebsiella aerogenes	6	–	3	2	5	–	16
Total							165
Potential spoilage:							
Sarcina lutea	–	–	–	8	3	–	11
Gaffkya tetragena	–	–	–	4	–	–	4
Brevibacterium linens	1	–	–	–	–	–	1
Brevibacterium fulvum	–	–	–	2	–	–	2
Lactobacillus brevis	–	2	–	3	2	–	7
Achromobacter guttatus	–	2	–	1	–	–	3
Total							28
Pathogenic:							
Staphyloccus aureus	19	8	13	5	9	7	61
Streptococcus faecalis	2	–	7	2	8	10	29
Clostridium perfringens	–	–	–	–	6	3	9
Bacillus cereus	8	7	6	1	6	9	37
Serratia marcescens	1	–	–	–	–	3	5
Escherichia coli	1	8	7	2	4	6	27
Proteus mirabilis	–	5	–	–	4	3	12
Klebsiella pneumoniae	–	–	–	–	–	4	4
Pseudomonas aeruginosa	–	–	–	–	1	–	1
Salmonella enteritidis	–	–	3	–	–	7	10
Total							195

* Figures in parenthesis indicate the total number of samples processed.

Table 7.10 Microbial load of carcasses at different stages of processing in a modern abattoir (Narasimha Rao and Ramesh, 1992)

Process stage	Total plate counts (log cfu/cm^2)
Skin of the live animal	7.5[a] (7.0–8.0)[b]
Carcass	
surface after skinning	3.5[a] (2.9–4.0)[b]
after evisceration	3.8[a] (3.0–4.2)[b]
after wash	3.6[a] (2.8–4.3)[b]

[a] Means followed by different letters differ significantly ($p < 0.05$).
[b] Values from ten sheep; figures in parenthesis range of counts SE $m = \pm$ 0.20 (36 df).

7.6 Microbiology of sheep carcasses processed in the CFTRI modern abattoir

Differences in the livestock handling practices and slaughter-house practices in different environmental conditions in various geographical regions have an impact on the microbial ecology of meats. A study (Narasimha Rao and Ramesh, 1992) conducted in the CFTRI modern abattoir in India on microbial profiles of sheep carcasses revealed interesting results (Tables 7.10 to 7.15).

The data revealed that careful handling at different stages of processing of sheep reduced the level of microbial contamination of carcasses. Indeed processing stages such as evisceration and washing did not increase contamination in the abattoir. Skin, floor washings, intestinal contents and gambrels were the major sources of microbial contamination. Seasonality did not have any effect on the microbial contamination of carcasses. The study revealed that total plate counts in 86.6% of the carcasses ranged between 3.0 and 4.9 log/cm^2. The counts of coliforms, staphylococci, enterococci and psychrotrophs were low. Pathogens such as *Salmonella* were not detected.

Table 7.11 Sources of microbial contamination in the CFTRI modern abattoir (Narasimha Rao and Ramesh, 1992)

	Total plate counts	
Source	Before operation	After operation
Skin (per cm^2)	5.8×10^5–6.7×10^6	–
Water (per ml)	1.0×10^1–2.0×10^1	–
Knives (per cm^2)	1.0×10^1–5.0×10^1	1.5×10^4–2.7×10^5
Floor washings (per ml)	5.0×10^1–2.0×10^2	6.4×10^5–2.5×10^6
Walls (per cm^2)	1.0×10^2–2.0×10^2	4.5×10^4–8.5×10^5
Hands (per cm^2)	1.0×10^1–2.0×10^1	2.5×10^4–8.8×10^4
Gambrel	1.0×10^2–2.0×10^2	4.6×10^5–5.8×10^6
Trolley (per cm^2)	1.0×10^1–5.0×10^1	1.8×10^4–2.6×10^5
Gum boots (per cm^2)	5.0×10^1–1.5×10^2	1.0×10^3–2.0×10^4
Intestines	–	4.5×10^9–7.8×10^{11}

Table 7.12 Microbial load of sheep carcasses processed in the modern abattoir in winter and summer (Narasimha Rao and Ramesh, 1992)

Microbial counts (log no. per cm^2)	Winter†	Summer†
Total plate counts	3.45 (2.60–4.90)	3.95 (2.60–4.50)
Coliforms	1.35 (1.00–2.50)	1.63 (1.00–2.00)
Staphylococcus	2.10 (1.20–2.80)	2.09 (1.20–3.00)
Enterococci	1.45 (1.00–2.00)	1.62 (1.00–2.00)
Psychrotrophs	2.30 (1.50–3.20)	2.64 (1.40–2.40)

† There was no significant difference between the two seasons for all the bacterial counts ($p > 0.05$).
Figures in parenthesis indicate range of counts.

The microbial counts were well within the generally acceptable levels. These findings demonstrated hygienic handling of carcasses. Shoulder and neck are obviously the critical points for microbiological sampling as these sites showed higher microbial counts (Narasimha Rao and Ramesh, 1992).

In comparison with other studies, the total plate counts recorded for the carcasses in this study were quite low. It was found that cattle and sheep carcasses in the UK had between 10^3 and 10^5 aerobic mesophiles per cm^2 on their surfaces (Ingram and Roberts, 1976). It was also reported (Nottingham and Wyborn, 1975) that, at the completion of dressing under good

Table 7.13 Carcasses grouped under different counts according to the mean counts of leg, rib, shoulder and neck. Data from 60 carcasses (Narasimha Rao and Ramesh, 1992)

Log no. of bacteria/cm^2	Carcasses	Percentage
2.0–2.9	8	13.3
3.0–3.9	25	41.7
4.0–4.9	27	45.0

Table 7.14 Microbial load from different sites of carcasses. Data from 20 carcasses (Narasimha Rao and Ramesh, 1992)

Site on carcass	Log cfu/cm^2 (total plate count)
Leg	3.43^{xy} (2.4–4.3)
Rib	3.28^{x} (2.0–4.2)
Shoulder	3.78^{y} (2.5–4.6)
Neck	3.86^{y} (2.6–4.7)
SE m	+/− 0.15 (76 df)

Means of the column followed by different letters differed significantly ($p < 0.05$).
Figures in parenthesis indicate range of counts.

Table 7.15 Types of micro-organisms isolated from sheep carcasses in a modern abattoir (Narasimha Rao and Ramesh, 1992)

Micro-organisms (genus)	Winter (%)	Summer (%)
Staphylococcus	22	25
Micrococcus	25	34
Acinetobacter	20	15
Pseudomonas	20	10
Escherichia	1	1
Enterobacter	1	1
Citrobacter	0	0
Enterococcus	5	7
Serratia	1	1
Bacillus	1	1
Proteus	2	1
Flavobacterium	1	2
Aerobacter	0	0
Total number of isolates tested	100	100

hygienic conditions, beef is likely to have surface aerobic bacterial counts of 10^2–10^4/cm^2. Sheep carcasses usually have a slightly higher level of contamination than beef, with bacterial counts of 10^2–10^5/cm^2 (Nottingham, 1982). Initial low levels of bacterial counts of 4.5–7.7 × 10^2/cm^2 on the beef carcass surface have also been reported (Nortje and Naude, 1981). Coliform counts were also very low in the study by Narasimha Rao and Ramesh (1992). Earlier investigations done elsewhere have shown that there were less than 10/cm^2 Enterobacteriaceae or *E. coli* (Ingram and Roberts, 1976; Grau, 1986). Coliforms tend to be more numerous on mutton than beef (Nottingham, 1982). The lower psychrotrophic counts compared to total plate counts recorded in the tropics are interesting (Narasimha Rao and Ramesh, 1992). Tropical soils contain fewer cold-tolerant bacteria than soil from temperate zones, a situation that was noticed on the skin of cattle and on meat (Ayres, 1955). Regarding *S. aureus*, the counts recorded in a recent study (Narashima Rao and Ramesh, 1992) are far lower than the 'pathogenic' dose. Counts of *Staph. aureus* of 10^5–10^6/g are needed to produce sufficient enterotoxin to cause disease (Evans, 1986).

Micrococcus and *Staphylococcus* predominated among different microorganisms in hygienically processed sheep meat (Narasimha Rao and Ramesh, 1992) both in winter and summer. It is interesting to note however, that staphylococci isolated in this study were nonpathogenic. Interestingly, salmonellas were not isolated from hygienically processed sheep carcasses in the CFTRI abattoir. The counts of *Escherichia* and *Enterobacter* indicated that carcasses were free from faecal contamination. Even contamination of carcasses from soil was low as evidenced by the presence of *Bacillus*. Studies on microbial profiles of carcasses processed in temperate zones have demonstrated the predominance of *Pseudomonas* and *Micrococcus* (Ayres, 1955;

Ingram and Dainty, 1971; Gill and Newton, 1977, 1978; Nottingham, 1982 – see Chapter 1 also). Thus there are differences in microbial types on carcasses processed in tropical and temperate climates. It was also observed that the percentage of *Staphylococcus* (within 2 h of death) was higher on the carcasses of sheep, beef and hog (Vanderzant and Nickleson, 1969). Food poisoning staphylococci are widely distributed and meat can become contaminated from both animal as well as human sources (Nottingham, 1982). The data generated in the CFTRI modern abattoir provided useful information for improving meat handling practices in tropical countries.

The Government of India has prescribed microbiological standards for raw meats (chilled/frozen); meats included were buffalo meat, veal, mutton and minced meat. A minimum of five samples should be drawn and tested: (1) Total plate count (TPC): out of five samples, three should have counts not exceeding 10^6/g and the remaining two samples can have up to 10^7/g; (2) *Escherichia coli*: out of five samples, three should have *E. coli* counts not exceeding 10/g and the remaining two samples can have *E. coli* counts up to 100/g; (3) *Salmonella* should be absent in all the samples (the Gazette of India, April, 1991; Thulasi, 1997).

7.7   Microbial growth and meat spoilage

Meat is considered as spoiled when it becomes unfit for human consumption. Spoilage is primarily due to metabolites formed from the utilization of nutrients present in the meat by micro-organisms; chemical change due to enzymatic process contributes to a limited extent (see Chapter 9). Spoilage of meat has been defined to signify that any single symptom or group of symptoms of overt microbial activity, are manifested by changes in meat odour, flavour or appearance – it does not take into account whether or not any particular consumer would find these changes objectionable although some almost certainly would (Gill, 1986). Reports have appeared on the deterioration of meat due to chemical changes caused by autolysis or proteolysis in the absence of micro-organisms (Lawrie, 1985; Narasimha Rao and Sreenivasa Murthy, 1986). In practice, however, the control of microbial spoilage is regarded as paramount in the preservation of meat.

The basic factors that affect the growth of micro-organisms in meat are pH, temperature, oxygen status, nutrients, water activity (a_w), autolytic enzymes such as cathepsins and types of bacteria. These factors will be discussed briefly in relation to spoilage of meat and to methods for preventing or retarding such spoilage.

Micro-organisms require a source of carbon and nitrogen, growth factors such as vitamins, minerals and water for their growth and survival. Meat contains many nutrients and is an ideal medium for the growth and survival

Table 7.16 The limits of pH allowing initiation of growth by various micro-organisms in laboratory media (ICMSF, 1980)

Organism	pH	
	Minimum	Maximum
Escherichia coli	4.4	9.0
Proteus	4.4	9.2
Pseudomonas aeruginosa	5.6	8.0
Salmonella	4.5	8.0
Bacillus cereus	4.9	9.3
Clostridium botulinum	4.7	8.5
Enterococcus spp.	4.8	10.6
Micrococcus	5.6	8.1
Staphylococcus aureus	4.0	9.8

of micro-organisms. It has been well documented (see Chapter 9) that micro-organisms on meat surface utilize initially glucose and low molecular substances and subsequently the amino acids (Gin, 1976).

Micro-organisms are affected by the level of free H^+ ions and the concentration of undissociated acid. The pH of meat along with other environmental factors will determine the types of micro-organisms that are able to grow and eventually cause spoilage or become a potential health hazard. Living animal tissue is near neutral pH (7.0–7.2). The ultimate pH of meat from healthy animals at the end of *post-mortem* glycolysis is 5.5–5.7 (Lawrie, 1985). Many micro-organisms grow optimally near pH 7.0 and grow well between pH 5 and 8 while others are favoured by an acid environment (Table 7.16). It has been found that meat with high ultimate pH (6.0) favours rapid microbial growth; that with a low ultimate pH (5.5–5.7) retards the rate of microbial growth. Meat from stressed animals reaching high ultimate pH spoils rapidly (described as DFD meat) while meat from well rested animals shows better keeping quality due to lower ultimate pH (Lawrie, 1985; Gill, 1986).

Redox potential implies the oxygen requirements of micro-organisms. Oxidation and reduction (OR) processes are defined in terms of electron migration between chemical compounds. Oxidation is the loss of electrons whereas reduction is the gain of electrons. Aerobic micro-organisms require positive redox values for growth while anaerobes often require negative redox values. Micro-aerophillic bacteria grow under slightly reduced conditions. Environmental redox potential is an important determinant of microbial growth. The OR potential of meat has been reported to range from –150 mV to +250 mV (Jay, 1992) and its level determines the types of spoilage of meat.

Water activity, the quantity of water available for microbial activity, is defined by the ratio of water vapour pressure of the food substrate to the vapour pressure of pure water at the same temperature (Scott, 1957). This concept is related to equilibrium relative humidity (R.H.) (R.H. = 100 ×

Table 7.17 Approximate minimal levels of water activity (a_w) permitting growth of micro-organisms at temperatures near optimal (ICMSF, 1980)

Organism	a_w
Bacteria	
Escherichia coli	0.95
Salmonella	0.95
Staphylococcus aureus	0.86
Micrococcus	0.93
Pseudomonas	0.97
Lactobacillus	0.94
Enterobacter	0.94
Bacillus	0.95
Clostridium perfringens	0.95
Clostridium botulinum	0.97
Moulds	
Aspergillus	0.75
Penicillium	0.81
Rhizopus	0.93
Yeasts	
Debaryomyces	0.83
Saccharomyces cerevisiae	0.90

a_w). The a_w of fresh meat is 0.99. Many micro-organisms (including spoilage and pathogenic ones) grow most rapidly at a_w in the range of 0.995–0.980. The minimum a_w values reported for the growth of some micro-organisms in foods are presented in Table 7.17.

Bacteria may be classified into four major groups by the temperature requirement for their growth (Table 7.18): (1) mesophiles; (2) thermophiles; (3) psychrophiles; and (4) psychrotrophs. The last mentioned group is of major importance to spoilage organisms of chilled meat in Europe and America, for example, whereas mesophiles are important on meats traded in traditional ways in India.

7.7.1 Spoilage of fresh meat at warm temperatures

Temperature plays a vital role in meat spoilage. Micro-organisms that grow in warm will be different from those that grow at chill temperatures. Thus

Table 7.18 Cardinal temperatures for micro-organisms (ICMSF, 1980)

Group	Temperature (°C)		
	Minimum	Optimum	Maximum
Thermophiles	40–45	55–65	60–90
Mesophiles	5–10	30–45	35–47
Psychrotrophs	−5 – +5	25–30	30–35
Psychrophiles	−5 – +5	12–15	15–20

Table 7.19 Growth of micro-organisms on the carcasses held at ambient temperature (28–30°C) (Narasimha Rao and Sreenivasa Murthy, 1985)

	Storage period (h)						
	0	4	8	12	16	18	20
Average total plate count (log cfu/cm^2)	5.7	5.6	5.7	6.0	6.5	7.0	8.5
Range of total counts on six carcasses	5.6–5.8	5.6–5.8	5.6–5.8	5.9–6.2	6.4–6.8	6.8–7.2	8.0–9.0

there will be differences in meat spoilage under warm and chill conditions (Ingram and Dainty, 1971). Spoilage under warm conditions was observed in whale carcasses. This was attributed to the proliferation of anaerobic bacteria – especially *Clostridium perfringens* present in the musculature (Robinson *et al.*, 1953; Ingram and Dainty, 1971). Now it is well established that deep muscle tissue from the healthy animals slaughtered under hygienic conditions is generally germ free (Gill, 1979). Therefore, there is only a remote chance for microbial spoilage of deeper meat tissue. Thus spoilage is a function of microbial activity on the surface of meat.

The practice of slaughter, holding and sale of meat under commercial conditions in India is different from that in Western countries. In many parts of India, the sheep carcasses, for example, are brought to the retail shops from the local slaughterhouse immediately after slaughter. The meat is offered for sale for 18–20 h in the retail shops. In such a situation, it is observed that some of the carcasses would remain in the shops until the end of sale. A study was made on the development of spoilage flora and shelf-life of sheep carcasses at 28–30 °C under commercial conditions (Narasimha Rao and Sreenivasa Murthy, 1985). Growth of micro-organisms on the carcasses held at ambient temperature is shown in Table 7.19.

There was an initial lag phase of 8 h before the microbial growth was initiated. The carcasses were not cold initially, their temperature varying from 35 to 37 °C. There was no warming up period during the study. There was surface drying of the carcasses – microbial growth is hindered by such drying (Scott, 1936). Hence surface drying can be one of the causes for the initial lag phase and slow growth of the micro-organisms on carcasses. An extended lag phase was observed on fresh meat during storage at high temperatures (Herbert and Smith, 1980). Development of off-odour was observed at 20 h in all the carcasses when the microbial count reached 8.5 log cfu/cm^2. Thus carcasses had a shelf-life of 19 h at ambient temperature (Narasimha Rao and Sreenivasa Murthy, 1985). Off-odours were evident by second or third day when beef was held at 25 °C (Ayres, 1955).

The proportions of different organisms in the initial and final microflora on the carcasses held at ambient temperature are shown in Table 7.20. *Staphylococcus* spp. predominated in the initial microflora. Meats such as

Table 7.20 Percent composition of aerobic flora on carcasses held at ambient temperatures (Narasimha Rao and Sreenivasa Murthy, 1985)

	Genera of bacteria
Initial	*Staphylococcus* (48), *Micrococcus* (19), *Acenitobacter*-like (4), *Pseudomonas* (3), *Escherichia* (12), *Enterobacter* (6), *Serratia* (1), *Flavobacterium* (1), *Bacillus* (1), *Proteus* (1), *Enterococcus* (1), *Brochothrix* (3)
Final	*Escherichia* (28), *Enterobacter* (16), *Acenitobacter*-like (22), *Staphylococcus* (18), *Pseudomonas* (16)

Figures in parenthesis indicate percent composition.

ham, lamb and beef were found to carry high percentage of staphylococci (Vanderzant and Nickleson, 1969; Narasimha Rao, 1982). Spoilage flora on the carcasses was dominated by mesophilic organisms such as *Escherichia* and *Acinetobacter*-like organisms (Narasimha Rao and Sreenivasa Murthy, 1985). *Enterobacter, Pseudomonas* and *Staphylococcus* spp. also formed a major part of the spoilage flora. The high percentage of staphylococci in the spoilage flora was mainly due to initial high contamination of the carcasses with these organisms. These findings suggest that spoilage of meat at ambient temperatures (28–30 °C) is mesophilic in nature. In a study on the spoilage of beef at 30 °C, it was reported that *Acinetobacter*, Enterobacteriaceae and *Pseudomonas* were predominant in the spoilage flora (Gill and Newton, 1980). Spoilage development at 30 °C and 40 °C of sheep carcasses (unwrapped) processed in the CFTRI modern abattoir was studied by Narasimha Rao (1982). Spoilage was noted at 24 h in carcasses held at 30 ± 1 °C, indicating a shelf-life of 22 h. Carcasses kept at 40 ± 1 °C, were spoiled

Table 7.21 Initial microbial counts of minced sheep meat (Narasimha Rao and Ramesh, 1988)

	Total plate count	Microbial counts (log cfu/g)			
		Coliforms	*Staphylococcus*	Enterococci	Psychrotrophs
Retail shop I					
Mean	6.11A	4.11A	4.73A	4.20A	4.18A
Range	5.2–7.3	3.0–4.5	3.8–5.0	3.7–4.6	3.5–4.8
Retail shop II					
Mean	5.99A	4.12A	4.62A	4.20A	4.18A
Range	5.0–7.4	3.4–4.5	4.0–5.4.	3.7–4.6	3.6–4.7
CFTRI modern training abattoir					
Mean	4.13B	1.41B	2.74B	1.72B	2.20B
Range	3.7–4.5	1.0–2.0	2.0–3.0	1.2–2.0	2.0–2.4
SEM	±0.113 2 (69 df)	±0.078 3 (69 df)	±0.069 5 (69 df)	±0.064 9 (69 df)	±0.053 0 (69 df)

Means of the same column followed by different letters differ significantly according to Duncan's New Multiple Range test ($p < 0.05$).

Table 7.22 Shelf-life of minced sheep meat kept at 30 ± 1°C (Narasimha Rao and Ramesh, 1988)

Storage period (h)	Total plate counts* (log cfu/g)		Psychrotrophs* (log cfu/g)	
	Retail shop	Modern abattoir	Retail shop	Modern abattoir
0	5.85 ± 0.43	4.18 ± 0.26***	4.27 ± 0.22	2.43 ± 0.26***
1	6.05 ± 0.42	4.65 ± 0.27***	4.42 ± 0.34	2.62 ± 0.34***
2	6.51 ± 0.67	5.22 ± 0.47**	4.85 ± 0.35	3.12 ± 0.32***
3	7.07 ± 0.49	5.75 ± 0.57**	5.42 ± 0.40	3.83 ± 0.47***
4	7.60 ± 0.39	6.47 ± 0.48**	5.78 ± 0.53	4.37 ± 0.50***
5	8.38 ± 0.35	6.98 ± 0.45***	6.62 ± 0.52	5.00 ± 0.33***
6	9.40 ± 0.22	7.48 ± 0.32**	7.60 ± 0.65	5.70 ± 0.38***
7	9.80	8.28 ± 0.20	7.50	6.66 ± 0.46
8	–	9.32 ± 0.44	0	7.60 ± 0.52

   * Mean of six experiments.
  ** Significant at 1% level.
*** Significant at 0.1% level.

Table 7.23 Shelf-life of minced sheep meat kept at 40 ± 1°C (Narasimha Rao and Ramesh, 1988)

Storage period (h)	Total plate counts (log cfu/g)†		Psychrotrophs (log cfu/g)†	
	Retail shop	Modern abattoir	Retail shop	Modern abattoir
0	5.90 ± 0.28NS	3.75 ± 0.33*	4.38 ± 0.26	2.32 ± 0.24NS
1	6.90 ± 0.46**	4.67 ± 0.67NS	4.98 ± 0.46*	2.93 ± 0.51NS
2	7.87 ± 0.12***	5.82 ± 0.68NS	6.22 ± 0.42***	4.08 ± 0.54**
3	8.92 ± 0.64***	6.75 ± 0.45**	7.03 ± 0.38***	5.08 ± 0.69**
4	9.65 ± 0.07***	7.79 ± 0.40***	7.20*	5.98 ± 0.53***
5	–	8.46 ± 0.46***	–	6.70 ± 0.53***
6	–	8.40	–	7.10
7	–	–	–	–

Statistical significance indicated as suffixes are both to be compared with the corresponding values at 30 ± 1°C.
† Mean of six experiments.
NS not significant.
   * Significant at 5% level.
  ** Significant at 1% level.
*** Significant at <0.001% level.

after about 12 h indicating a shelf-life of 10 hours. A lag phase or slow growth of micro-organisms was observed on the carcasses.

7.7.2 Microbial spoilage in minced meat at ambient temperatures

Minced meat was examined for microbiological quality and for shelf-life at high temperatures – 30° C and 40° C (Narasimha Rao and Ramesh, 1988). It was observed (Table 7.21) that minced meat obtained from local retail shops showed significantly higher microbial counts than that processed under hygienic conditions. It is seen also that *Staph. aureus, Micrococcus* and *Escherichia* were predominant in fresh meat. It may be observed that there was a significant difference in TPC at 30 ± 1 °C between retail shops and the modern abattoir in respect of all the different storage periods considered (Table 7.22) and, in all cases, it was observed that the counts were lowest in the case of meat from the modern abattoir. Similar results were observed for psychrotrophs. In respect of the shelf-life of minced meat at 40 ± 1 °C (Table 7.23), it was again observed that there were highly significant ($p < 0.001$) differences between the retail shop and the modern abattoir regarding the total and psychrotrophs counts for all the time periods considered. Invariably the counts were lower in meat from the modern abattoir than in that from retail shops. The shelf-life of minced meat at 30 ± 1 °C (obtained from retail shops) was 4 h. Incipient spoilage of meat was observed when the total plate counts reached 8.38 ± 0.20 and psychrotrophic counts were 6.62 ± 0.52 log cfu/g, at the end of 5 h. Minced meat from the modern abattoir had a shelf-life of 6 h. Samples were spoilt by the seventh hour of storage. Total plate counts had reached 8.28 ± 0.20 log cfu/g and psychrotrophic counts reached 6.66 ± 0.46/g at the time of the incipient spoilage. The shelf-life of meat from retail shops held at 40 ± 1 °C was only 2 h; off-odours had developed by the third hour. The shelf-life of minced meat from the modern abattoir was 4 h. The rate of spoilage plotted (Fig. 7.1) showed the same rate of change of

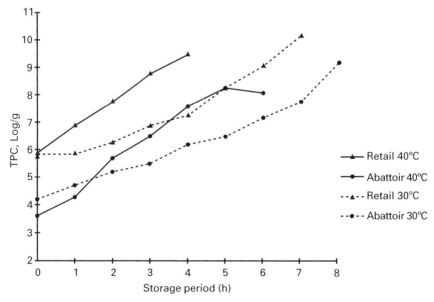

Figure 7.1 Rate of change of bacterial numbers in samples of minced meat obtained from a retail shop or modern abattoir and stored at 30 °C or 40 °C. (Narasimha Rao and Ramesh, 1988).

numbers in the samples from the retail shops and the abattoir. The curves at 30 °C and 40 °C were displaced by the differences in initial counts. These data revealed the differences in counts between retail and abattoir at any given time. It was observed that both increase in temperature and the initial microbial counts influenced the shelf-life of the product. Similar findings have been reported by other workers also (Ayres, 1955).

Micro-organisms associated with spoilage of meat in the study discussed above are listed in Table 7.24. *Escherichia* were predominant followed by *Acinetobacter, Staphylococcus* and *Micrococcus* in meat from retail shops stored at 30 ± 1 °C whereas *Staphylococcus* was predominant followed by *Micrococcus* in minced meat from the modern abattoir. The presence of small percentages of *Escherichia, Acinetobacter* and *Enterobacter* at the end of spoilage was due to the low initial contamination of minced meat with these organisms in the modern abattoir. Even at 40 ± 1 °C *Escherichia* was predominant, followed by *Staphylococcus aureus*, in minced meat from the retail shop whereas *Staphylococcus* and *Micrococcus* were predominant in the minced meat from the modern abattoir at the end of the spoilage period. It is interesting to note that the percentage of *Pseudomonas* organisms was low by the end of spoilage of minced meat. Thus the spoilage of meat at high temperatures (up to 40 °C) was due mainly to mesophilic organisms. It has been reported that meat stored at 15 °C or higher temperature showed an approximately equal number of *Pseudomonas* and *Micrococcus* (Ayres, 1955).

It is pertinent to note that the data obtained under ambient temperature in India have given important information on meat handling practices under

Table 7.24 Micro-organisms associated with meat spoilage (Narasimha Rao and Ramesh, 1988)

Storage temperature (°C)	Type of micro-organisms	Retail shop (%)	Modern abattoir (%)
30 ± 1	*Staphylococcus*	19.0	32.0
	Escherichia	25.0	8.0
	Acinetobacter-like	20.0	10.0
	Micrococcus	16.0	23.0
	Pseudomonas	13.0	12.0
	Enterococcus	5.0	9.0
	Enterobacter	12.0	6.0
40 ± 1	*Staphylococcus*	20.0	30.0
	Escherichia	35.0	6.0
	Acinetobacter	11.0	14.0
	Micrococcus	14.0	27.0
	Pseudomonas	9.0	6.0
	Streptococcus	11.0	15.0

tropical conditions. The results revealed that minced meat from local retail shops was a product with a large microbial load and varied microbial profile which frequently included a number of potentially pathogenic bacteria and resulted in rapid spoilage. This short shelf-life led to economic losses. Minced meat from carcasses processed in the modern abattoir demonstrated an initially low level of bacterial contamination and smaller numbers of pathogenic bacteria. Furthermore, spoilage of the product was delayed thereby giving shelf-life longer than meat from the shops. These data are relevant as a baseline in order to identify those changes in practice needed to improve the hygiene, safety and storage life of the product in the local developing situation.

7.7.3   Meat spoilage at chill temperatures

Chilling is a well recognized method of meat preservation commonly practised for many years in the western part of the world. Thus Western consumers are accustomed to the purchase of cold meat.

The primary purpose of chilling is to inhibit the growth of micro-organisms thereby extending the shelf-life of meat. The other purposes are to remove the heat from the carcasses, to firm the flesh, to delay the chemical changes and to prevent shrinkage (Ayres, 1955; Lawrie, 1985; Gill, 1986; Cassins, 1994). Hot carcasses are transferred to cool rooms (2–4 °C) to remove heat. Chilling of hot carcasses at too fast a rate immediately after slaughter leads to toughness (Lawrie, 1985) due to cold shortening (shortening of muscle fibres). Cold-shortening can be prevented by cooling the meat to about 15 °C and keeping it at this temperature to allow the onset of *rigor mortis*. Electrical stimulation is a recent innovative method to minimize cold shortening (Lawrie, 1985). Carcasses which have been held at chill temperatures till the interior temperature of carcasses reaches 7 °C are then fabricated into wholesale cuts, retail cuts or converted into mince. Packed meat is stored at 2–4 °C. Though chilling is an ideal method of preservation, it does favour the growth of psychrotrophs on meat surfaces eventually resulting in meat spoilage. Aerobically packed meat will have a shelf-life of 3 d at 4 °C. Microbial spoilage of meat stored at chill temperatures has been well documented (Ayres, 1955; Mossel, 1971; Ingram and Dainty, 1971; McMeekin, 1975; Patterson and Gibbs, 1977; Gill and Newton, 1977, 1978; Narasimha Rao, 1987; Huis in't Veld, 1996; Borch *et al.*, 1996; Dainty, 1996). Spoilage occurs when the microbial population reaches 10^8 on the surface and the meat has an off-odour and slime formation is evident. In aerobically stored fresh meat at chill temperatures, the predominant spoilage micro-organism is *Pseudomonas* though other micro-organisms such as *Acinetobacter, Moraxella, Enterobacter* spp. and *Brochothrix thermosphacta* contribute to the microbial association. In anaerobically packed meat, lactobacilli are predominant at the time of

Table 7.25 Microbial growth and shelf-life of meat cuts at different chill temperatures for market meat (Narasimha Rao, 1986)

Storage period in days	Leg cuts						Shoulder cuts					
	1° C		4° C		10° C		1° C		4° C		10° C	
	Total plate count*	Psychrotrophs	Total plate count*	Psychrotrophs	Total plate count*	Psychrotrophs	Total plate count*	Psychr9o-trophs	Total plate count*	Psychrotrophs	Total plate count*	Psychrotrophs
						(Log cfu/cm²)						
0	5.6	3.8	6.0	4.2	5.8	4.3	6.1	4.3	5.8	4.0	6.1	3.9
1	–	–	–	–	6.4	5.0	–	–	–	–	–	–
2	4.4	3.0	4.5	3.7	7.6	6.0	4.6	4.0	4.8	4.1	7.4	6.1
3	5.0	5.2	5.5	5.8	–	–	5.3	4.4	5.2	5.0	–	–
4	5.5	5.9	7.2	7.8	–	–	–	–	–	8.0	–	–
5	6.0	6.4	–	–	–	–	6.3	5.9	–	–	–	–
6	7.4	7.6	–	–	–	–	7.3	7.8	–	–	–	–

• Mean of six trials.

Table 7.26 Microbial growth and shelf-life of meat cuts (from modern abattoir) at different chill temperatures (Narasimha Rao, 1986)

Storage period in days	Leg cuts						Shoulder cuts					
	1° C		4° C		10° C		1° C		4° C		10° C	
	Total plate count*	Psychro-trophs	Total plate count*	Psychro-trophs	Total plate count*	Psychro-trophs	Total plate count*	Psychro-trophs	Total plate count*	Psychro-trophs	Total plate count*	Psychro-trophs
					($Log\ cfu/cm^2$)							
0	3.8	2.1	3.6	1.8	3.7	2.0	4.0	2.5	3.8	2.2	3.9	2.1
2	2.6	1.5	2.8	1.5	4.9	4.0	3.0	2.0	3.2	2.0	4.9	3.9
4	3.0	2.2	3.7	2.6	7.6	6.0	3.6	2.6	3.8	2.8	7.4	6.4
6	3.8	3.0	5.2	5.4	–	–	4.0	3.8	5.2	4.8	–	–
8	4.7	3.9	7.2	7.9	–	–	4.5	4.8	7.2	8.0	–	–
10	5.1	5.0	–	–	–	–	5.4	5.4	–	–	–	–
12	5.7	5.6	–	–	–	–	5.8	5.9	–	–	–	–
14	7.4	7.8	–	–	–	–	7.6	8.0	–	–	–	–

* Mean of six trials.

spoilage. The mechanism of meat spoilage due to micro-organisms is reviewed in Chapter 9 (Gill and Newton, 1977, 1978; Gill, 1986).

Shelf-life of sheep meat at 1 °C, 4 °C and 10 °C was studied in India by Narasimha Rao (1987). Sheep carcasses were processed in the CFTRI modern abattoir and chilled at 4 ± 1 °C for 18 h and retail cuts produced. Market cuts of meat (unchilled) for the market were also used in the study. Leg and shoulder cuts were packed in polyethylene pouches (200 gauge) and kept at 1 °C, 4 °C and 10 °C. The microbiological data are presented (Tables 7.25 to 7.27).

There was a reduction in the microbial counts during the first 1–2 d in meat cuts stored at 1 °C or 4 °C; microbial growth increased subsequently. Spoilage was detected on fifth day with market meat cuts and on the eighth day in hygienically processed meat cuts obtained from the abattoir and stored at 4 °C. Spoilage was manifested when the total plate and psychrotrophic counts reached 7.2–8.0 log/cm^2. The results showed that the shelf-life of meat cuts from the market at 4 °C was 4 d and that of meat cuts processed a modern abattoir was 7 d. The shelf-life of market meat and that from a modern abattoir stored at 1 °C were 6 d and 12 d respectively. Growth of micro-organisms took place slowly from the beginning of storage period on meat cuts stored at 10 °C and off-odour was detected on the fourth day in meat from a modern abattoir whereas market meat cuts spoiled on the second day. Thus shelf-life of meat cuts from the market and from a modern abattoir was 1 d and 3 d respectively at 10 °C. Microflora at the time of spoilage at different temperatures were isolated (Table 7.27). At 1 °C and 4 °C, the initial heterogeneous microflora was replaced by Gram-negative rods, i.e. *Pseudomonas* [*P. fragi* (90%) and *P. geniculata* (10%)] on market meat and *P. fragi* (100%) in meat cuts prepared in a modern abattoir. At 10 °C, both Gram-negative rods and Gram-positive cocci were present. The results indicate that both mesophiles and psychrotrophs grew at 10 °C.

Table 7.27 Microflora of meat cuts at the time of spoilage (Narasimha Rao, 1986)

Storage temperature (°C)	Types of microflora	Market meat cuts (% of microflora)	Meat cuts from modern abattoir (% of microflora)
1	*Pseudomonas fragi*	10	100
	Pseudomonas geniculata	10	–
4	*Pseudomonas fragi*	100	100
10	*Staphylococcus*	30	24
	Escherichia	18	1
	Enterobacter	9	3
	Micrococci	4	23
	Enterococci	10	6
	Acinetobacter	5	9
	Pseudomonas	24	30

In a recent study, it was reported (Borch, 1996) that the predominant bacteria associated with spoilage of refrigerated beef and pork was *Brochothrix thermosphacta, Carnobacterium* spp., Enterobacteriaceae, *Lactobacillus* spp., *Leuconostoc* spp., *Pseudomonas* spp. and *Shewanella putrefaciens*. The main defects in meat are off-odours and off-flavours, but discoloration and gas production also occur (Borch, 1996).

7.8   Control of spoilage of meat: possible approaches

Many attempts have been made all over the world to reduce the levels of initial microbial contamination and pathogenic bacteria of carcasses by certain treatments. Possible approaches that help in the control of spoilage of carcasses have been briefly discussed, by Gill in Chapter 4.

7.8.1   Lowering water activity

Surface drying of carcasses reduces the water activity and inhibits microbial growth (Scott, 1936; Scott and Vickery, 1939; Hicks *et al.*, 1955; Gill, 1986; Cassens, 1994). Desiccation delays for several hours microbial growth on the surface of carcasses held at ambient temperatures, by extension of the lag phase. Atmospheric humidity and air speed can maintain the drying process. If surface drying is prevented by covering the carcasses, bacterial growth increases rapidly. Surface drying extended the shelf-life of sheep and buffalo carcasses at ambient temperature (Nottingham, 1982; Narasimha Rao and Sreenivara Murthy, 1985; Syed Ziauddin *et al.*, 1995).

7.8.2   Reduction of surface pH

pH reduction has been attempted to reduce the levels of micro-organisms on meat surfaces. Normal pH of meat may inhibit the growth of certain bacteria but encourages the growth of others. A low pH ranging 4.0–4.5 inhibits the growth of both spoilage and pathogenic micro-organisms (Cassens, 1994). pH reductions have often been achieved by treating meat with organic acids such as acetic or lactic acids.

7.8.3   Treatment with organic acids

The use of many organic acids has been proposed as a general method of extending shelf-life. These acids cause a transient drop in surface pH and affect the micro-organisms thereon. The lactic acid in meat is itself inhibitive of some groups of organisms. It is the undissociated molecule of organic

acid that is responsible for antimicrobial activity. It has been suggested that
lactic acid treatment results in the reduction of spoilage and pathogenic
micro-organisms on carcasses (Smulders *et al.*, 1987). *In vitro* studies were
conducted on the effect of lactic acid alone or in combination with sodium
chloride on *Staph. aureus, Salm. newport, Ent. faecalis, Bc. cereus, Ps. fragi*
and *E. coli* (Syed Ziauddin *et al.*, 1993). Considerable reduction in the
counts of these organisms was observed in nutrient broth adjusted to pH 4.0,
4.5 or 4.8 with lactic acid (Fig. 7.2). Their inhibitory action on spoilage and
pathogenic micro-organisms was enhanced when lactic acid was used in
combination with sodium chloride (Fig. 7.3). Studies conducted at CFTRI
showed that sheep carcasses processed under hygienic conditions had a
shelf-life of 18 h when wrapped and 24 h when unwrapped and stored at
ambient temperatures (Fig. 7.4). *Staphylococcus* and *Micrococcus* were

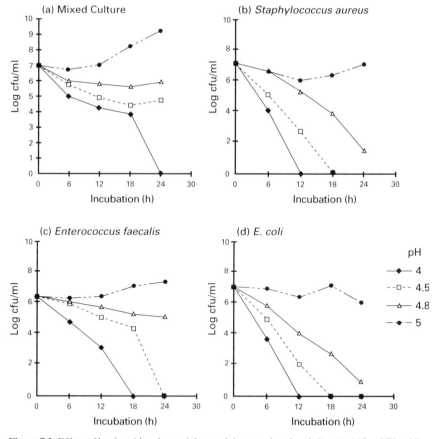

Figure 7.2 Effect of lactic acid on bacterial growth in a nutrient broth (*in vitro*) (Syed Ziauddin *et al.*, 1993).

predominant in the spoilage flora. Treatment with a spray of solution containing 2% lactic acid extended the shelf-life to 24 h in wrapped carcasses and 30 h with unwrapped ones (Fig. 7.5), while treatment with spray of a solution containing 2% lactic acid and 20% sodium chloride showed (Fig. 7.5) a shelf-life of 28 h in wrapped and 36 h in unwrapped carcasses stored at ambient temperature (Narasimha Rao and Nair, 1990). Extended shelf-life was observed (Figs. 7.6 and 7.7) with cuts of buffalo meat treated with lactic acid or a combination of lactic acid and sodium chloride and stored at ambient temperature (26 ± 1 °C) (Syed Ziauddin *et al.*, 1995a,b). Substantial reductions occurred in the microbial counts on carcass surface by these treatments. These studies suggest the possibilities of controlling the microbial spoilage of carcasses. The data generated in these studies will prove useful in programmes planned for the hygienic handling systems and for

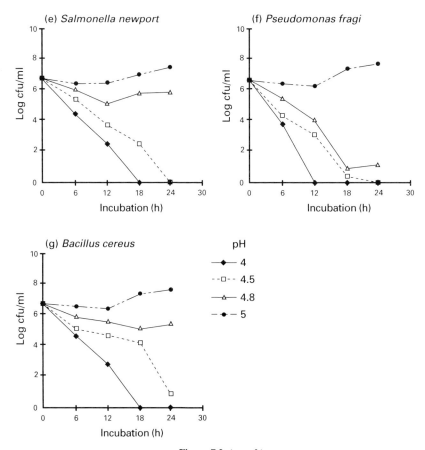

Figure 7.2 (contd.)

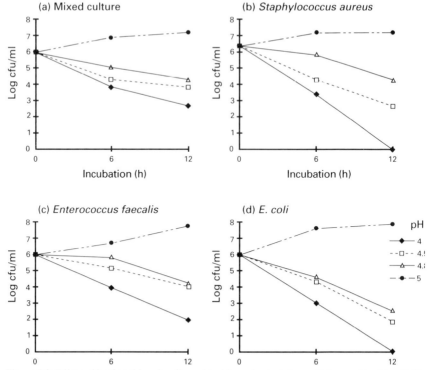

Figure 7.3 Effect of lactic acid and sodium chloride on bacterial growth in a nutrient broth (*in vitro*) (Syed Ziauddin *et al.*, 1993).

transportation and marketing of meat in the tropical regions where refrigeration facilities are inadequate.

7.8.4 Treatment with chlorine and hot water

Chlorination of wash water and hot water washing have both been examined as methods for reducing initial contamination. Although there is a real decline in numbers after chlorination it does not have a significant effect on shelf-life (Emswiler *et al.*, 1976; Marshall *et al.*, 1977). Hot water treatment appears to be more successful provided the surface reaches a high enough temperature for sufficient period (Smith and Graham, 1978; Graham *et al.*, 1978). Raising the surface temperature to 60 °C for 10 s can give a 3 log reduction in microbial counts. There will be slight discoloration of meat, which may be regained during the holding period. The treatment may have a significant effect on the shelf-life of carcasses.

7.8.5 Sodium chloride treatment

Sodium chloride has been used to flavour and preserve a variety of meats. Sodium chloride lowers water activity (a_w) and inhibits microbial growth.

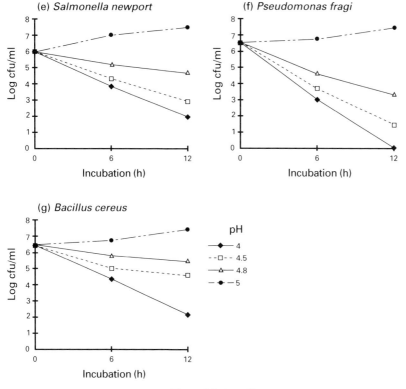

Figure 7.3 (contd.)

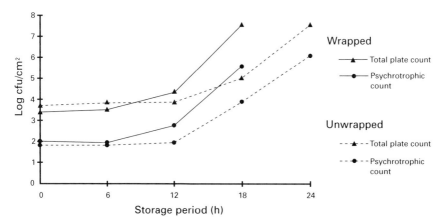

Figure 7.4 Microbial growth on carcasses held at 25–30 °C (Narasimha Rao and Nair, 1990).

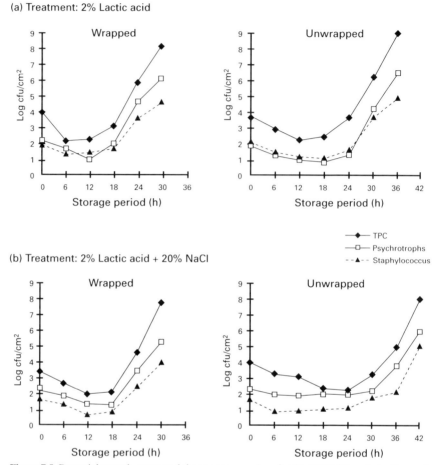

Figure 7.5 Bacterial growth on treated sheep carcasses stored at 25–30 °C (Narasimha Rao and Nair, 1990).

Sodium chloride was used along with organic acids for reducing microbial levels and for controlling microbial growth on carcass surfaces (Narasimha Rao and Nair, 1990).

7.8.6   Sorbate treatment

The primary inhibitory action of sorbate is against yeasts and moulds. Sorbate inhibits many bacteria including *Salmonella, Escherichia, Staphylococcus* and *Clostridium* (Sofos and Busta, 1983). Sorbate treatments are in use for controlling microbial growth in beef carcasses held at temperatures of 15 °C (Leistner, 1983). Chemical dips containing potassium sorbate substantially reduced the counts of *E. coli, Staph. aureus, Strep. faecalis* and *Cl. perfringens* on unchilled beef and on beef stored at 30 °C and 20 °C and extended the shelf-life up to 32 h at 30 °C and up to 68 h at 2 °C (Kondaiah *et al.*, 1985a,b,c).

(a) Total plate count

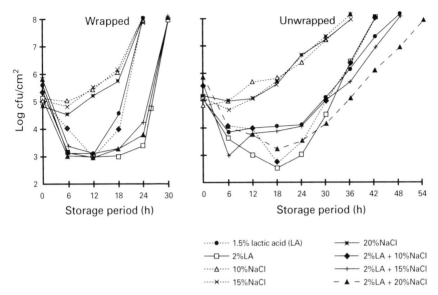

···•··· 1.5% lactic acid (LA)   —✳— 20%NaCl
—□— 2%LA   —◆— 2%LA + 10%NaCl
···△··· 10%NaCl   —+— 2%LA + 15%NaCl
···✕··· 15%NaCl   − ▲ − 2%LA + 20%NaCl

(b) *Staphylococcus*

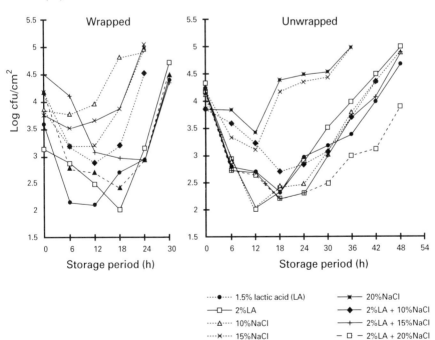

···•··· 1.5% lactic acid (LA)   —✳— 20%NaCl
—□— 2%LA   —◆— 2%LA + 10%NaCl
···△··· 10%NaCl   —+— 2%LA + 15%NaCl
···✕··· 15%NaCl   − □ − 2%LA + 20%NaCl

Figure 7.6 Bacterial growth on treated cuts of buffalo meat at 26 ± 1 °C (Syed Ziauddin, 1988).

(c) *Enterococcus*

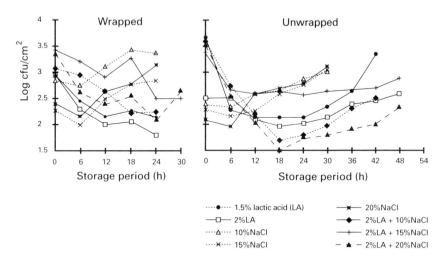

(d) Coliforms

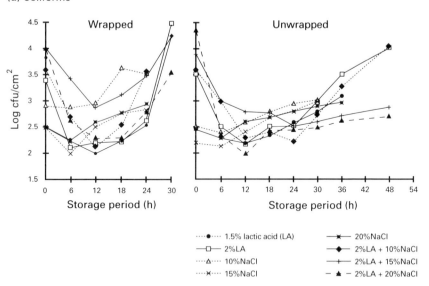

Figure 7.6 (contd.)

(e) Psychrotrophs

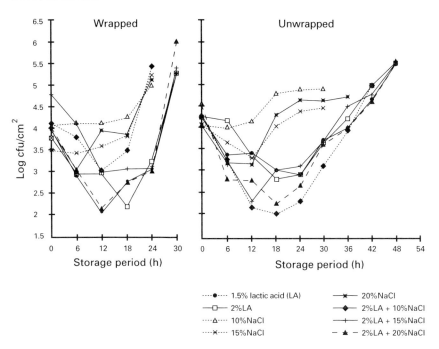

(f) Yeast and Moulds

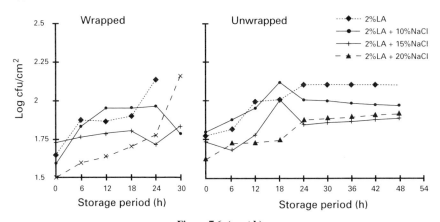

Figure 7.6 (contd.)

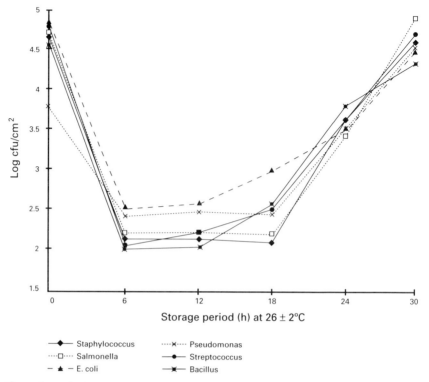

Figure 7.7 Behaviour of micro-organisms on meat slices treated with 2% lactic acid + 20% NaCl (Syed Ziauddin, 1988).

7.8.7 Enzyme inhibitors

It was demonstrated that ante-mortem administration of epinephrine controlled the *post-mortem* autolysis of meat by inhibiting catheptic activity at ambient temperatures (Lawrie, 1995). It was suggested that this process could be useful for long-term storage and for more efficient meat distribution at ambient temperatures (Radouco-Thomas *et al.*, 1959). Anti-autolytic activity of urea was demonstrated in meat kept at ambient temperatures (Rao and Murthy, 1986).

7.8.8 Cooling

Though hot meat consumption is common in tropical countries, especially in developing ones, traditional meat handling systems need to be replaced slowly by at least applying chilling to limit microbial growth. Motivation and education of people on the benefits of cooling systems and encouragement from the governments for financial support need to be addressed in developing countries. Refrigeration and freezing are the two most commonly used cooling systems for preservation of fresh meat.

Refrigeration.   Refrigeration is the most commonly used method for carcasses immediately after slaughter, during transport and storage and for packed meat (cuts and ground meat). At refrigeration temperature (4 °C) the shelf-life of properly packaged retail meat is 72 h, after which some discoloration can be expected to appear (Cassens, 1994), while the shelf-life of ground meat is for only one day. Initial microbial contamination, as affected by production practices, will have a major influence on the shelf-life of refrigerated meat. Modern methods, such as vacuum packaging and modified atmosphere packaging, are in use in advanced countries for increasing the shelf-life of refrigerated meat for extended periods (see Chapter 6).

Temperature, air flow and humidity will play a vital role in the efficient management of chilling rooms. In modern practices, carcass chilling rooms are normally operated in the temperature range of –2 ° to –4 °C (28–25 °F) with relative humidity of 88–92%. Shrinkage loss from the carcass is in the range of 1–2% (Cassens, 1994). The faster the air movement, the more rapid is the cooling. Rapid cooling enhances dehydration of the surface of the carcasses. Nevertheless, rapid cooling is essential in order to control microbial growth. In European community standards, the deep temperature of beef carcasses should reach 7 °C before cutting.

There is an increased interest in advanced countries in accelerated cooling (an expensive technique) (Cassens, 1984). Accelerated cooling is achieved by using extremely low temperatures (–15 °C to –35 °C) or by spraying with or immersion in cryogenic liquids. Liquid nitrogen is the ideal cryogenic agent. Solid carbon dioxide (dry ice) can be utilized.

Freezing.   Freezing is an effective method of storing cuts of large carcasses, whole small carcasses, retail cuts in fresh state for extended periods. Marketing of frozen meat in retail outlets is unsuccessful in the West due to the resistance from the consumers (Cassens, 1994). Consumers prefer to see the appearance of the product. Frozen meat will not give the appearance of fresh meat due to the ice crystal formation on the meat surface.

Meat freezes between –1.5 ° and –7 °C. The recommended storage temperature for frozen meat is –18 °C (0 °F). A lower temperature maintains quality better but this should be balanced against cost.

In India, meat from buffaloes and cattle is exported in frozen form to Malaysia and the Gulf countries. Sheep and goat carcasses are chilled, packed and subjected to freezing and frozen storage. Deboned meat is prepared from chilled buffalo carcasses. It is packed in small volumes and is subjected to freezing and frozen storage. Plate freezers and blast freezers are used. A worthwhile study is necessary on the quality of meat exported from the point of microbiological considerations. This would broaden our knowledge for applying modern methods of freezing to retain the freshness of meat to be exported.

In meat freezing, there are four stages: pre-freezing, freezing, frozen storage and thawing. Quality of meat is important in pre-freezing. It depends on both intrinsic (*post mortem* changes, tenderness and flavour) and extrinsic (microbiological status) factors. Films impermeable to oxygen and water need to be used for packaging of meat. Freezing must be rapid. Blast freezers and contact plate freezers are commonly used. Individual quick freezing (IQF) is gaining importance for certain speciality small sized meat cuts. Most rapid freezing can be achieved by cryogenic agents (liquid nitrogen or solid carbon dioxide). Rapid freezing produces smaller ice crystals on the surface of meat and damage to the meat tissues is very much less. Frozen storage of meat is very critical. Due to fluctuations in temperature, ice crystals grow in size and damage the meat tissue during frozen storage. Freezer burn is another interesting condition. It is nothing but surface dehydration due to moisture loss from meat surface which appears dry. It occurs if proper packaging material is not used. Enzymatic and chemical reactions proceed at a slower rate even in frozen meat. Lipid oxidation is common. Thawing is the final stage in freezing. Thawing is nothing but reverse freezing. Immersion in water (meat must be in an impermeable bag) and microwave methods are advocated for thawing. The problems need to be addressed in thawing are microbiological and drip loss. Microbial growth occurs as the temperature of the surface is higher than that of interior meat during thawing. Therefore thawing methods are to be standardized from the point of microbiological safety. Drip is a watery, red exudate. It contains proteins, vitamins and minerals. It was stated that drip is more a problem in frozen whole carcasses or quarter for shipment (Cassens, 1994). The lesser is the damage to the tissues during freezing and frozen storage, the lesser is the drip loss.

7.8.9   Lactic fermentation

Application of lactic fermentation is one of the traditional ways of preserving meat. Meat preservation is attributed to the combined effect of several substances (lactic acid, volatile acids such as acetic acid, antibiotics and bacteriocins) produced by lactic acid bacteria (LAB) though lactic acid plays a vital role (Reddy *et al.*, 1970; Gilliland and Speck, 1975; Fetlinski *et al.*, 1979; Narasimha Rao, 1995 – and Chapter 2 also). Lactic fermentation is a simple, 'low-tech' and inexpensive method that can be practised at ambient temperatures. Recent studies in India revealed that meat chunks or minced meat can be kept for 36 h at 37 °C when inoculated with a combination of *L. plantarum, L. casei* and *Lactococcus lactis* (Narasimha Rao, 1995). More studies are needed in this direction to develop a system for nonrefrigerated preservation of meat for areas where there are no refrigeration facilities particularly in developing/tropical countries. Lactic fermentation would find a place to achieve this target.

7.8.10 Irradiation

Irradiation has good potential in the elimination of pathogenic and spoilage micro-organisms from carcasses, cuts and minced meat and in the preservation of meat. It has emerged as a cost-effective method and it finds a place in developing countries. WHO clarified in 1980 the medical acceptability of irradiated foods and said "no health hazard results from consuming any food irradiated up to a dose of one mega rad (1 Mrad)" (Dempster, 1985). Irradiation reduces microbial levels and pathogenic micro-organisms and eliminates parasites like *Trichinella spiralis*. The USA permitted irradiation in pork and poultry (Cassens, 1994). The UK has permitted this in poultry. South Africa has permitted irradiation in cold meats, poultry, chicken, sausages, bacon and smoked salami. Several other countries have also permitted irradiation in meat, fish and poultry.

7.8.11 Packaging

Packaging plays a significant role in meat handling practices irrespective of warmer or cooler climates and deserves the merit of mention for the benefit of improving meat handling practices in developing countries (see Chapter 6).

Packaging protects the meat from moisture loss, contamination by micro-organisms, changes in colour and physical damage. Colour plays a vital role in consumer selection of meat offered for sale. Colour of meat can be displayed in a very appealing way through packaging to the consumers. Packaging imparts attractiveness to the product. Packaging of fresh meat varies from simple wrapping to advanced systems like vacuum packing (VP) and modified atmosphere packaging (MAP). Carcasses and large size meat cuts are wrapped in simple polyethylene films to protect them from contamination during handling. Fresh retail meat cuts are packed in pouches (polyethylene or polyvinyl chloride). These pouches allow oxygen transmission which maintains the bright red colour of meat and reduces moisture loss. Shelf-life of these meat cuts varies between 3 and 5 d at 4 °C (Ayres, 1955; Narasimha Rao, 1982; Cassens, 1994). Recent advances in the extension of shelf-life of fresh meat cut or mince are the use of vacuum packing and modified atmosphere storage (Hintallian and Hotchkiss, 1986; Brody, 1989; Bruce, 1990; Wei *et al.*, 1991; Taylor, Down and Shaw, 1990; Cassens, 1994). Exclusion of oxygen prior to packing is the principle in the vacuum packing of meat. Oxygen impermeable films are used. The growth of aerobic micro-organisms is prevented in meat under vacuum. Vacuum packaging provides at least three weeks shelf-life for the product under adequate refrigeration but the product looks dark. When the pack is opened and exposed to air (oxygen), the meat regains its bright red colour. Vacuum packing can be used in larger wholesale meat cuts, smaller retail cuts and for minced meat.

Modified atmosphere packing (MAP) is a new innovation in controlling the microbial spoilage and extending the shelf-life of fresh meat. MAP means alteration of atmospheric gas concentrations in the pack prior to packing. The three principal gases used in MAP are carbon dioxide, nitrogen and oxygen. Carbon dioxide exerts inhibitory effects on bacterial and mould growth. Nitrogen can inhibit oxidation of fats, rancidity and pack collapse and reduce the possibility of mould growth. Oxygen prevents anaerobic spoilage. MAP is very effective when used with a combination of carbon dioxide, nitrogen and oxygen. A gas mixture containing 10% carbon dioxide, 5% oxygen and 85% nitrogen was very effective for extending the shelf-life of meats in zero barrier film (Wei *et al.*, 1991). Expected shelf-life of fresh meat in MAP is ten days (Cassens, 1994).

In India, there is no proper packing of meat in the traditional system of meat handling. The common practice at retail level is to keep the meat in a special leaf (tamra) to hand over to the customers. There is a general resistance from the butcher community against the use of polyethylene films for meat packing. In organized meat processing plants, bacon factories and poultry processing plants, meat or chicken is packed in polyethylene films for storage and distribution in the local market. In export houses, low and high-density polyethylene (LDPE) films are used for packing meat and sheep and goat carcasses. In metro-cities, there is a shift from traditional meat handling to real packing of meat. This is mainly because large numbers of educated people live in metro-cities. Simple wrapping of carcasses with low cost films needs to be encouraged in the traditional system of meat handling from the viewpoint of controlling microbial spoilage to some extent for retaining the freshness of meat at ambient temperatures for a certain period. Beef packed in LDPE, in two-layer LD/ID under vacuum, could be stored up to 4 d at 4 °C (Venkataramanujam, 1997). Mutton packed by ordinary methods in LDPE and stored up to 120 h at 4 ± 1 °C had the highest Munsell colour value (Dushyanthan, 1997).

7.9 Summary

Quick consumption of meat without much in the way of storage of the carcass is the unique feature in India. Indeed the consumers continue to prefer 'hot meat' due to inbuilt traditional food habits. This underlines the need to improve the hygienic practices during meat processing in the abattoirs, during transport, in the meat markets and in the retail outlets. Available data have been presented with regard to the sources of microbial contamination, the microbiology of market meat and microbial spoilage of meat at ambient temperatures. Hygienic practices in a modern abattoir and suggested approaches for control of spoilage can minimize initial microbial contamination, reduce microbiological hazards and extend the shelf-life of meat. Consumers' reluctance and the high costs involved limit the use of refrigeration in meat handling but people need to be educated on the benefits of the cold chain. It is essential to introduce refrigeration for short periods at critical points during meat handling. Freezing is used for export meat. A detailed study on microbiological status of frozen meat is essential.

The data presented revealed that microbial ecology of meat is different in tropical climates from that of temperate ones. Hence, more understanding is needed with regard to the physiological characteristics of micro-organisms associated with tropical meat, characteristics of spoilage flora, the interaction of spoilage micro-organisms with meat tissue, development of aerobic and anaerobic spoilage of meat at warm temperatures and survival of pathogens in competitive microbial growth. Basic knowledge derived along these lines would form a sound basis for devising methods for microbiological quality control during meat handling in hot climates. Application of modern techniques such as Hazard Analysis of Critical Control Points concept (HACCP) would bring tremendous improvements in microbiological status during meat handling and marketing.

Acknowledgement

The author expresses gratitude and thanks to Dr V. Prakash, Director, Central Food Technological Research Institute, Mysore 570013, for his valuable suggestions and encouragement in the preparation of the chapter.

References

Ayres, J.C. (1955) Microbial implications in the handling, slaughtering and dressing of meat animals. *Advances Food Research*, **6**, 109–61.

Bachhil, V.N. and Jaiswal, T.N. (1988a) Occurrence of *Salmonella* in meats. *Journal of Food Science and Technology*, **25**, 310–12.

Bachhil, V.N. and Jaiswal, T.N. (1988b) *Bacillus cereus* in meats: Incidence, prevalence and enterotoxigenicity. *Journal of Food Science and Technology*, **25**, 371–2.

Bergdoll, M.S. (1967) The staphylococcal enterotoxins. In: *Biochemistry of some Food borne Microbial Toxins*. (eds. R.I. Mateles and G.N. Wogan), M.I.T. Press, Cambridge.

Borch, E., Marie-Louise Kant-Muermans, Blixt, Y. (1996) Bacterial spoilage of meat and cured meat products. *International Journal Food Microbiology*, **33**, 103–20.

Brody, A.L. (1989) Controlled or Modified Atmosphere or Vacuum Packaging of Foods. *Food and Nutrition Press*, New York.

Bruce, J. (1990) In 'International Conference on Modified Atmosphere Packaging' Stratford-upon-Avon, U.K, 15–17 October, 1990, Chipping Campden, GLOCS, 6 LD, U.K. Campden Food and Drink Research Association.

Callow, E. H. (1949) Benjamin Ward Richardson Lecture. Part-1. Hygiene and storage. *Journal of Royal Sanitation Institute*, **69**, 35.

Carl, K.E. (1975) Oregon's experience with microbiological standards for meat. *Journal of Milk and Food Technology*, **38**, 483–6.

Cassens, R.G. (1994) *Meat Preservation*, Food and Nutrition Press, Connecticut.

Childers, A.B., Keahey, E.E. and Vincent, P.G. (1973) Sources of salmonellae contamination of meat following approved livestock slaughtering procedures. II. *Journal of Milk and Food Technology*, **36**, 635–8.

Corlett, D.A. and Brown, M.H. (1980) In *"Microbial Ecology of Foods,"* Vol. 1, ICMSF, Academic Press, New York.

Dainty, R.H. (1996) Chemical/biochemical detection of spoilage. *Int. J. Food Microbiol.*, **33**, 19–33.

Das Guptha, P. (1974) Incidence of *Salmonella* in beef and goat in West Bengal and its public health significance. *Indian J. Animal Health*, **13**, 161–3.

Das Guptha, P. (1975) Incidence of *Salmonella* in pig meat in West Bengal and its public health significance. *Indian J. Animal Health*, **14**, 63–5.

Demster, J.F. (1985) Radiation preservation of meat products: A Review. *Meat Sci.*, **12**, 61–89.

Dushyanthan, K. (1997) In *Training Manual on Meat Processing and Packaging Technology*, Department of Meat Science and Technology, Madras Veterinary College, Chennai, India.

Emswiler, B.S., Kotula, A.W. and Rough, D.K. (1976) Bactericidal effectiveness of three chlorine sources used in beef carcass washing. *J. Animal Sci.*, **42**, 1445.

Evans, J.B. (1986) Staphylococci, in *Advances in Meat Research., Vol 2. Meat and Poultry Microbiology*, (eds A.M. Pearson and T.R. Dutson.), AVI Publishing, Connecticut.

Fetlinski, A., Knaut, T. and Kornacki, K. (1979) Insatz Von milchasanrebakterien also starter kulturen zur Haltbareiverlangerung von Hackfleish. *Fleischwirtschaft*, **59**, 1729–30.

Foegeding, P.M. (1986) Detection and quantitation of spore forming pathogens and their toxins, in Proceedings of the 1985 IFT-IUFOST Basic Symposium, *Food Microorganisms and their Toxins – Developing Methodology*, (eds M.D. Fierson and N.J. Stern), Marcel Dekker, New York.

Fung, D.Y.C. (1987) Types of microorganisms, in *Microbiology of Poultry Meat Products*, (eds F.E. Cunningham and N.A. Cox), Academic Press, London.

Gilbert, R.J. (1979) *Bacillus cereus* gastro-enteritis, in *Food-borne Infections and Intoxications*. 2nd edn. (eds H. Riemann and F.L. Bryan), Academic Press, New York, p. 495.

Gill, C.O. and Newton, K.G. (1977) The development of aerobic spoilage flora on meat stored at chill temperatures. *J. Appl. Bacteriol.*, **43**, 189–95.

Gill, C.O. (1976) Substrate limitation of bacterial growth at meat surfaces. *J. Appl. Bacteriol.*, **41**, 401–10.

Gill, C.O. (1979) Intrinsic bacteria in meat. *J. Appl. Bacteriol.*, **47**, 367–78.

Gill, C.O. (1986) The control of microbial spoilage in fresh meats, in *Advances in Meat Research*, Vol. 2, *Meat and Poultry Microbiology*, (eds A.M. Pearson and T.R. Dutson), AVI Publishing, Connecticut.

Gill, C.O. and Newton, K.G. (1978) The ecology of bacterial spoilage of fresh meat at chill temperatures. *Meat Sci.*, 207–17.

Gill, C.O. and Newton, K.G. (1980) Growth of bacteria on meat at room temperatures. *J. Appl. Bacteriol.*, **49**, 315–23.

Gilliland, S.E. and Speck, M.L. (1975) Inhibition of psychrotrophic bacteria by lactobacilli and pediococci in non-fermented refrigerated foods. *J. Food Sci.*, **40**, 903–5.

Graham, A., Cain, B.P. and Eustace, I.J. (1978) An enclosed hot water spray cabinet for improved hygiene of carcass meat. *CSIRO Meat Res. Lab. Rep. 11/78*, Brisbane, Australia.

Grua, F.H. (1986) Microbial ecology of meat and poultry, in *Advances in Meat Research*, Vol. 2, *Meat and Poultry Microbiology*, (eds A.M. Pearson and T.R. Dutson), AVI Publishing, Connecticut.

Guptha, P.K., Chauhan, G.S. and Bains, G.S. (1987) Bacteriological quality of fresh pork collected from different abattoirs and retail shops. *J. Food Sci. Technol.*, **24**, 270–2.

Herbert, L.S. and Smith, M.G. (1980) Hot boning of meat. Refrigeration requirements to meet microbiological demands. *CSIRO, Food Research Qt.*, **40**, 65–70.

Hicks, B.W., Scott, W.J. and Vickery, J.R. (1955) cited by Lawrie, R.A. (1985).

Hintlian, C.B. and Hotchkiss, J.H. (1986) The safety of modified atmosphere packaging: a review. *Food Technol.*, **40**, 70–7.

Hobbs, B.C. (1974) *Food Poisoning and Food Hygiene*. Edward Arnold, London.

Huis in't Veld, J.H.J. (1996) Microbial and biochemical spoilage of foods: an overview. *Int. J. Food Microbiol.*, **33**, 1–18.

ICMSF (1980) *Microbial Ecology of Foods*, Vol. 1, *Food Commodities*, Academic Press, London.

Ingram, M. and Dainty, R.H. (1971) Changes caused by microbes in spoilage of meats. *J. Appl. Bacteriol.*, **34**, 21–39.

Ingram, M. and Roberts, T.A. (1976) The microbiology of the red meat carcass and slaughter house. *R. Soc. Health J.*, **96**, 270–6.

Ingram, M. and Simonsen, B. (1980) Meat and meat products, in *Microbial Ecology of Foods:* Vol. 2, Academic Press, New York, p. 333.

Jay, J.M. (1962) Further studies on staphylococci in meats. *Appl. Microbiol.*, **10**, 247–51.

Jay, J.M. (1992) *Modern Food Microbiology*, 4th edn. Van Nostrand Reinhold, New York.

Jensen, L.B. (1945) *Microbiology of Meats*. The Gar Press, Illinois.

Kondiah, N., Zeuthen, P. and Jul, M. (1985a) Effect of chemical dips on unchilled fresh beef inoculated with *E. coli, S. aureus, S. faecalis* and *Cl. perfringens* and stored at 30 °C and 20 °C. *Meat Sci.*, **12**, 17–30.

Kondiah, N., Zeuthen, P. and Jul, M. (1985b) Chemical dips for extending the shelf-life of unchilled fresh beef at ambient temperature. *Journal of Food Science and Technology*, **22**, 142–5.

Kondiah, N., Zeuthen, P. and Jul, M. (1985c) Chemical dips for extending the shelf-life of unchilled fresh beef at ambient temperature. *J. Food Sci. Technol.*, **22**, 142–5.

Krishnaswamy, M.A. and Lahiry, N.L. (1964) Microbiological examination of market meat. *Indian J. Public Health*, **8**, 105–6.

Lawrie, R.A. (1985) *Meat Science*, Pergamon Press, New York.

Leistner, L. (1983) Paper presented at *Prospects of the Preservation and Processing of Meat*, in Proceedings 5th World Conference on Animal Production, Aug. 14–19, 1983, Tokyo, Japan, Vol. 1, 1255–62.

Mandokhot, U., Mayura, K., Chaterjee, A.K. and Dwarkanath, C.T. (1972) Microbiological examination of market meat for the presence of *Salmonella*, staphylococci and streptococci. *Indian J. Public Health*, **16**, 131–5.

Manickam, R. and Victor, D.A. (1975) A study on the occurrence of *Salmonella* in market meat. *Indian Vet. J.*, **52**, 209–11.

Marshall, R.J., Anderson, M.E., Naumann, H.D. and Stringer, W.C. (1977) Experiments in sanitising beef with sodium hypochlorite. *J. Food Protect.*, **40**, 246–9.

McMeekin, T.A. (1975) Spoilage association of chicken breast muscle. *Appl. Environ. Microbiol.*, **33**, 1244–6.

Minor, T.E. and Marth, E.H. (1971) *Staphylococcus aureus* and staphylococcal food intoxications: A review. *J. Milk Food Technol.*, **34**, 557–64.

Mossel, D.A.A. (1971) Physiological and metabolic attributes of microbial groups associated with foods. *J. Appl. Bacteriol.*, **34**, 95–118.

Nair, K.K.S., Madhwaraj, M.S., Kadkol, S.B. and Baliga, B.R. (1984) Bacteriological quality of pig carcasses. *Indian J. Animal Sci.*, **54**, 547–8.

Nair, K.K.S., Madhwaraj, M.S., Kadkol, S.B., Baliga, B.R. and Moorjani, M.N. (1983) Microbial examination of meat of buffalo (*Bubalus Bubalis* Linn) for the presence of salmonella and staphylococci. *Indian Vet. J.*, **60**, 189–92.

Nair, R.B., Rao, D.N., Mahendrakar, N.S. and Nair, K.K.S. (1991) The progressive changes observed in the quality of fresh goat and sheep meat during post-slaughter handling and marketing, in *Meat and Slaughterhouse by-products Handling Systems*. Proceedings of workshop held at CLRI, Madras, India, 15–17 July 1991 (eds K. Seshagiri Rao, K. Sarjuna Rao, A. Subbarama Naidu and D. Chandramouli).

Narasimha, Rao, D. (1982) Studies on microflora of prepackaged meat cuts with special reference to pathogenic staphylococci. Ph.D. Thesis, University of Mysore, Mysore, India.

Narasimha Rao D. (1983) Salmonella in meat. *J. Food Science Technol.*, **20**, 156–60.

Narasimha Rao, D. (1987) Shelf-life of meat at chill temperatures, in *Advances in Meat Research*, (eds J.B. Khot, A.T. Sherikar, B.M. Jaya Rao and S.R. Pillai), Red and Blue Cross Publishers, Bombay, India.

Narasimha Rao, D. (1995) Lactic fermentation in the preservation of animal products, in *Microbes for Better Living*, (eds R. Sankaran and K.S. Manja), Bangalore Printing and Publishing Company, Bangalore-560 018, India.

Narasimha Rao, D. and Nair, R.B. (1990) Preservation of hot carcass meat at ambient temperatures. Paper presented in the FAO Seminar on *Meat Development and Traditional and Low cost Meat Preservation Methods in Asia and the Pacific* held in Bangkok, Thailand, 24–28 October, 1990.

Narasimha Rao, D. and Ramesh, B.S. (1988) Microbial profiles of minced meat. *Meat Sci.,* **23**, 279–91.

Narasimha Rao, D. and Ramesh, B.S. (1992) The microbiology of sheep carcasses processed in modern Indian abattoir. *Meat Sci.,* **32**, 425–36.

Narasimha Rao, D. and Sreenivasa Murthy, V. (1985) A note on microbial spoilage of sheep meat at ambient temperature. *J. Appl. Bacteriol.,* **58**, 457–60.

Nartje, G.L. and Naude, R.T. (1981) Microbiology of beef carcass surface. *J. Food Prot.,* **44**, 355–8.

Nazer, A.H.K. and Osborne, A.D. (1976) *Salmonella* infection and contamination of veal calves. A slaughter house survey. *Br. Vet. J.,* **132**, 192.

Newton, K.G. and Gill, C.O. (1978a) The development of the anaerobic spoilage flora of meat stored at chill temperatures. *J. Appl. Bacteriol.,* **44**, 91–5.

Newton, K.G. and Gill, C.O. (1978b) Storage quality of dark, firm, dry meat. *Appl. Environm. Microbiol.,* **36**, 375–6.

Newton, K.G., Harrison, J.C.L. and Wauters, A.M. (1978) Sources of psychrotrophic bacteria on meat at the abattoir. *J. Appl. Bacteriol.,* **45**, 75–82.

Nottingham, P.M. and Wyborn, R. (1975) Microbiology of beef processing. Chilling and aging. *NZ J. Agric. Res.,* **18**, 23–7.

Nottingham, P.M. (1982) Microbiology of carcass meats, in *Meat Microbiology* (ed. M.H. Brown), Applied Science, London.

Panda, P.C. (1971) Bacteriological study of dressed chicken during the process of retailing. *Indian Vet. J.,* **48**, 927–31.

Panduranga Rao, C.C. (1977) Enterotoxigenic staphylococci in raw meats. *J. Food Sci. Technol.,* **14**, 224–5.

Patterson, J.T. and Gibbs, P.A. (1977) Incidence and spoilage potential of isolates from vacuum-packaged meat of high pH value. *J. Appl. Bacteriol.,* **43**, 25–38.

Paturkar, A.M., Sherikar, A.A. and Jaya Rao (1992) Prevalence of *Salmonella* in meats and sea foods of Bombay City. *J. Food Sci. Technol.,* **29**, 242–3.

Prasad, M., Chandiramani, N.K. and Mandokhot, U. (1983) Studies on the hygienic quality of pork products sold in the Indian market. *J. Food Sci. Technol.,* **20**, 272–6.

Radouco-Thomas, C.A., Lataste-Dorolle, C., Zender, R., Burset, R., Meyer, H.M. and Monton, R.F. (1959) cited by Narasimha Rao, D. (1982).

Randhwa, A.S. and Kalara, D.S. (1970) Human pathogens from goat meat with special reference to source of contamination. Part II – Salmonella. *Indian J. Med. Res.,* **88**, 283–7.

Rao, D.N. and Murthy, V.S. (1986) A preliminary study of the effect of urea in the preservation of meat. *Meat Sci.,* **17**, 251–65.

Rao, D.N. and Panda, P.C. (1976) Salmonellosis through meat and meat products. *Livestock Advisor,* **1**, 5.

Reddy, S.G., Hendrickson, R.L. and Olsen, H.I.C. (1970) The influence of lactic cultures on ground beef quality. *J. Food Sci.,* **35**, 787–91.

Roberts, T.A. (1976) *Microbiological Guidelines for Meat.* Proceedings of 22nd European Meeting of Meat Research Workers, Malmo, Sweden.

Robinson, R.H.M., Ingram, M., Case, R.A.M. and Benstead, J.G. (1953) *Whale Meat Bacteriology and Hygiene.* Department of Scientific and Industrial Research., London, Fd Inves. Spec. Rep. No. 59.

Scott, W.J. (1936) The growth of microorganisms on ox muscle. I. The influence of water content of substrate on rate of growth at –1 °C. *J. Council Ind. Res. Australia,* **9**, 177.

Scott, W.J. (1957) Water relations of food spoilage microorganisms. *Adv. Food Res.,* **7**, 83–127.

Scott, W.J. and Vickery, J.R. (1939) cited by Lawrie (1985).

Sherikar, A.A., A Jinkya, S.M., Khot, J.B. and Sherikar, A.T. (1979) The microbial flora of ready-to-cook pork products – A public health point of view. *J. Food Sci. Technol.,* **16**, 228–32.

Smith, M.G. and Graham, A. (1978) Destruction of *Escherichia coli* and salmonellae on mutton carcasses by treatment with hot water. *Meat Sci.,* **2**, 119.

Smulders, F.J.M. (1987) In *Elimination of Pathogenic Organisms from Meat and Poultry* (ed. F.J.M. Smulders), Elsevier, New York.

Sofas, J.N. and Busta, F.F. (1983) In *Antimicrobials in Foods* (eds A.L. Branen and P. Davidson), Marcel Dekker, New York.

Stiles, M.E. and Ng, L.K. (1981) Use of Baird-Parker's medium to enumerate *Staphylococcus aureus* in meats. *J. Food Protect.*, **44**, 583–7.

Stringer, W.C., Bilskie, M.F. and Naumann, H.D. (1969) Microbial profiles of fresh beef. *Food Technol.*, **23**, 97–102.

Syed Ziauddin, K. (1988) Ph.D Thesis. Studies on physico-chemical characteristics and shelf-life of buffalo meat. University of Mysore, Mysore, India.

Syed Ziauddin, K., Narasimha Rao, D. and Amla, B. L. (1993) *In vitro* study on the effect of lactic acid and sodium chloride on spoilage and pathogenic bacteria of meat. *J. Food Sci. Technol.*, **30**, 204–7.

Syed Ziauddin, K., Rao, D.N. and Amla, B.L. (1994) A study of microbial profiles of buffalo carcasses processed in local slaughter units. *Indian J. Comp. Microbiol. Immunol. Infect. Dis.*, **15**, 61–2.

Syed Ziauddin, K., Rao, D.N. and Amla, B.L. (1995a) Effect of lactic acid, ginger extract and sodium chloride on quality and shelf-life of refrigerated buffalo meat. *J. Food Sci. Technol.*, **32**, 126–8.

Syed Ziauddin, K., Rao, D.N., Ramesh, B.S. and Amla, B.L. (1995b) Extension of shelf-life of buffalo meat by pre-treatment-lactic acid and sodium chloride. *Current Res.*, **24**, 62–5.

Tarwate, B.G., Sherikar, A.T. and Murugkar, H.V. (1993) Microbiological analysis of environmental sources of contamination in Deonar abattoir. *J. Food Sci. Technol.*, **30**, 127–9.

Taylor, A.A., Down, N.F. and Shaw, B.G. (1990) A comparison of modified atmosphere and vacuum skin packing for the storage of red meats. *Int. J. Food Sci. Technol.*, **25**, 98–109.

The Gazette of India (1993) Schedule T; Specifications for Bacteriological Standards, January 15, 1993. Ministry of Commerce, New Delhi, India.

Thulasi, G. (1997) Microbial quality control standards for meat and meat products, in *Training Manual on Meat Processing and Packaging Technology*, Department of Meat Science and Technology, Madras Veterinary College, Chennai, India.

Valin, C., Pinkas, A., Dragnev, G., Boikovski, S. and Polikronov, D. (1984) Comparative study of buffalo meat and beef. *Meat Sci.*, **10**, 321–3.

Vanderzant, C. and Nickleson, R. (1969) A microbiological examination of muscle tissue of beef, pork, lamb carcasses. *J. Milk Food Technol.*, **32**, 357–61.

Varadarajulu, P. and Rao, D.N. (1975) *Microbial Levels of Pork as Affected by Processing*. Proceedings of the Conference of Microbiologists of India, Tirupati, 22–24 Dec, 1975.

Venkataramanujam, V. (1997) Packaging aspects of meat and meat products, in *Training Manual on Meat Science and Technology*, Madras Veterinary College, Chennai, India.

Wei, C.I., Balaban, M.O., Fenando, S.Y. and Replow, A.J. (1991) Bacterial effect of high pressure carbon dioxide treatment on foods spiked with *Listeria* or *Salmonella*. *J. Food Protect.*, **54**, 189–93.

8 The microbiology of stored poultry

N.A. COX, S.M. RUSSELL AND J.S. BAILEY

8.1 Introduction

The microbiology of the carcasses and meat of broiler chicken has been extensively studied. This review of the literature has been limited to information dealing with the multiplication of bacteria responsible for the reduction of shelf-life of fresh chicken under a variety of environmental conditions and the multiplication of pathogenic bacteria under conditions of temperature abuse. Microbiological methods for detection of these bacteria are also considered. In some cases, research done on beef, fish, or other types of poultry, such as turkey, that can be directly applied to fresh broiler chicken has been reviewed.

In the early 1990s, Pennington (1910) summarized the knowledge concerning the preferred methods for handling fresh poultry. As early as 1910, producers recognized the need to maintain poultry carcasses at cold temperatures and to minimize handling as well as hauling times. Pennington noted that fresh meat was stored on ice because mechanical refrigeration systems were either unavailable or extremely expensive to purchase and/or operate. It was recommended that for chicken, 'low temperature' should be maintained until the product was consumed. Pennington observed also that fluctuating temperatures caused condensation of moisture, which enhanced bacterial growth and hastened decomposition of the carcass.

8.2 Factors affecting shelf-life of fresh poultry

Temperature is the most important factor affecting the growth of psychrotrophic bacteria and hence the shelf-life of fresh poultry. Pooni and Mead (1984) stressed that poultry products were subjected to variations in (holding) temperature during processing, storage, distribution, and whilst being displayed for retail sale. Ayres et al. (1950) studied the effect of storage temperature on the shelf-life of fresh poultry. Their study revealed that the average shelf-life of fresh, eviscerated and cut-up birds under commercial conditions was 2–3, 6–8 and 15–18 d, when held at temperatures of 10.6, 4.4 and 0 °C respectively. Barnes (1976) observed that turkey carcasses stored at –2, 0, 2, and 5 °C, developed off odours in 38, 22.6, 13.9, and 7.2 d respectively. Other investigators (Baker *et al.*, 1956) have reported that temperature and time of storage are important because increases in counts of aerobic bacteria on ready-to-cook broiler carcasses stored for more than 7 d at 1.7 °C or 7.2 °C, were much larger than those on carcasses stored for shorter periods of time. In summary, storage temperatures extend the shelf-life of poultry carcasses.

Daud *et al.* (1978) reported that broiler carcasses maintained under optimal conditions should have a shelf-life of 7 d when stored at 5 °C. The rate of spoilage was twice as fast at 10 °C and three times at 15 °C, than for carcasses stored at 5 °C.

Spencer *et al.* (1954) identified factors that affect spoilage of chicken meat and reported that the following are important: (1) scalding temperature, (2) chlorinated chiller water, and (3) storage temperature. Under simulated commercial conditions, carcass halves scalded for 40 s at 53.3 °C had on average shelf-life of 1 d longer than carcass halves scalded at 60 °C. Carcass halves scalded at 53.3 °C and cooled for 2 h in ice water containing 10 ppm of residual chlorine had a shelf-life of 15.2 d, compared with 12.8 d for halves chilled in nonchlorinated water. Spencer *et al.* (1954) reported that chlorinated carcass halves stored at –0.6 °C had a shelf-life of 18 d whereas those stored at 3.3 °C had a shelf-life of 10 d.

Conflicting results on the influence of chilling methods on the microbiology of stored chicken have been published. Lockhead and Landerkin (1935) observed that chickens suspended in a refrigerator at –1.1 °C did not develop spoilage odours as rapidly as those surrounded by ice or ice water at the same temperature. In contrast, Naden *et al.* (1953) reported that there were significant advantages to packing poultry on ice, viz: (1) fresh quality was maintained for longer, (2) drying out was prevented, and (3) the carcasses in display cases were more attractive. Baker *et al.* (1956) demonstrated that the bacterial counts on ready-to-cook poultry stored on ice for 9 d were similar to those stored in a refrigerator for 5 d at 1.7 °C or 4 d at 7.2 °C. Spencer *et al.* (1954) observed, however, that carcasses stored in crushed ice had the same shelf-life as those stored in a refrigerator at –0.6 °C. Thus it is interesting to note that three possible conclusions can be drawn from four separate studies: i.e. refrigeration is best; ice is best; and there is no difference. This may be due to the parameters, such as odour or slime production, used for judging spoilage.

Evisceration is another factor purported to affect the shelf-life of fresh poultry. Lockhead and Landerkin (1935) demonstrated that eviscerated chickens developed spoilage odours more quickly than New York dressed (uneviscerated) chickens held under similar conditions. Baker *et al.* (1956) reported that bacterial counts on ready-to-cook poultry increased much more rapidly than on New York dressed poultry during 4 d of storage in ice at 1.7 or 7.2 °C. They attributed the increased rate of spoilage of ready-to-cook poultry to the fact that the abdominal region of the carcass was vulnerable to contamination and that the water used for washing the carcasses might be a means of spreading spoilage bacteria (Baker *et al.*, 1956).

The bacterial load present immediately after processing has also been shown to affect shelf-life. Thus Brown (1957) demonstrated that as the initial bacterial load increased, shelf-life decreased dramatically. Less time was required for bacterial populations to reach numbers high enough to produce spoilage defects when large numbers of bacteria were present initially.

8.3   Multiplication of psychrotrophic spoilage bacteria

When broiler carcasses are held at low temperatures, conditions for the growth of most species of bacteria are no longer optimal. Ayres *et al.* (1950) noted that the total number of bacteria on poultry stored at 0 °C decreased during the first few days of storage. They attributed this decrease to: (a) the unsuitability of the temperature for reproduction and survival of chromogenic (pigment producing) bacteria and mesophilic bacteria, and (b) insufficient time for psychrotrophic bacteria to enter the exponential phase of growth.

8.4   Effect of cold storage on generation times of bacteria found on broiler carcasses

Although spoilage bacteria – mainly psychrotrophs – grow at refrigeration temperatures, their rate of multiplication is much reduced. Most mesophilic bacteria are unable to grow at refrigeration temperatures. Olsen and Jezeski (1963) found that generation times for mesophiles and psychrotrophs did not increase proportionally when incubation temperatures were lowered progressively from the temperature ranges for optimum growth. When near to the lower temperature limit for growth, not only is the doubling time of *Escherichia coli*, a mesophile commonly found on broiler carcasses, much slower, but there is also a longer lag period before growth begins (Barnes, 1976). The authors observed that the generation time of *Escherichia coli* at –2, 1, 5, 10, 15, 20, 25, and 30 °C, was 0, 0, 0, 20, 6, 2.83, 1.37, and 0.55 h, respectively. Elliott and Michener (1965) reported that, at storage temperatures below 0 °C, the generation time for mesophilic bacteria might exceed 100 h.

8.5   Effect of elevated storage temperature on generation times

Firstenberg-Eden and Tricarico (1983) determined that at temperatures slightly above commonly used refrigeration temperatures, such as 10 °C, the generation times of psychrotrophic bacteria are much shorter than those of mesophiles. At about 18 °C, the multiplication rate of psychrotrophic and mesophilic bacteria was equal. At temperatures above 18 °C, mesophiles multiplied more rapidly than psychrotrophs.

8.6   Identification of spoilage flora

Studies to identify the spoilage flora of fresh chicken and other muscle foods date back to the late 1800s. Forster (1887) observed that most foods were exposed to saprophytic bacteria abounding in the air, soil, and water. He reported that, when cold storage was to be used to preserve foods, it was important to be able to predict the behaviour of these saprophytes over a given range of temperature.

Glage (as reported by Ayres, 1960) isolated slime-forming spoilage bacteria from the surfaces of meat stored at low temperature and high humidity. He named them '*Aromobakterien*' and observed seven species of spoilage bacteria, one of which predominated. These bacteria were oval to rod-shaped, with rounded ends, and occurred occasionally in chains. Glage (as reported by Ayres, 1960) demonstrated that '*Aromobakterien*' grew well at 2 °C but poorly at 37 °C. Their optimum growth temperature was 10–12 °C.

Later, Haines (1937) demonstrated that Glage's '*Aromobakterien*' were similar to bacteria that produced slime on meat stored at low temperatures. Haines (1933) reported that, with the exception of some members of the *Pseudomonas* group and a few strains of *Proteus*, bacteria found on lean meat stored at 0–4 °C belonged mainly to the *Achromobacter* group. Empey and Vickery (1933) found that 95% of the bacteria on beef immediately after processing that were capable of growth at –1 °C, were *Achromobacter* with a few *Pseudomonas* and *Micrococcus*. During cold storage, populations of *Achromobacter* and *Pseudomonas* increased while those of *Micrococcus* decreased.

Studies by Haines (1937), Empey and Scott (1939) and Lockhead and Landerkin (1935) all found that species of *Achromobacter* were the predominant spoilage organisms of fresh meat. Later, Ayres *et al.* (1950), Kirsch *et al.* (1952), and Wolin *et al.* (1957) reported that species of *Pseudomonas* were the predominant spoilage bacteria thus contradicting the results of these earlier studies. These three groups of workers attributed this discord to changes in nomenclature used in the sixth edition of *Bergey's Manual of Determinative Bacteriology* (Breed *et al.*, 1948) and in the third edition (Bergey, 1930).

Brown and Weidemann (1958) reassessed the taxonomy of the 129 psychrotrophic spoilage bacteria isolated by Empey and Scott (1939) from meat and concluded that most of these were in fact pseudomonads. Empey and Scott (1939) previously classified meat spoilage bacteria as *Pseudomonas* largely on the basis of the production of a water-soluble green pigment. Brown and Weidemann (1958) determined that 21 strains originally classified as pseudomonads, on the basis of this attribute, failed to produce pigment. Ayres *et al.* (1950), using Bergey's sixth edition (Breed *et al.*, 1948) as their taxonomic guide, reported that isolates from slimy carcasses were closely related to the following species of *Pseudomonas*: *ochracea, geniculata, mephitica, putrefaciens, sinuosa, segnis, fragi, multistriata, pellucida, rathonis, desmolytica (um)* and *pictorum*. Ayres *et al.* (1950) emphasized that, because of changes in *Bergey's Manual* between the third (Bergey, 1930) and sixth (Breed *et al.*, 1948) editions, many organisms that were previously assigned to *Achromobacter* should be classified as members of the genus *Pseudomonas* because they moved by means of polar flagella. Kirsch *et al.* (1952) came to the same conclusion.

Halleck *et al.* (1957) found that non-pigmented *Achromobacter–Pseudomonas* bacteria formed about 85% of the bacterial population on fresh meats during the first 2 weeks of storage at 1.1–3.3 °C and during the first week of storage at 4.4–6.7 °C. *Pseudomonas fluorescens* was dominant (upwards of 80% of the bacteria) on the meat in the latter stages of storage. At the beginning of the storage period, *Pseudomonas fluorescens* seldom exceeded 5% of the population on meat (Halleck *et al.*, 1957).

Ayres *et al.* (1950) reported that *Pseudomonas putrefaciens*, a common spoilage bacterium on meat, had lateral and polar flagella and hence should not be included in the genus *Pseudomonas*. This species is characterized by forming brownish colonies on nutrient agar and it differs from other pseudomonads by its production of proteases and hydrogen sulfide.

Barnes and Impey (1968) found that *Pseudomonas, Acinetobacter,* and *Pseudomonas putrefaciens* were the bacteria isolated most commonly from spoiled poultry. Fluorescent (pigmented) and non-pigmented strains of *Pseudomonas* dominate the microflora of spoiled poultry.

Since the report (Barnes and Impey, 1968) in which *Ps. putrefaciens* was found to be the principal spoilage bacterium of fresh poultry, this species has been reclassified as *Alteromonas putrefaciens* (MacDonell and Colwell, 1985). The genus was then changed from *Alteromonas* to *Achromobacter*, the latter organisms being transferred to the genus *Pseudomonas* in the seventh edition of *Bergey's Manual* (Breed *et al.*, 1957). MacDonell and Colwell (1985) retained the specific epithet but placed *P. putrefaciens* into a new genus, *Shewanella*. According to Thornley (1960), *Acinetobacter* was also part of the genus *Achromobacter* until the mid-1960s. A detailed account of the contemporary classification of Gram-negative spoilage bacteria on meat is given in Chapter 1.

8.7   The origin of psychrotrophic spoilage bacteria on broiler carcasses

Barnes (1960) reported that psychrotrophic bacteria present on the carcass immediately after processing, were present also on the feathers and feet of the live bird, in the water used in the processing plant, especially the chill tanks, and on equipment. These bacteria are rarely found in the intestines of the live bird. Schefferle (1965) found large populations of *Acinetobacter* (10^8/g) on the feathers of birds and she suggested that these probably originated from the deep litter on which the birds were recorded. Other psychrotrophs, such as *Cytophaga* and *Flavobacterium*, were often found in the chill tanks. They were rarely found on carcasses (Barnes and Impey, 1968).

Immediately after slaughter the psychrotrophic species of bacteria on chicken carcasses are *Acinetobacter* and pigmented pseudomonads occur (Barnes and Impey, 1968). Although strains or nonpigmented *Pseudomonas* produce off-odours and off-flavours on spoiled poultry they are less prevalent initially and *P. putrefaciens* is rarely found (Barnes and Impey, 1968).

8.8   Identification of spoilage flora on broilers held at elevated temperatures

The bacterial genera responsible for off-odours and slime on spoiled chicken at chill temperature are not as prevalent as the storage temperature increases. The spoilage bacteria on chickens stored at various temperatures, as reported by Barnes and Thornley (1966), are listed in Table 8.1. Initially, the predominant bacterial species on a broiler carcass are mesophilic, such as micrococci, Gram-positive rods and flavobacteria. If carcasses are held at

Table 8.1 The spoilage flora of eviscerated chickens initially, and after storage at 1, 10 and 15°C until spoiled (Barnes and Thornley, 1966)

	Number of strains			
	Initial	1°C	10°C	15°C
Total strains	58	40	80	69
Gram-positive rods	14	0	4	6
Enterobacteriaceae (lactose pos.)	8	0	3	10
Enterobacteriaceae (lactose neg.)	0	3	12	17
Micrococci	50	0	4	0
Streptococci	0	0	6	8
Flavobacteria	14	0	0	0
Aeromonas	0	0	4	6
Acinetobacter	7	7	26	34
Pigmented *Pseudomonas*	2	51	21	9
Nonpigmented *Pseudomonas*	0	20	12	2
Pseudomonas putrefaciens	0	19	4	4
Unidentified	5	0	4	4

10 °C, *Acinetobacter*, pseudomonads, and Enterobacteriaceae multiply. At 15 °C, *Acinetobacter* and Enterobacteriaceae, whose optimum growth temperatures are higher than those of the pseudomonads, predominate.

The presence of these spoilage bacteria is not the main concern when considering the shelf-life of fresh poultry. If allowed to increase in number, these bacteria produce by-products that result in off-odours and slime formation.

Spoilage bacteria must increase in number to levels above 10^5 cfu/cm^2 before off-flavours, off-odours, and visual defects are evident. Lockhead and Landerkin (1935) detected off-odours in New York dressed poultry (uneviscerated) once the number of bacteria had reached 2.5×10^6 to 1×10^8 cfu/cm^2. Ayres *et al.* (1950) observed that odour and slime were not present until the bacterial population exceeded 1×10^8 cfu/cm^2. Elliott and Michener (1961) determined that odour was produced at 1.6×10^5 to 1×10^8 cfu/cm^2. The authors reported that slime was produced at 3.2×10^7 to 1×10^9 cfu/cm^2.

8.9   Major causes of spoilage defects

Pooni and Mead (1984) determined that initial off-odours did not result from breakdown of the protein in skin and muscle of broilers, as previously thought, but from the direct microbial utilization of low-molecular-weight nitrogenous compounds. The spoilage is caused by the accumulation of metabolic by-products of the psychrotrophic bacteria that multiply at the expense of single compounds in carcasses at chill temperatures. As noted in Chapter 9, these by-products eventually become detectable as off-odours and slime as bacteria utilize the substrates available to them. With small population size, the spoilage bacteria utilize glucose for energy. The by-products of glucose metabolism do not contribute substantially to spoilage. As glucose availability diminishes, the bacteria switch to other substrates, especially amino acids, from the metabolism of which odorous end products make the carcass unacceptable (Pooni and Mead, 1984).

8.10   The development of off-odour and slime formation

Spoilage defects have been described by a number of workers. Glage (as reported by Ayres, 1960) observed that spoilage organisms produce a grey coating, which later becomes yellow on red meat. An aromatic odour accompanied the growth of these bacteria. The meat became covered with tiny drop-like colonies, which increased in size and coalesced to form a slimy covering. Glage (as reported by Ayres, 1960) reported that micro-organisms appeared first in damp pockets, such as folds between the foreleg and breast of a carcass, and their spread was promoted by condensation, which occurs when a cold carcass is exposed to warm, damp air.

Ayres *et al.* (1950) identified a characteristic ester-like odour described as a 'dirty dishrag' odour, developing on cut-up chickens. Off-odour preceded the production of slime and was considered to be the first sign of spoilage. Shortly after off-odours were produced, many small, translucent, moist colonies appeared on the cut surfaces and skin of the carcass. Initially, these colonies appeared as droplets of moisture; however, they eventually became large, white or creamy in colour, and often coalesced to form a uniform sticky or slimy layer. In the final stage of spoilage, the meat had a pungent ammoniacal odour in addition to the 'dirty dishrag' odour (Ayres *et al.*, 1950).

8.11   Metabolic adaptation of spoilage bacteria to refrigeration temperatures

At cold temperatures, psychrotrophic bacterial populations are able to multiply on broiler carcasses and produce spoilage defects; however, the numbers of mesophilic bacteria that were predominant on the carcass initially will remain the same or diminish (Barnes and Thornley, 1966). This phenomenon may be explained by examining metabolic changes that occur when these two groups of bacteria are incubated at refrigerator temperatures.

Wells *et al.* (1963) determined that the minimum growth temperature of a bacterium exists because, as incubation temperatures decrease, so does the absorption of nutrients. Also, as incubation temperatures decrease, bacteria increase the amount of lipid in their cell membranes. Graughran (as reported by Wells *et al.*, 1963) observed that, when mesophilic bacteria are incubated at progressively lower temperatures, the degree of saturation and the quantity of cellular lipids increase. As lipids in the cell membrane increase, nutrient absorption is inhibited. Eklund (1962) observed that *Brevibacterium linens* contained 7.2% fat when growing rapidly at 25 °C, but 16.7%, when growing poorly, at 4 °C. Eklund (1962) also determined that bacteria produced more fat at 4 °C, than at 9.4 or 22 °C. However, two typical psychrotrophic bacteria displayed no such temperature-induced differences when grown at 4 °C (Wells *et al.*, 1963).

8.12   Effect of cold storage on lipase production

Studies have determined that the amount of lipase produced by psychrotrophic bacteria increases as a result of incubation at low temperatures. Nashif and Nelson (1953) reported that lipase production in *P. fragi* was high at incubation temperatures between 8 and 15 °C, but almost absent at 30 °C and above. Alford and Elliott (1960) reported that lipase production in *P. fluorescens* was the same at 5–20 °C but slight at 30 °C.

8.13 Effect of cold storage on proteolytic activity

Changes in proteolytic activity of bacteria at low temperatures have also been studied. Peterson and Gunderson (1960) determined that production of proteolytic enzymes by a psychrotrophic strain of *Pseudomonas fluorescens* was highest when this bacterium was grown at low temperatures.

8.14 Effect of cold storage on carbohydrate metabolism

Metabolism of carbohydrates by bacteria is reduced at low temperatures. Brown (1957), Ingraham and Bailey (1959), and Sultzer (1961) observed that, at reduced incubation temperatures, carbohydrate oxidation rates of psychrotrophic bacteria decreased less than carbohydrate oxidation rates of mesophilic bacteria. Ingraham and Bailey (1959) determined that temperature coefficient differences between mesophiles and psychrotrophs had been identified for the catabolic processes, glucose oxidation, acetate oxidation, and formate oxidation by resting cells. Maintenance of a high rate of carbohydrate metabolism for psychrotrophs at low temperatures may be an indication of their ability to maintain their metabolism under adverse temperature conditions.

8.15 Bacterial 'conditioning'

Culturing psychrotrophic bacteria at low temperatures increases their ability to grow at cold temperatures. Hess (1934) determined that, by culturing psychrotrophs (*P. fluorescens*) at 5 °C, strains could be produced that were more active at 0 °C and –3 °C than other *P. fluorescens* strains that had been cultured at 20 °C. Chistyakov and Noskova (1955) successfully adapted a variety of bacterial strains to grow at –2 °C by growing them at 0–8 °C for two years. Ingraham and Bailey (1959) and Wells *et al.* (1963) suggest that this 'adaptation' may be a result of cellular reorganization. A greater comprehension of how psychrotrophic bacteria react to their environment is important when exploring the spoilage of fresh foods. This 'adaptation' is also important for understanding how bacteria react to very low temperatures, such as freezing.

MacFadyen and Rowland (1902) reported that bacteria were unique because they were able to survive freezing and thawing, by stating:

> It is difficult to form a conception of living matter under this new condition, which is neither life nor death, or to select a term which will accurately describe it. It is a new and hitherto unobtained state of living matter – a veritable condition of suspended animation.

In the late 19th century Burden-Sanderson (1871) observed that not all bacteria were destroyed by freezing. Forster (1887) and Muller (1903) isolated bacterial cultures that grew at 0 °C from sausage and fish. Fischer (1888) isolated 14 different bacteria that grew at 0 °C. Not only were these micro-organisms widely distributed, but their growth characteristics were the same at 0 °C as at higher temperatures; the rate of growth however was decreased (Muller, 1903). Rubentshik (1925) observed uro-bacteria that multiplied at –2.5 to –1.3 °C. Bedford (1933) found that strains of *Achromobacter* were able to grow at temperatures as low as –7.5 °C. Berry and Magoon (1934) stated that –10 °C was the lowest temperature at which bacteria would multiply.

8.16   Survival of bacteria during storage

Berry and Magoon (1934) reported that, under certain conditions, moderately cold storage temperatures (–2 to –4 °C) resulted in greater destruction of bacteria than storage at –20 °C. When cells are frozen rapidly, both intra- and extracellular fluid freezes. Slow freezing, however, causes an intra- and extracellular osmotic gradient, which can result in cellular disruption (Mazur, 1984). Upon freezing various types of bacteria at –190 °C for 6 months, MacFadyen and Rowland (1902) reported no appreciable difference in the vitality of the organisms and that the ordinary life functions ceased at –190 °C; they hypothesized that intracellular metabolism must also cease as a result of withdrawal of heat and moisture.

A proportion of the microbial population is killed or sublethally injured during the freezing process (Elliott and Michener, 1960). During frozen storage, the population of bacteria that survives on food can range from 1 to 100%, but averages 50% depending on the type of food (Elliott and Michener, 1960). Straka and Stokes (1959) concluded that some nutrients, required by bacteria for growth, were rendered inaccessible by the freezing process, thereby preventing bacterial multiplication.

In contrast, other studies have indicated that freezing and thawing might enhance recovery of bacteria. Hartsell (1951) observed that *E. coli* that survive freezing and thawing grow more rapidly than *E. coli* that are not frozen. Sair and Cook (1938) indicated that one reason for accelerated bacterial growth on frozen and thawed foods was that tissue damage due to freezing might result in nutrient release and increased moisture, providing a more suitable growth medium.

8.17   Effects of freezing on shelf-life

The effect of freezing on the shelf-life of chicken has been extensively studied. Spencer *et al.* (1955) determined that carcasses that were frozen and held for two months and then thawed had the same shelf-life as unfrozen controls. Similar observations were made by Spencer *et al.* (1961) and Newell *et al.* (1948), who reported no major increases or decreases in shelf-life of carcasses as a result of freezing and thawing. Elliott and Straka (1964) reported that chicken meat that was frozen for 168 days at –18 °C and thawed spoiled at the same rate as unfrozen controls.

8.18   Effect of elevated storage temperature on bacterial multiplication growth temperature classification

Since holding temperature is the most important factor that affects the growth of both spoilage and pathogenic bacteria, the holding temperature of fresh poultry is of great concern to the poultry industry. When considering the relationship of temperature to microbial life, two things must be considered: the holding temperature of the micro-organism and the length of time for which the micro-organism is exposed to that temperature (Olsen, 1947). All living cells respond in various ways to temperature, and bacteria, being living cells, are no exception. Their metabolism, physical appearance, or morphology may be altered and their growth may be stimulated or retarded depending upon the particular combination of temperature and time of exposure. Olsen (1947) reported that there is a minimum growth temperature below which growth ceases, an optimum growth temperature, which is the most favourable for rapid growth, and a maximum growth temperature, above which growth ceases. Bacteria vary not only with regard to their growth temperature range, but also in their minimum, optimum, and maximum growth temperatures (Olsen, 1947). Greene and Jezeski (1954) reported that the two criteria that are used to determine optimum growth conditions are generation time and maximum cell population. Generation time is an indicator of speed of cell division, whereas maximum cell population takes into account cell destruction as well as cell production.

The minimum, optimum, and maximum growth temperatures for psychrotrophic and mesophilic bacteria are listed in Table 8.2. Olsen (1947) placed bacteria now considered psychrotrophic in the psychrophilic category. Muller (1903), Zobell and Conn (1940), and Ingraham (1958) objected to the term 'psychrophiles' as many spoilage bacteria are able to

Table 8.2 Minimum, optimum, and maximum growth temperature ranges for psychrophilic and mesophilic bacteria (Olsen, 1947)

	Minimum (°C)	Optimum (°C)	Maximum (°C)
Psychrophilic	−5 – 0	10–20	25–30
Mesophilic	10–25	20–40	40–45

Table 8.3 Minimum, optimum, and maximum growth temperatures for psychrophiles, low-temperature mesophiles, psychrotrophic and psychroduric micro-organisms, nonfastidious high-temperature mesophiles, and fastidious high-temperature mesophiles (Ayres *et al.*, 1980)

	Minimum	Optimum	Maximum
Psychrophiles		≤ 0 5 to 15	± 20
Low-temperature mesophiles, psychrotrophic and psychroduric micro-organisms	± 10 – +8	20–27	32–43
Nonfastidious high-temperature mesophiles	± 8	35–43	43–45
Fastidious high-temperature mesophiles	20–25	37	?

survive and grow at low temperatures, but their optimal growth temperatures are well above freezing. Ayres *et al.* (1980) reported that the optimum growth temperature for psychrophilic bacteria was between 5 and 15 °C. Muller (1903) reported that the psychrotrophic bacteria were a group of mesophiles that were able to multiply relatively slowly at a lower temperature range than for most other organisms.

A more current grouping of bacteria based on their growth temperatures is given in Table 8.3. Many of these bacteria cannot be placed into any single category because their growth temperature range is broad (Ayres *et al.*, 1980). Some bacteria are able to grow well at both refrigerator temperatures and high temperatures. These bacteria, however, do not represent the average groups of bacteria, which can be separated based on their minimum, optimum, and maximum growth temperatures.

8.19 Enumeration of psychrotrophic bacteria

Elliott and Michener (1961) reported that total bacterial populations are used as a measure of sanitation, adequacy of refrigeration, or handling speed. Determining which of these factors were responsible for a high count was thought to be impossible, with only a total count on the product as a guide (Elliott and Michener, 1961). Elliott and Straka (1964) emphasized the importance of incubating plates at or near the temperature at which the product spoiled in order to obtain a true indication of bacteriological changes during spoilage and to enumerate the bacteria responsible for spoilage. To enumerate psychrotrophs, the incubation temperature must be low enough to preclude the multiplication of mesophiles. Ayres (1951) determined that, at temperatures of 0 °C and 4.4 °C, many mesophilic bacteria did not multiply. Senyk *et al.* (1988) observed that, in raw milk samples held at 1.7, 4.4, 7.2, and 10.0 °C, mesophilic bacteria increased by 0.12, 0.13, 0.40, and 1.12 $\log_{10}$, respectively, after 48 h. When milk is held above 4.4 °C, mesophiles are able to multiply more rapidly (Senyk *et al.*, 1988). Barnes (1976) determined that mesophilic bacteria, such as *E. coli*, did not multiply at storage temperatures below 5 °C. The temperature at which mesophilic microflora are able to grow seems to be between 4 and 7.2 °C, and, if aerobic plate counts are performed at temperatures below 4 °C, mesophilic bacteria should not contribute to the total plate count.

8.20 Enumeration of mesophilic bacteria

To enumerate mesophilic bacteria, a temperature must be used that is high enough to retard the growth of the psychrotrophic populations of bacteria. Enumeration of mesophilic bacteria is more difficult than that of psychrotrophs because some psychrotrophs are able to multiply at elevated temperatures. Studies have been conducted to determine the maximum growth temperature of psychrotrophs. Greene and Jezeski (1954) determined that *Pseudomonas* and *Aerobacter* were capable of normal development at 0–30 °C. At 35 °C none of the pseudomonads grew and the *Aerobacter* was inhibited. Changes in temperature influence every stage in the growth of bacteria. Lower temperatures result in extended lag phases, longer and slower lag phases, and maximum stationary phases of higher population and longer duration (Greene and Jezeski, 1954). Higher temperatures (35 °C) inactivate certain essential enzyme systems associated with cell division (Greene and Jezeski, 1954). An understanding of the maximum growth temperature of psychrotrophic bacteria is important for determining the incubation temperature that should be used to enumerate mesophilic bacteria in mixed samples without interference from psychrotrophs.

8.21 Determination of temperature abuse

Bacteria isolated from temperature-abused broiler carcasses are different from those isolated from carcasses held at correct storage temperatures. For carcasses severely temperature-abused (at 20 °C), the spoilage flora consists mainly of *Acinetobacter* spp. and Enterobacteriaceae. Pooni and Mead (1984) determined that, at 20–22 °C, only 20% of the bacteria were pseudomonads and 70% of the bacteria were *Proteus* spp. which are mesophilic. Regez *et al.* (1988) observed that pseudomonads made up less than 2% of the carcass flora after 1.5 d of storage at 20–22 °C.

One approach to differentiate mesophiles from psychrotrophs and measure their growth on chicken involves using selective temperatures above 37 °C because psychrotrophic spoilage bacteria isolated from chickens fail to grow at 37 °C (Barnes and Impey, 1968).

Buchanan *et al.* (1991) used selective temperatures of incubation to differentiate samples of raw ground chicken stored at elevated temperatures, 12 or 19 °C, from controls held under refrigeration (5 °C) for 10 d. For aerobic plate counts (APC) at 42 °C, bacterial populations in the samples stored at 5 °C remained below 10^6 cfu/g for the 10-d storage period, whereas those samples stored at 12 or 19 °C contained bacterial populations in the range of 10^8–10^{10} cfu/g. Mesophilic bacteria in the samples held at 12 or 19 °C increased, but not if held at 5 °C.

Another study was conducted to determine the storage temperature at which mesophilic bacteria were able to initiate growth. Russell *et al.* (1992a) found that mesophiles increased on carcass rinse samples held for 7 d above 5 °C as determined by APC, and above 6 °C as determined by impedance detection time (DT), using an incubation temperature of 42 °C. These results indicate that 5 °C, using APC, and 6 °C, using DT, are the 'abuse' temperatures. If carcasses are held above these temperatures, mesophilic populations of bacteria will be able to multiply.

Another study (Russell *et al.*, 1992b) revealed that whole broiler chicken carcasses that were temperature-abused at 15 °C for 12 h or 20 °C for as little as 4 h could be microbiologically distinguished from those that were held at proper cold storage temperature (4 °C). Temperature-abused carcasses were microbiologically distinguished from unabused controls by monitoring the growth of mesophiles as 'indicators' of temperature abuse.

8.22   Use of different microbiological methods to determine temperature abuse

When considering the disposition of a product that may have been temperature-abused, such as fresh broiler carcasses, the time required for microbiological evaluations to be completed becomes critical.

Since freshly processed carcasses should be shipped to market within 24 h of processing (Anonymous, 1988) and traditional plate count methods require 48–72 h to conduct (Busta *et al.*, 1984), poultry products may be shipped to the consumer and consumed before bacterial counts are obtained. If carcasses are temperature-abused, the time required for analysis using traditional plate counts may be sufficient to allow the carcasses to spoil completely.

The food industry is continually searching for novel rapid methods to identify pathogens and to determine the total populations of bacteria in food products. Detection of specific pathogenic bacteria such as *Salmonella*, using traditional techniques, may require up to 7 days (Andrews *et al.*, 1984). Psychrotrophic plate counts require an incubation period of 10 d at $7 \pm 1 \, °C$ (Gilliland *et al.*, 1984). Impedance microbiological techniques have been shown to be an effective means of rapidly enumerating bacteria from a variety of foods.

Electrical methods to measure the growth of bacteria date back to the late 1800s (Stewart, 1899). Parsons (as reported by Strauss *et al.*, 1984) demonstrated that conductivity could be useful to measure the ammonia produced by clostridia in various environments. Allison *et al.* (1938) used conductance to measure proteolysis induced by bacteria. Ur and Brown (1975) proposed the use of impedance as a tool for enumeration of micro-organisms. Cady *et al.* (1978) investigated the ability of a variety of micro-organisms to produce impedance changes when cultured in different culture media.

Impedance is defined as the opposition to flow of an alternating electrical current in a conducting material. As bacteria multiply, they convert large molecules into smaller, more mobile metabolites, which change the impedance of the medium. These metabolites increase the conductance and decrease the impedance of the medium. When the microbial population reaches a level of 10^6–10^7 cells/ml, a change in the impedance of the medium is observed. The time required for this exponential change to occur is known as the impedance detection time (DT) (Firstenberg-Eden, 1985).

DT can be obtained in very short periods of time, 12 h or less, compared with aerobic plate counts (APC). There are however several fundamental differences between impedance microbiological techniques and APC (Firstenberg-Eden, 1985). When performing APC, all bacteria that are able to reach a visible biomass are counted (Firstenberg-Eden, 1983), whereas the impedance technique relies on the measurement of metabolic changes (Firstenberg-Eden, 1985). Since impedance measurements depend on metabolic change produced by the fastest-growing bacterium or group of bacteria in a sample, factors such as media, time, and temperature, become critical parameters in the assay because specific bacteria use different metabolic pathways depending on the media in which they multiply. Some end-products of metabolism produce stronger impedance signals than others when bacteria are allowed to multiply and utilize different substrates in the media (Firstenberg-Eden, 1985). The substrates that bacteria are grown on will therefore determine the by-products they produce, and hence are an important consideration when performing impedance assays.

Another consideration when selecting a medium for conducting impedance measurements is that some bacteria will multiply in a given medium, produce a detection time, exhaust the nutrients necessary for growth, and stop growing. Subsequently, another group of bacteria will use the remaining nutrients and begin to multiply, creating a 'double hump' in the impedance curve. The impedance curve represents the relationship of impedance change to incubation time (Firstenberg-Eden and Eden, 1984). The DT of these bimodal curves is difficult to assess.

When enumerating total numbers of bacteria in a sample, it is essential that the media and temperature are selected so that differences in generation times between the different genera of bacteria in the sample are minimized (Firstenberg-Eden, 1985). Firstenberg-Eden and Tricarico (1983) revealed that 18 °C is the appropriate temperature for impedance monitoring of mixed flora samples (mesophilic and psychrotrophic) when enumerating total populations of bacteria, because this is the temperature at which generation times for the mesophiles (1–5 h) and psychrotrophs (1.2 h) are most similar. Minimization of differences in generation time allows most of the genera in the sample to multiply at a similar rate, and hence gives a more accurate indication of the total population present.

If a selective medium and temperature are used, the bacteria or group of bacteria able to multiply most rapidly and reach the threshold level of 10^6 will be responsible for the DT. This feature of impedance microbiology makes it a useful tool in that, for mixed samples, a particular bacterium or group of bacteria can be measured by selecting for its growth over the other competing microflora in the sample. For example, if a mixed sample contained 100 000 pseudomonads and 1 coryneform, and the sample was incubated at 30 °C, the coryneform would be the bacteria responsible for the DT (Firstenberg-Eden and Eden, 1984). At 30 °C, the generation time of the pseudomonads is four times that of the coryneform, which allows the coryneform to multiply and reach 10^6 before the pseudomonads. Thus, selective media can be useful for enumerating one species of bacteria in mixed samples by selecting for its growth over that of other species present in the sample.

The most commonly used application of impedance microbiological methods is to determine whether samples contain above or below a certain concentration of bacteria. This is determined by comparing the results from a given analysis with a calibration curve. DT and APC are determined for 100 samples and a calibration curve is generated in which DT (h) is regressed against APC ($\log_{10}$ cfu/ml). This curve defines the relationship between impedance and the APC method for a given product (Firstenberg-Eden, 1985). After the calibration curve has been generated, future samples can be analysed using the impedance method and APC can be estimated. This approach has been used to determine the total number of bacteria in a variety of foods including meats (Firstenberg-Eden, 1983), raw milk (Firstenberg-Eden and Tricario, 1983), frozen vegetables (Hadley et al., 1977) and fresh fish (Gibson et al., 1984; Gibson, 1985; Ogden, 1986; van Spreekins and Stekelenburg, 1986). Firstenberg-Eden (1985) found that, frequently, the medium, temperature, and pH routinely used in the standard method are not appropriate for the impedance method. Only the development of suitable media that optimize the impedance signal and the appropriate incubation temperature that minimizes difference in generation times in a mixed sample will yield an accurate and consistent impedimetric estimation of microbial counts.

Using selective media and selective incubation temperatures, researchers have developed many procedures for enumerating specific bacteria or groups of bacteria. Impedance has been used as a means of enumerating coliforms (Martins and Selby, 1980; Firstenberg-Eden and Klein, 1983; Firstenberg-Eden et al., 1984; Strauss et al., 1984; Tenpenny et al., 1984), faecal coliforms (Mischak, et al., 1976; Silverman and Munoz, 1978, 1979; Rowley et al., 1979), and Enterobacteriaceae (Cousins and Marlatt, 1989) from samples of wastewater, dairy products, and meat.

Many media and incubation temperatures have been analysed to determine which medium and temperature produces the most accurate and consistent DT when enumerating coliforms or Enterobacteriaceae. Firstenberg-Eden and Klein (1983) define the optimal medium for rapid enumeration of coliforms or E. coli as a medium that allows these organisms present in a food sample to produce a strong, consistent, and early impedance change, which can be uniformly interpreted by the computerized data analysis system.

Silverman and Munoz (1978, 1979) used a general lactose-based broth medium and a selective incubation temperature of 44.5 °C to enumerate E. coli in pure cultures. Mischak et al. (1976) were able to quantify faecal coliforms from mixed samples containing initial noncoliform:coliform ratios of 10^4:1, by conducting impedance determinations using a highly selective faecal coliform broth and incubating samples at 44.5 °C.

Martins and Selby (1980) developed a novel medium that selected for the growth of coliforms over that of other bacteria present in the sample. The medium consisted of tryptone (20 g), lactose (5 g), L-asparagine (1 g), Triton X-100 (4 ml), sodium dihydrogen phosphate (7 g), K_2SO_3 (0.35 g), novobiocin (3 mg), and distilled water (1000 ml). The asparagine and phosphates preferentially enhance the impedimetric responses of Gram-negative bacteria and inhibit the impedimetric responses of Gram-positive bacteria. The K_2SO_3 and novobiocin, at pH 6.3, are inhibitory to Proteus, Pseudomonas, and other non-coliform organisms.

EC broth was shown to be a suitable medium for enumerating coliforms (Rowley et al., 1979). These researchers observed that, in samples containing Enterobacter aerogenes, Streptococcus faecalis, and E. coli, only the E. coli produced detectable impedance changes in the medium when incubated in EC broth at 45.5 °C.

Strauss et al. (1984) described a procedure for enumerating coliforms using m-Endo broth. Temperatures of 40–42 °C produced a good compromise between rapid reaction rates and thermal inactivation. Thermal inactivation refers to an inability to recover bacteria as a result of heat stress at temperatures above 42 °C. Hence, using a selective medium and temperature of incubation, specific species of bacteria may be enumerated from mixed samples by creating optimal conditions for their multiplication while inhibiting the growth of other competing microflora.

In another study, Firstenberg-Eden and Klein (1983) used CM broth, lauryl tryptose broth (LTB), MacConkey (MAC), brilliant green bile broth (BGB), and EC broth (EC). CM broth was shown to have the highest slopes and an average maximum change value more than twice as large as the maxima for the other media. Detection time was delayed in MAC, BGB, and EC compared with CM and LTB (Firstenberg-Eden and Klein, 1983). They concluded that CM provided better impedance signals than conventional media and was more selective than LTB and violet red bile agar (Firstenberg-Eden et al., 1984).

If a particular species of bacteria can be identified that increases on broiler carcasses when carcasses are temperature-abused and which do not increase on carcasses if the carcasses are held properly, this species may be used as an 'indicator' of temperature abuse instead of measuring the growth of all mesophilic bacteria in general. This may provide a more accurate and sensitive method for determination of temperature abuse.

References

Alford, J.A. and Elliott, L.E. (1960) Lipolytic activity of microorganisms at low and intermediate temperatures. I. Action of *Pseudomonas fluorescens* on lard. *Food Res.*, **25**, 296–303.

Allison, J.B., Anderson, J.A. and Cole, W.H. (1938) The method of electrical conductivity in studies on bacterial metabolism. *J. Bacteriol.*, **36**, 571–86.

Andrews, W.H., Poelma, P.L. and Wilson, C.R. (1984) Isolation and identification of *Salmonella* species, in *Bacteriological Analytical Manual*, 6th edn, Association of Official Analytical Chemists, Arlington, Virginia, pp. 7.01–7.17.

Anonymous (1988) Poultry processor predicts product perishability. *Prepared Foods*, **157**, 118.

Ayres, J.C. (1951) Some bacterial aspects of spoilage of self-service meats. *Iowa St. J. Sci.*, **26**, 31–48.

Ayres, J.C. (1960) Temperature relationships and some other characteristics of the microbial flora developing on refrigerated beef. *Food Res.*, **25**, 1–18.

Ayres, J.C., Mundt, J.O. and Sandline, W.E. (1980) *Microbiology of Foods*. W.H. Freeman, San Francisco, California, p. 55.

Ayres, J.C., Ogilvy, W.S. and Stewart, G.F. (1950) Post mortem changes in stored meats. I. Micro-organisms associated with development of slime on eviscerated cut-up poultry. *Food Technol.*, **4**, 199–205.

Baker, R.C., Naylor, H.B., Pfund, M.C., Einset, E. and Staempfli, W. (1956) Keeping quality of ready-to-cook and dressed poultry. *Poultry Sci.*, **35**, 398–406.

Barnes, E.M. (1960) Bacteriological problems in broiler preparations and storage. *R. Soc. Health J.*, **80**, 145–8.

Barnes, E.M. (1976) Microbiological problems of poultry at refrigerator temperatures – A review. *J. Sci. Food Agric.*, **27**, 777–82.

Barnes, E.M. and Impey, C.S. (1968) Psychrophilic spoilage bacteria of poultry. *J. Appl. Bacteriol.*, **31**, 97–107.

Barnes, E.M. and Thornley, M.J. (1966) The spoilage flora of eviscerated chickens stored at different temperatures. *J. Food Technol.*, **1**, 113–19.

Bedford, R.H. (1933) Marine bacteria of the Northern Pacific Ocean. The temperature range of growth. *Contr. Can. Biol. Fish.*, **8**, 433–8.

Bergey, D.H. (1930) *Bergey's Manual of Determinative Bacteriology*, 3rd edn, Williams and Wilkins, Baltimore, Maryland.

Berry, J.A. and Magoon, C.A. (1934) Growth of microorganisms at and below 0 °C. *Phytopathology*, **24**, 780–96.

Breed, R.S., Murray, E.G.D. and Smith, N.R. (1948) *Bergey's Manual of Determinative Bacteriology*, 6th edn, Williams and Wilkins, Baltimore, Maryland.

Breed, R.S., Murray, E.G.D. and Smith, N.R. (1957) *Bergey's Manual of Determinative Bacteriology*, 7th edn, Williams and Wilkins, Baltimore, Maryland.

Brown, A.D. (1957) Some general properties of a psychrophilic pseudomonad: the effects of temperature on some of these properties and the utilization of glucose by this organism and *Pseudomonas aeruginosa. J. Gen. Microbiol.*, **17**, 640–8.

Brown, A.D. and Weidemann, J.F. (1958) The taxonomy of the psychrophilic meat-spoilage bacteria: a reassessment. *J. Appl. Bacteriol.*, **21**, 11–17.

Buchanan, R.L., Schultz, F.J. and Klawitter, L.A. (1991) Effectiveness of various indicators for detection of temperature abuse in refrigerated meat, poultry, and seafood. Presented at Institute of Food Technologies, Abstract 135.

Burdon-Sanderson, J.S. (1871) The origin and distribution of microzymes (bacteria) in water, and the circumstances which determine their existence in the tissues and liquids of the living body. *Quart. J. Microbiol. Sci. n.s.*, **11**, 323–52.

Busta, F.F., Peterson, E.H., Adams, D.M. and Johnson, M.G. (1984) Colony count methods, in *Compendium of Methods for the Microbiological Examination of Foods*, (eds C. Vanderzant and D.F. Splittstoesser). American Public Health Association, Washington, DC, pp. 62–76.

Cady, P., Dufour, S.W., Shaw, J. and Kraeger, S.J. (1978) Electrical impedance measurements: rapid method for detecting and monitoring microorganisms. *J. Clin. Microbiol.*, **7**, 265.

Chistyakov, F.M. and Noskova, G. (1955) The adaptations of micro-organisms to low temperatures. Proceedings of the 9th International Congress on Refrigeration, Paris, France, 1955, **2**, 4.230–4.235.

Cousins, D.L. and Marlatt, F. (1989) The use of conductance microbiology to monitor Enterobacteriaceae levels. *Dairy Food Environ. Sanit.*, **9**, 599.

Daud, H.B., McMeekin, T.A. and Olley, J. (1978) Temperature function integration and the development and metabolism of poultry spoilage bacteria. *Appl. Environ. Microbiol.*, **36**, 650–4.

Eklund, M.W. (1962) Biosynthetic responses of poultry meat organisms under stress. Ph.D. thesis, Purdue University, Indiana.

Elliott, R.P. and Michener, H.D. (1960) Review of the microbiology of frozen foods, in *Conference on Frozen Food Quality*. US Department of Agriculture, ARS-74-21, pp. 40–61.

Elliott, R.P. and Michener, H.D. (1961) Microbiological standards and handling codes for chilled and frozen foods. A review. *Appl. Microbiol.*, **9**, 452–68.

Elliott, R.P. and Michener, H.D. (1965) Factors affecting the growth of psychrophilic microorganisms in foods, a review. *Technical Bulletin No. 1320*, Agricultural Research Service, US Department of Agriculture.

Elliott, R.P. and Straka, R.P. (1964) Rate of microbial deterioration of chicken meat at 2 °C after freezing and thawing. *Poultry Sci.*, **43**, 81–6.

Empey, W.A. and Scott, W.J. (1939) Investigations on chilled beef. I. Microbial contamination acquired in the meatworks. *Commonwealth Aust. Coun. Sci. Ind. Res. Bull.*, 126.

Empey, W.A. and Vickery, J.R. (1933) The use of carbon dioxide in the storage of chilled beef. *Aust. Coun. Sci. Ind. Res. J.*, **6**, 233–43.

Firstenberg-Eden, R. (1983) Rapid estimation of the number of microorganisms in raw meat by impedance measurement. *Food Technol.*, **37**, 64–70.

Firstenberg-Eden, R. (1985) Electrical impedance method for determining microbial quality of foods, in *Rapid Methods and Automation in Microbiology and Immunology*, (ed. K.O. Habermehl), Springer-Verlag, Berlin, p. 679–87.

Firstenberg-Eden, R. and Eden, G. (1984) *Impedance Microbiology*. Research Studies Press, Letchworth, Hertfordshire, England, pp. 48–9.

Firstenberg-Eden, R. and Klein, C.S. (1983) Evaluation of a rapid impedimetric procedure for the quantitative estimation of coliforms. *J. Food Sci.*, **48**, 1307–11.

Firstenberg-Eden, R. and Tricarico, M.K. (1983) Impedimetric determination of total, mesophilic and psychrotrophic counts in raw milk. *J. Food Sci.*, **48**, 1750–4.

Firstenberg-Eden, R., Van Sise, M.L., Zindulis, J. and Kahn, P. (1984) Impedimetric estimation of coliforms in dairy products. *J. Food Sci.*, **49**, 1449–52.

Fischer, B. (1888) Bakterienwachsthum bei 0 °C. sowie über das Photographiren von Kulturen leuchtender Bakterien in ihrem eigenen Lichte. *Cent. Bakt.*, **4**, 89–92.

Forster, J. (1887) Ueber einige Eigenschaften leuchtender Bakterien. *Cent. Bakt.*, **2**, 337–40.

Gibson, D.M. (1985) Predicting the shelf-life of packaged fish from conductance measurements. *J. Appl. Bacteriol.*, **58**, 465–70.

Gibson, D.M., Ogden, I.D. and Hobbs, G. (1984) Estimation of the bacteriological quality of fish by automated conductance measurements. *Int. J. Food Microbiol.*, **1**, 127–34.

Gilliland, S.E., Michener, H.D. and Kraft, A.A. (1984) Psychrotrophic microorganisms, in *Compendium of Methods for the Microbiological Examination of Foods*. American Public Health Association, Washington, DC, p. 136.

Greene, V.W. and Jezeski, J.J. (1954) Influence of temperature on the development of several psychrophilic bacteria of dairy origin. *Appl. Microbiol.*, **2**, 110–17.

Hadley, D., Kraeger, S.J., Dufour, S.W. and Cady, P. (1977) Rapid detection of microbial contamination in frozen vegetables by automated impedance measurements. *Appl. Environ. Microbiol.*, **34**, 14–17.

Haines, R.B. (1993) The bacterial flora developing on stored lean meat, especially with regard to 'slimy' meat. *J. Hyg.*, **33**, 175–82.

Haines, R.B. (1937) Microbiology in the preservation of animal tissues. Department of Science Industrial Research Food Invest. Board, Special Report No. 45, HMSO, London, 269 pp.

Halleck, F.E., Ball, C.O. and Stier, E.F. (1957) Factors affecting quality of prepackaged meat. IV. Microbiological studies. A. Cultural studies on bacterial flora of fresh meat; classification by genera. *Food Technol.*, **12**, 197–203.

Hartsell, S.E. (1951) The growth initiation of bacteria in defrosted eggs. *Food Res.*, **16**, 97–106.

Hess, E. (1934) Cultural characteristics of marine bacteria in relation to low temperatures and freezing. *Can. Biol. Fisheries Contrib.*, **8**, 459–74.

Ingraham, J.L. (1958) Growth of psychrophilic bacteria. *J. Bacteriol.*, **76**, 75–80.

Ingraham, J.L. and Bailey, G.F. (1959) Comparative study of effect of temperature on metabolism of psychrophilic and mesophilic bacteria. *J. Bacteriol.*, **77**, 609–13.

Kirsch, R.H., Berry, F.E., Baldwin, G.L. and Foster, E.M. (1952) The bacteriology of refrigerated ground beef. *Food Res.*, **17**, 495–503.

Lockhead, A.G. and Landerkin, G.B. (1935) Bacterial studies of dressed poultry. I. Preliminary investigations of bacterial action at chill temperatures. *Sci. Agric.*, **15**, 765–70.

MacDonell, M.T. and Colwell, R.R. (1985) Phylogeny of the Vibrionaceae and recommendation for two new genera. *Listonella* and *Shewanella*. *System Appl. Microbiol.*, **6**, 171–82.

MacFadyen, A. and Rowland, S. (1902) On the suspension of life at low temperatures. *Ann. Bot.*, **16**, 589–90.

Martins, S.B. and Selby, M.J. (1980) Evaluation of a rapid method for the quantitative estimation of coliforms in meat by impedimetric procedures. *Appl. Environ. Microbiol.*, **39**, 518–24.

Mazur, P. (1984) Freezing of living cells: mechanisms and implications. *Am. J. Physiol.*, **247**, C125–42.

Mischak, R.P., Shaw, J. and Cady, P. (1976) Specific detection of fecal coli by impedance measurement. *Abst. Ann. Meet. Am. Soc. Microbiol.*, **P3**, 187.

Muller, M. (1903) Ueber das Wachstum und die Lebenstatigkeit von Bakterien sowie den Ablauf fermentativer Prozesse bei niederer Temperaturen unter spezieller Berucksichtigung des Fleisches als Nahrungsmittel. *Arch. Hyg.*, **47**, 127–93.

Naden, K.D. and Jackson, Jr., G.A. (1953) Some economic aspects of retailing chicken meat. *California Agric. Expt. Stn. Bull.*, **734**, 107–8.

Nashif, S.A. and Nelson, F.E. (1953) The lipase of *Pseudomonas fragi*. II. Factors affecting lipase production. *J. Dairy Sci.*, **36**, 471–80.

Newell, G.W., Gwin, J.M. and Jull, M.A. (1948) The effect of certain holding conditions on the quality of dressed poultry. *Poultry Sci.*, **27**, 251–6.

Ogden, I.D. (1986) Use of conductance methods to predict bacterial counts in fish. *J. Appl. Bacteriol.*, **61**, 263–8.

Olsen, R.H. and Jezeski, J.J. (1963) Some effects of carbon source, aeration, and temperature on growth of a psychrophilic strain of *Pseudomonas fluorescens*. *J. Bacteriol.*, **86**, 429–33.

Olson, J.C., Jr. (1947) Psychrophiles, mesophiles, thermophiles and thermodurics. What are we talking about? *Milk Plant Monthly*, **36**, 32–6.

Penfold, W.J. (1914) On the nature of bacterial lag. *J. Hyg.*, **14**, 215–41.

Pennington, M.E. (1910) Studies of poultry from the farm to the consumer. *US Dept. Agric. Bur. Chem. Cir. 64*, Government printing office, Washington.

Peterson, A.C. and Gunderson, M.F. (1960) Some characteristics of proteolytic enzymes from *Pseudomonas fluorescens. Appl. Microbiol.*, **8**, 98–104.

Pooni, G.S. and Mead, G.C. (1984) Prospective use of temperature function integration for predicting the shelf-life of non-frozen poultry-meat products. *Food Microbiol.*, **1**, 67–78.

Regez, P., Gallo, L., Schmitt, R.E. and Schmidt-Lorenz, W. (1988) Microbial spoilage of refrigerated fresh broilers. III. Effect of storage temperature on the microbial association of poultry carcasses. *Lebensm.-Wiss. Technol.*, **21**, 229–33.

Rowley, D.B., Vandemark, P., Johnson, D. and Shattuck, E. (1979) Resuscitation of stressed fecal coliforms and their subsequent detection by radiometric and impedance techniques. *J.Food Protect.*, **42**, 335–41.

Rubentshick, L. (1925) Ueber die Lebenstatigkeit der Urobakterien bei einer Temperatur unter 0 °C. *Centr. Bakt.*, **64**, 166–74.

Russell, S.M., Fletcher, D.L. and Cox, N.A. (1992a) A model for determining differential growth at 18 and 42 °C of bacteria removed from broiler chicken carcasses. *J. Food Protect.*, **55**, 167–70.

Russell, S.M., Fletcher, D.L. and Cox, N.A. (1992b) A rapid method for the determination of temperature abuse of fresh broiler chicken. *Poultry Sci.*, **71**, 1391–5.

Sair, L. and Cook, W.H. (1938) Effect of precooling and rate of freezing on the quality of dressed poultry. *Can. J. Res.*, **D16**, 139–52.

Schefferle, H.E. (1965) The microbiology of built up poultry litter. *J. Appl. Bacteriol.*, **28**, 403–11.

Senyk, G.F., Goodall, C., Kozlowski, S.M. and Bandler, D.K. (1988) Selection of tests for monitoring the bacteriological quality of refrigerated raw milk samples. *J. Dairy Sci.*, **71**, 613–19.

Silverman, M.P. and Munoz, E.F. (1978) Enumeration of fecal coliforms by electrical impedance. *Abst. Ann. Meet. Am. Soc. Microbiol.*, **Q69**, 206.

Silverman, M.P. and Munoz, E.F. (1979) Automated electrical impedance technique for rapid enumeration of fecal coliforms in effluents from sewage treatment plants. *Appl. Environ. Microbiol.*, **37**, 521–6.

Spencer, J.V., Ziegler, F. and Stadelman, W.J. (1954) Recent studies of factors affecting the shelf-life of chicken meat. *Wash. Agr. Exp. Stat. Inst. Agr. Sci. Stat. Circ. 254.*

Spencer, J.V., Sauter, E.A. and Stadelman, W.J. (1955) Shelf life of frozen poultry meat after thawing. *Poultry Sci.*, **34**, 1222–3.

Spencer, J.V., Sauter, E.A. and Stadelman, W.J. (1961) Effect of freezing, thawing, and storing broilers on spoilage, flavor, and bone darkening. *Poultry Sci.*, **40**, 918–20.

Stewart, G.N. (1899) The changes produced by the growth of bacteria in the molecular concentration and electrical conductivity of culture media. *J. Exp. Med.*, **4**, 235–47.

Straka, R.P. and Stokes, J.L. (1959) Metabolic injury to bacteria at low temperatures. *J. Bacteriol.*, **78**, 181–5.

Strauss, W.M., Malaney, G.W. and Tanner, R.D. (1984) The impedance method for monitoring total coliforms in wastewaters. *Folia Microbiol.*, **29**, 162–9.

Sultzer, B.M. (1961) Oxidative activity of psychrophilic and mesophilic bacteria on saturated fatty acids. *J. Bacteriol.*, **82**, 492–7.

Tenpenny, J.R., Tanner, R.D. and Malaney, G.W. (1984) The impedance method for monitoring total coliforms in wastewaters. *Folia Microbiol.*, **29**, 170–80.

Thornley, M.J. (1960) Computation of similarities between strains of *Pseudomonas* and *Achromobacter* isolated from chicken meat. *J. Appl. Bacteriol.*, **23**, 395–7.

Ur, A. and Brown, D.F.J. (1975) Impedance monitoring of bacterial activity. *J. Med. Microbiol.*, **8**, 19.

van Spreekins, K.J.A. and Stekelenburg, F.K. (1986) Rapid estimation of the bacteriological quality of fresh fish by impedance measurements. *Appl. Microbiol. Biotechnol.*, **24**, 95–6.

Wells, F.E., Hartsell, S.E. and Stadelman, W.J. (1963) Growth of psychrophiles. I. Lipid changes in relation to growth-temperature reductions. *J. Food Sci.*, **28**, 140–4.

Wolin, E.F., Evans, J.B. and Niven, C.F. (1957) The microbiology of fresh and irradiated beef. *Food Res.*, **22**, 268–88.

Zobell, C.E. and Conn, J.E. (1940) Studies on the thermal sensitivity of marine bacteria. *J. Bacteriol.*, **40**, 223–38.

9 Chemical changes in stored meat

G.-J.E. NYCHAS, E.H. DROSINOS AND R.G. BOARD

9.1 Introduction

In 1971, The Society for Applied Bacteriology organized a symposium on Microbial Changes in Foods. Among the contributions, those by Ingram (1971) and Ingram and Dainty (1971) laid a conceptual foundation for our current understanding of chemical changes in meat. In the following decades review chapters (e.g. Dainty *et al.*, 1983) and articles (e.g. Dainty and Mackey, 1992) on the spoilage phenomena and chemical changes in meat have appeared in the literature. The rationale of why physico-chemical and chemical changes have attracted the interest of researchers is based on their efforts to correlate these changes with the loss of quality of fresh meat. Are observed changes objective indices that reflect the quality of meat? An answer will be presented here.

9.1.1 The meat ecosystem

The development of a microbial association in a meat ecosystem depends not only on the imposed environmental conditions but also on microbial competition (Mossel and Ingram, 1955; Fredickson and Stephanopoulos, 1981). As a consequence the observed chemical changes are essentially an expression of the development of a food ecosystem. Two distinct situations are possible. One where competition between facultatively anaerobic Gram-positive floras determines the changes in an ecosystem and the other where competition is between aerobic Gram-negative floras. The physiological attributes of the organisms under the ecological determinants result in the ecophysiological ones that are examined in detail in the following sections.

9.2 The status of substrates

While the microbiological changes are well established (see Chapters 1, 2 and 3), the physico-chemical ones accompanying the growth of bacteria on meat during storage, either aerobically or anaerobically, have not been

Table 9.1 Nutritional properties of lamb chop meat. Based on Passmore and Eastwood (1986)

Component		
Moisture	49	%
Energy	1558	kJ
	377	kcal
Carbohydrate	0	g
Protein	15	g
Fat	35	g
Calcium	7	mg
Iron	1.2	mg
Retinol equivalent	trace	μg
Ascorbic acid	0	mg
Thiamin	0.09	mg
Riboflavin	0.16	mg
Nicotinic acid equivalent	7.1	mg
Vitamin B12	1	μg

studied in equal detail (Jay, 1986b; Dainty and Mackey, 1992). The changes which occur during spoilage take place in the aqueous phase of meat (Nychas *et al.*, 1994). This phase contains glucose, lactic acid, certain amino acids, nucleotides, urea and water soluble proteins which are catabolized by almost all members of the meat microflora (Gill, 1976; Nychas *et al.*, 1988; Drosinos, 1994). For this reason meat is considered to be a very rich nutritive substratum for the growth of organisms (Table 9.1). Although the

Table 9.2 Chemical composition of typical adult mammalian muscle after *rigor mortis* and before commencement of decomposition *post mortem*. Based on Lawrie (1985) and Gill (1986)

Components	Wet weight (%)	
1. Water	75.0	
2. Protein	19.0	
(a) Myofibrillar		11.5
(b) Sarcoplasmic		5.5
(c) Connective tissue and organelle		2.0
3. Lipid	2.5	
4. Carbohydrate and lactic acid	1.2	
lactic acid		0.90
glucose-6-phosphate		0.05
glycogen		0.10
glucose, traces of other glycolytic intermediates		0.15
5. Miscellaneous soluble nonprotein substances	2.3	
(a) Nitrogenous	1.65	
creatine		0.55
inosine		0.30
monophosphate ATP, AMP		0.10
amino acids		0.40
carnosine, anserine		0.30
(b) Inorganic (dipeptides)	0.65	
6. Vitamins	Traces	

Table 9.3 Substrates used for growth by major meat spoilage micro-organisms[a]

Micro-organism	Substrates used for growth[b]	
	Aerobic	Anaerobic[c]
Pseudomonas spp.	Glucose[1], glucose-6-P, D, L-lactic acid[2], pyruvate, gluconate, gluconate 6-P, amino acids[2], creatine, creatinine, citrate, aspartate, glutamate	Glucose[1], actic acid[1], pyruvate, gluconate, amino acids (glutamate)
Acinetobacter/ Moraxella	Amino acids[1], lactic acid[2], glucose[1], amino acids[1,2]	Glucose[1], amino acids[1,2]
Shewanella putrefaciens	Glucose, lactic acid, pyruvate, gluconate, propionate, ethanol, acetate, amino acids (serine)	Formate
Brochothrix thermosphacta	Glucose[1], amino acids[2] (glutamate, L-valine, L-leucine, *iso*-leucine), ribose, glycerol	Glucose[1]
Enterobacter spp.	Glucose[1], glucose-6-P[2], amino acids[3] (lysine, arginine, threonine), lactic acid[4]	Glucose[1], glucose-6-P[2], amino acids[3]
Lactobacillus spp.	Glucose[1]	Glucose[1], lactic acid[2], amino acids[2]

[a] Adapted from Hitchener *et al.* (1979), Dainty and Hibbard (1980), McMeekin (1982), Molin (1985), Dainty *et al.* (1985), Nychas *et al.* (1988), Borch and Agerhem (1992), Nychas (1994), Kakouri and Nychas (1994), Drosinos and Board (1994, 1995a, b), Scott and Nealson (1994).
[b] The number in superscript indicates the order of utilization of this substrate according to Gill (1986).
[c] Under oxygen limitation and/or carbon dioxide inhibition.

concentrations of low molecular weight compounds, especially carbohydrates such as glycogen and the intermediate glycolytic products – glucose, glucose-6-phosphate, lactate etc. (Table 9.2) – are low in comparison with those of protein and lipids the former are sufficient to support a massive development of the microcosm on the meat (Gill, 1986; Nychas *et al.*, 1988). Indeed several studies have shown that bacteria grow on meat at the expense of one or more of the low molecular weight soluble components (Gill, 1976; Drosinos, 1994). The order in which these substrates are attacked by the various groups of spoilage bacteria, under aerobic or anaerobic conditions, is shown in Table 9.3. Under aerobic conditions none of the bacteria ceased growth because of substrate exhaustion at the meat surface and oxygen availability was suggested to be the limiting factor. During growth of mixed cultures it was demonstrated that there were no interactions until one organism had attained its maximum cell density. It was suggested that the pseudomonads predominated because of their faster growth rates and their greater affinity for oxygen, and as a consequence greater catabolism of glucose, over the other meat bacteria (Gill and Newton, 1977;

Gill and Molin, 1991). Molin (1985) and Drosinos and Board (unpublished results) showed that this could not be the reason. Their studies showed that *Ps. fragi* grown under oxygen limitation can use alternative carbon sources but the order of utilization remains the same with the exception of glucose/lactate. They concluded that there is no convincing evidence that specific affinity for glucose is the determinant which results in the preponderance of *Ps. fragi* on chilled meat. Indeed, in recent studies (Drosinos and Board, unpublished results) it was shown that the glucose uptake by *Ps. fragi* makes only a minor contribution to its domination over other pseudomonads on meat. Indeed they found that the transport capacity of *Ps. fragi* was lower than that of the other pseudomonads tested (lower V_{max}). The strains of *Ps. fluorescens* which were unable to oxidize D-glucose appeared to have a high transport capacity. *Pseudomonas lundensis* had an intermediate capacity for transport equal to that of the fluorescent pseudomonad that oxidized D-glucose. When values of V_{max}/K_T, corresponding to the specific affinity of a micro-organism for a substrate (Button, 1983) are considered, *Ps. fragi* and the fluorescent strain unable to oxidize D-glucose gave the higher values.

The observation that a strain of *Ps. fluorescens* with the highest apparent V_{max} when correlated with that of its inability to form gluconate, provides evidence that D-glucose oxidation is not a prerequisite for the uptake mechanism. Indeed, Eisenberg *et al.* (1974), who studied *Ps. fluorescens*, demonstrated that glucose oxidation extracellularly was not an obligatory first step for glucose uptake. *Pseudomonas aeruginosa* possesses two distinct and inducible transport systems for the uptake of glucose, one with a low and another with a high affinity (Midgley and Dawes, 1973). In the former, glucose is oxidized extracellularly prior to uptake and in the latter, a phosphorylative pathway, the substrate is transported directly into the cytoplasm. Baldwin and Henderson (1989) stated that: 'The mechanisms used by cells to take up sugars reflect the sugar concentrations in their normal environments' and pointed out the homologies between sugar transporters of eukaryotes and prokaryotes.

Under anaerobic conditions, Newton and Gill (1978) found that none of the bacteria studied utilized more than two of the compounds listed in Table 9.3. They reported that the affinity among micro-organisms for the common substrate, glucose, occurred in the sequence *Enterobacter* > *Br. thermosphacta* > *Lactobacillus*. However the ability of the last named bacterium to utilize other low molecular compounds (e.g. lactate, arginine – Newton and Gill, 1978; Drosinos and Board, 1995b,c) or to produce antimicrobial substances (Stiles, 1996) overcomes its deficiency in respect of glucose affinity. Drosinos and Board (1995b) showed that in a mixed culture, *Br. thermosphacta* grew faster than *Ps. fragi* under both aerobic or modified atmosphere conditions. This was attributed to acetic acid produced by *Br. thermosphacta* inhibiting *Ps. fragi*.

In both aerobic and anaerobic conditions, the key role of glucose in meat and meat products has been well documented (Nychas *et al.*, 1988). There is criticism, however, that this role has been over-emphasized (Molin, 1985). Nychas *et al.* (1992) and Drosinos (1994) correlated good microbiological quality (low bacterial numbers) of retail beef and lamb stored under different conditions with glucose concentration. Boers *et al.* (1994) found that there was a relationship between glucose depletion in wild boar meat and the onset of spoilage during storage under vacuum. They observed that the glucose concentration had become very low with the first signs of spoilage. It has been concluded also that glucose limitation caused a switch from a saccharolytic to an amino acid degrading metabolism in at least some bacterial species (Borch *et al.*, 1991). Glucose has been found to be an important intrinsic factor, among others, for describing or predicting the degree of spoilage (Borch, personal communication; Seymour *et al.*, 1994). For this reason the preferential utilization of low molecular weight compounds, in particular glucose, has been proposed as a potential and important 'hurdle' for the keeping quality of meat which may be used to extend the shelf-life of the product. Indeed the addition of carbohydrates, particularly glucose, has been suggested by Shelef (1977) and Barua and Shelef (1980) as a factor that can be used to delay spoilage particularly in dark, firm, dry (DFD) meat (pH > 6.0). This is due to the fact that the glucose content affects not only the cell density attained at the onset of spoilage (Gill, 1986; Nychas *et al.*, 1988; Drosinos and Board, 1995a) but also the metabolic products produced by the flora (Dainty and Hibbard, 1980, 1983; Nychas and Arkoudelos, 1990; Lambropoulou *et al.*, 1996). Meat with DFD characteristics spoils more rapidly than meat of normal pH (pH 5.5–5.8). It is well known that the accumulation of L-lactic acid, as a product of the glycolysis in the tissues, reflects the final pH of meat and affects also the selection of the microbial flora (Gill, 1986; Nychas *et al.*, 1988). Indeed the growth of *Br. thermosphacta* and *Sh. putrefaciens* is influenced by the pH value of meat. *Brochothrix thermosphacta* needs a pH > 5.8 while *Sh. putrefaciens* fails to grow on meat with pH less than 6.0. Such values occur with DFD meat which, if vacuum packed, results in the growth of the latter organism and, through hydrogen sulfide production from cysteine and serine, and green discoloration of the meat by sulfmyoglobin formation.

Lactate is another low molecular weight component utilized by the meat microflora under both aerobic and anaerobic conditions (Gill and Newton, 1977; Molin, 1985; Drosinos, 1994). The preferential use of glucose to lactate and amino acids is an observation that has attracted considerable attention by researchers of this field (Gill, 1976; Farber and Idziak, 1982; Nychas, 1984, 1994; Drosinos, 1994; Lasta *et al.*, 1995; Lambropoulou *et al.*, 1996). Indeed so far all the available data suggest a sequential use, glucose > lactate. Molin (1985) observed however that lactate was being utilized by *Ps. fragi* in a broth culture in the presence of glucose under both aerobic and oxygen limiting conditions. Similar results were reported with meat samples (broth, beef strip loins, poultry), naturally contaminated or inoculated with *Lactobacillus* sp., and stored aerobically, with oxygen limitation or under anaerobic conditions (Nychas and Arkoudelos, 1991; Borch and Agerhem, 1992; Drosinos and Board, 1994; Kakouri and Nychas, 1994; Nychas *et al.*, 1994). The decrease of lactate concentration (Table 9.4) followed glucose utilization (Table 9.5) in meat samples stored aerobically or under vacuum-modified atmosphere packaging (VP/MAP) conditions. It was evident also that the rate of glucose and lactate utilization in VP/MAP samples was less than that in samples stored aerobically (Tables 9.4 and 9.5).

Similar conclusions can be drawn for amino acids. Many workers have used the changes of free amino acids in meat in attempts to determine whether protein degradation has occurred. One would expect that the content of amino acids would remain constant until shortly before the onset of spoilage – due to glucose exhaustion (Gill, 1986; Dainty, 1996), or when the bacterial numbers reach about $10^7–10^8$ cfu/g (Dainty *et al.*, 1975) – and then presumably decline before again rising sharply when proteolysis begins. This scenario contrasts with the results reported by Newton and

Table 9.4 Changes[a] in L,D-lactic acid content in beef, minced lamb and poultry during storage at 3°C under aerobic or in different modified atmospheres

Meat	Atmosphere	\multicolumn{5}{c}{Days of storage}				
		0	2	4	7	9
Beef	Air[b]	824.0	953.4	684.3	639.4	215.3
	O_2/CO_2[c]	824.0	752.8	859.2	647.1	798.3
	O_2/CO_2[d]	4.3	9.0	7.0	56.4	100.0
		0	2	3	4	5
Minced lamb	Air	464.0	521.7	429.7	427.6	275.9
	O_2/CO_2	536.6[c]	–[f]	–	–	503.0
	O_2/CO_2	4.0[c]	–	–	–	83.9
		0	3	7	9	11
Poultry	Air	260	190	160	110	70
	CO_2	260	260	230	225	230
	CO_2	0	–	138	225	249

[a] mg/100 g.
[b] Data obtained during a study on the effect of glucose supplementation in the minced beef ecosystem.
[c] Modified atmosphere was composed of oxygen and carbon dioxide (80/20% v/v). Values of L-lactic acid.
[d] Values of the concentration of the D- lactic acid. This stereo-isomer was not detected under aerobic conditions.
[e] Data obtained during a survey of minced lamb packaged in a high oxygen modified atmosphere available in major supermarkets in the UK. Samples were analyzed on the purchase and expiry date printed on the product label (Drosinos and Board, 1995a).
[f] –, no data.

Table 9.5 Changes (mg/100 g) in glucose content in beef, pork and chicken during storage at 3 °C under aerobic or different modified atmospheres. Based on Kakouri and Nychas (1994), Nychas *et al.* (1994)

Beef:

Pack	Days of storage				
	0	2	4	7	9
Aerobic	78.0	66.2	40.6	6.2	6.3
20% CO_2/80% O_2	78.0	56.7	64.9	17.2	14.2
100% CO_2	78.0	75.0	80.0	72.0	66.0

Pork:

Pack	Days of storage				
	1	2	4	8	10	12
Aerobic	110.0	85.0	45.0	25.0	10.0	0.0
100% CO_2	110.0	120.0	100.0	90.0	85.0	45.0
100% N_2	110.0	122.0	86.0	56.0	34.0	7.0

Chicken:

Pack	Days of storage				
	1	3	7	10	14
Vacuum pack	22.2	20.8	10.4	15	9.3
100% CO_2	22.2	17.8	16.0	18.0	17.6
100% N_2	22.2	11.6	15.5	10.0	12.5

Rigg (1979), Nychas and Arkoudelos (1990), Schmitt and Schmidt-Lorenz (1992b) and Nychas and Tassou (1997) for beef, pork, poultry skin and poultry fillets respectively. These workers reported that under aerobic conditions the sum of the free amino acids and the water soluble proteins increased during storage and it corresponded well with colony counts. Nychas and Arkoudelos (1991) and Nychas and Tassou (1997) showed that this increase occurred in meat samples with a relatively high concentration of glucose. Moreover the rate of increase of free amino acids under aerobic was higher than that occurring under modified atmosphere conditions. These observations could be of great importance commercially since spoilage is most frequently associated only with post-glucose utilization of amino acids by pseudomonads (Gill, 1986).

Indeed according to Gill (1976), as long as low molecular weight components – especially glucose – are available, meat proteolysis is inhibited. This view is not fully supported by recent findings from Schmitt and

Table 9.6 Changes (mg/100g) of glucose, L-lactate and water soluble proteins (WSP) in poultry fillets inoculated with *Pseudomonas fragi* and stored under VP, 100% CO_2 or aerobic conditions at 3 °C. Water soluble proteins I: estimated with the Coomassie blue method. Each number is the mean of two samples taken from different experiments. Each sample was analysed in duplicate (coefficient variation < 5%). Reproduced with permission from Nychas and Tassou (1997)

Pack		\multicolumn{4}{c}{Days of sampling}			
		0	4	8	12
Vacuum	Glucose	10.6	8.0	7.5	6.7
	L-lactate	396	nd	332	nd
	WSP	9.3	11.6	8.0	11.4
100% CO_2	Glucose	10.6	10.9	7.1	9.1
	L-lactate	396	432	291	343
	WSP	9.3	11.5	9.9	10.1
Aerobic	Glucose	10.6	6.6	6.2	4.5
	L-lactate	396	378	22	12
	WSP	9.3	10.5	12.3	23.7

Schmidt-Lorenz (1992a,b) and Nychas and Tassou (1997). For example, Nychas and Tassou (1997) reported that there were always significant amounts of glucose and lactate present (Table 9.6) when proteolysis was evident with changes of Coomassie blue compounds (nitrogen compounds with MW higher than >3000 Da) and the HPLC analysis of water soluble proteins. The amount of these compounds increased progressively in most samples stored under aerobic, vacuum and modified atmospheres. Furthermore the HPLC profile of water soluble proteins changed significantly (Table 9.7). Indeed it was evident that new hydrophilic and hydrophobic peaks appeared progressively or at the end of storage in all samples stored at both temperatures (Table 9.7). Moreover the final concentration of the peaks present initially varied significantly among all the samples tested at the end of storage. These changes, which were evident even during the earlier stages of storage regardless of microbial size (Schmitt and Schmidt-Lorenz, 1992a,b), could be attributed not only to the indigenous proteolytic meat enzymes (autolysis) but also to the microbial proteolytic activity. If only autolysis occurred during storage, a similar pattern of protein breakdown would be expected in all samples, irrespective of the manner of their storage or of the contribution of the various spoilage groups in the final composition of the microbial flora. This was not the case (Table 9.7). The concept of microbial proteinase involvement must take into account also factors controlling the regulation of such enzymes in proteolytic bacteria. Catabolite repression, feedback inhibition, end-product repression and induction operate in the synthesis of extracellular proteases (Venugopal, 1990). For example, in some bacteria control is *via* end product inhibition, synthesis being inhibited by specific amino acids. Others show inhibition by glucose in a manner superficially similar to catabolite repression of intracellular catabolic enzymes or induction by specific amino acids. In particular it is reported that glucose inhibited proteinase production by a milk isolate of *Ps. fluorescens* (Juffs, 1976). *Pseudomonas fragi*, an active spoiler of meat, does not produce extracellular protease when grown in medium devoid of amino acids and proteins, but produces the enzyme when grown in meat (Tarrant *et al.*, 1973). Fairburn and Law (1986) found that, in continuous culture, a proteinase was produced by *Ps. fluorescens* under carbon but not under nitrogen limiting conditions. This was taken as evidence that proteinase induction ensured that an energy rather than a nitrogen source was available to the organisms. In general it is accepted that (i) organisms produce very low basal levels of extracellular enzymes in the absence of an inducer, and (ii) that the regulation and extracellular production of proteinases are based on induction and end product and/or catabolite repression (Harder, 1979).

It is also a common observation that the production of proteolytic enzymes is delayed until the late logarithmic phase of growth even when conditions appear to be favourable for enzyme production at an earlier

Table 9.7 The contribution (area %) of trifluoroacetic acid soluble nitrogen compounds eluted from a 250×4 mm MZ-SIL 300 C_{18} 7 µm column ($\lambda = 215$ nm), found to be present initially and at the end of the storage of poultry fillets inoculated or not with *Pseudomonas fragi*, under aerobic (air), vacuum pack (VP), and modified atmosphere (CO_2) at 3°C and 10°C. Reproduced with permission from Nychas and Tassou (1997)

Peak	RT	Day 0	12 days at 3 °C			7 days at 10 °C		
			Air	VP	CO_2	Air	VP	CO_2
Inoculated samples:								
1	4.33	–	4.8	–	–	6.3	2.7	2.5
2	4.71	8.7	13.8	13.0	11.6	11.1	10.9	11.9
3	5.03	5.1	8.1	–	1.9	10.7	6.2	5.1
4	5.45	3.9	10.2	11.7	8.8	16.1	11.4	15.0
5	6.71	17.3	8.6	15.6	17.3	1.65	8.5	9.6
6	7.27	–	11.1	5.5	5.0	17.2	8.0	8.4
7	9.75	38.3	25.0	33.5	33.2	18.4	19.4	27.3
8	12.36	–	1.4	3.0	1.2	2.5	2.3	2.5
9	13.70	–	–	–	–	2.1	–	–
Uninoculated samples:								
1	4.33	–	–	–	–	2.9	2.2	0.2
2	4.71	8.7	8.0	6.5	9.8	5.9	8.1	7.1
3	5.03	5.1	2.0	3.0	–	10.3	2.0	2.1
4	5.45	3.9	4.8	5.5	4.5	8.7	9.5	9.4
5	6.71	17.3	19.1	17.1	29.6	9.0	19.0	21.6
6	7.27	–	5.2	5.4	–	5.5	–	–
7	9.75	38.3	43.5	49.5	41.0	42.9	45.9	44.1
8	12.36	–	0.8	3.3	–	1.4	–	2.3
9	13.70	–	–	–	–	2.1	–	–

–, peak was not present; RT, Retention Time of eluted peak.

stage of growth (Pollock, 1963; Glenn, 1976) although some protease synthesis occurs in many organisms even during exponential growth (Boethling, 1975; Venugopal, 1990).

Whether the amino acid content will ultimately increase or decrease as a result of microbial action will depend upon the composition of the bacterial flora. A number of workers have examined the effects of individual species of bacteria on muscle protein by comparing the free amino acids or protein composition of sterile or antibiotic treated meat with that of inoculated samples. For example Adamcic *et al.* (1970) using chicken skin found that individual species had either little effect on or decreased the amount of free amino acids in the substrate during the early stages of growth. Only those species capable of a high level of proteolytic activity produced a significant increase in free amino acids, and then only during the final stage of spoilage. Nonproteolytic species caused a decrease in free amino acids. Mixed cultures gave intermediate results, producing little changes in free amino acid content at any time. Amino acid analysis will give only the net result of amino acid consumption by microbes and production by all mechanisms and the balance of these processes may well alter in an unknown fashion during development of the flora.

The soluble sarcoplasmic proteins are probably the initial substrate for proteolytic attack (Hasegawa *et al.*, 1970a,b; Jay and Shelef, 1976). Bacteria appear to affect changes in solubility but protein breakdown by bacteria cannot be demonstrated from increases in the nonprotein nitrogen fraction until after prolonged storage. The proportion of sarcoplasmic proteins (water soluble proteins, consisting of several nutrients including amino acids and vitamins, form the ideal medium for growth of micro-organisms) decreased, whereas the proportion of myofibrial (salt-soluble) and stromae (insoluble) proteins tends to increase initially, probably because of solubility changes in some of the muscle proteins (Ockerman *et al.*, 1969; Borton *et al.*, 1970a,b).

The production of proteases by psychrotrophic bacteria (e.g. pseudomonads) increased as the temperature decreased from 30 to 0 °C (Peterson and Gunderson, 1960). In general Gram-negative bacteria in chill meat secrete aminopeptidases. These could be measured for a rapid estimation of bacteriological quality of meat (de Castro *et al.*, 1988). It needs to be noted however that although lactic acid bacteria are considered to be weakly proteolytic (Law and Kolstad, 1983) when compared with many other bacteria, such as *Pseudomonas* spp., their limited proteolytic activity can lead to their penetration into meat (Gill and Penney, 1977; Gupta and Nagamohini, 1992). The proteolytic bacteria may gain an ecological advantage through penetration because they have access to a new environment with a 'fresh' supply of nutrients which would not be available to the nonproteolytic bacteria. In conclusion, there is no doubt that many bacteria can secrete proteases (endoproteases, proteinases, aminopeptidases and carboxypeptidases) which can degrade sarcoplasmic and myofibrial proteins.

9.3   Chemical changes in aerobic ecosystem

Microbial associations developing on meat stored aerobically at chill temperatures are characterized by an oxidative metabolism. The Gram-negative bacteria that spoil meat are either aerobes or facultative anaerobes (see pp. 3–6). The aerobic Gram-negative bacteria are the common cause of spoilage of meat joints and broilers stored under aerobic conditions at 4 °C (Nychas, 1994; Davies, 1995). *Pseudomonas fragi, Ps. fluorescens* and *Ps. lundensis* were found to be the dominant species on meat (beef, lamb, pork and poultry) stored at chill temperatures (Nychas *et al.*, 1994; Chapter 1, pp. 10–14). *Brochothrix thermosphacta* and cold-tolerant Enterobacteriaceae (e.g. *Hafnia alvei, Serratia liquefaciens, Enterobacter agglomerans*) also occur on chilled meat stored aerobically (Nychas, 1984) but in terms of numbers they do not contribute to the microbial associations. The contribution of these last mentioned organisms to the chemical changes of meat is discussed below.

9.3.1   Chemical changes by Gram-negative bacteria

(a)   Pseudomonads. The key chemical changes associated with the metabolic attributes of pseudomonads have been studied extensively in broth and in model systems such as meat juice (Gill, 1976; Molin, 1985; Drosinos and Board, 1994). A synopsis of metabolic activities of pseudomonads studied in a meat juice is shown in Table 9.8. Drosinos and Board (1994) investigated the attributes of pseudomonads isolated from minced lamb in meat juice. They observed that D-glucose and L- and D-lactic acid were used sequentially. D-glucose was used preferentially to DL-lactate (Fig. 9.1). This diauxie can be observed in the meat ecosystem discussed by Nychas *et al.* (1988). The oxidization of glucose and glucose-6-phosphate *via* the extracellular pathway that caused a transient accumulation of D-gluconate and an increase in the concentration of 6-phosphogluconate was an important observation (Fig. 9.2). The transient appearance of a peak in the concentration of gluconate is a phenomenon reported repeatedly in the literature (Farber and Idziak, 1982; Nychas and Arkoudelos, 1991; Drosinos and Board, 1994). It is influenced by the concentration of glucose as well as by the storage conditions (Lambropoulou *et al.*, 1996). The delay in the production of gluconate in meat (beef, pork, poultry) stored in carbon dioxide was probably due to high pCO_2 or low pO_2 inhibiting the activity of the glucose-dehydrogenase of pseudomonads (Mitchell and Dawes, 1982; Nychas *et al.*, 1988; Drosinos, 1994). The increase in the concentration of D-

Table 9.8 Metabolic activity of pseudomonads in meat juice at 4°C. Adapted from Drosinos and Board (1994)

	Pseudomonas		
Substrate	fragi[a]	lundensis	fluorescens
D-glucose[b]	+	+	+
D-glucose 6-P[c]	+	+	−
D-gluconate[b]	+	+	+
D-gluconate-6-P[c]	+	+	−
L-lactic acid[d]	+	+	+
D-lactic acid[d]	+	+	+
Pyruvate[d]	+	+	+
Acetic acid[d]	+	nd	nd
Amino acids[e]	+	+	+
Creatine[f]	+	−	−
Creatinine[f]	+	−	−
Ammonia[f]	+	+	+

[a] +, the substrate was catabolized or formed during growth; −, neither catabolized nor formed; nd, no data.
[b] D-glucose and L- and D-lactic acid were used sequentially. D-glucose was used preferentially to DL-lactate. All strains but one were able to oxidize this substrate *via* the extracellular pathway and cause a transient accumulation of D-gluconate. With one exception, a *Ps. fluorescens* (glc-) deficient in glucose dehydrogenase and an obtuse peak in gluconate concentration with considerable delay were observed.
[c] D-Glucose 6-phosphate was oxidized to 6-phosphogluconate during late stationary phase by *Ps. fragi* and *Ps. lundensis* growing aerobically. The former species was unable to do so under an atmosphere enriched with carbon dioxide.
[d] L- and D-lactic acid were used after depletion of D-glucose. A transient accumulation of pyruvate during catabolism was observed. With *Ps. fragi* the rate of catabolism under an atmosphere enriched with carbon dioxide was less than that under aerobic conditions. Acetic acid, formed by the fermentation of carbohydrates in meat by a facultatively anaerobic flora, was catabolized in a later phase by this taxon (Drosinos and Board, 1995b).
[e] A slight decrease in the concentration of amino acids by the end of the exponential growth was observed. Thereafter, a drastic increase in their concentration under aerobic conditions was noted.
[f] *Pseudomonas fragi* was able to catabolize creatine and creatinine under aerobic conditions but not with an atmosphere enriched with carbon dioxide. The phenomenal release of ammonia and the increase in pH was inextricably linked with the catabolism of these substrates.

gluconate led investigators to propose a method for the control of the microbial activity in meat, namely the addition of glucose to meat and its transformation to gluconate (Shelef, 1977; Gill, 1986; Lambropoulou *et al.*, 1996). The rationale for this is the fall in pH through the accumulation of the oxidative products. The transient pool of gluconate and the inability of the taxa of the association to catabolize it may offer a selective determinant on the meat ecosystem (Nychas *et al.*, 1988). In their study, Drosinos and Board found that with strains of *Ps. fluorescens* (glc-) deficient in glucose dehydrogenase, an obtuse peak in gluconate concentration after a considerable delay was observed. L- and D-lactic acid were used after depletion of D-glucose. A transient accumulation of pyruvate during catabolism

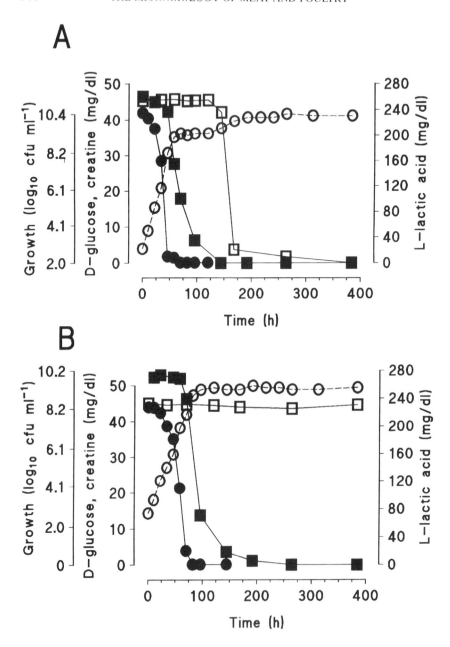

Figure 9.1 Sequential catabolism of glucose (●), lactic acid (■) and creatine (□) by *Pseudomonas fragi* growing in a meat juice under (A) aerobic or (B) carbon dioxide-enriched atmosphere at 4 °C. (○), Growth curve. (Reprinted from Drosinos, E.H., and Board, R.G. (1994) Metabolic activities of pseudomonads in batch cultures in extract of minced lamb. *J. Appl. Bacteriol.*, **77**, 613–20, with kind permission from Blackwell Science, Osney Mead, Oxford, OX2 0EL, UK).

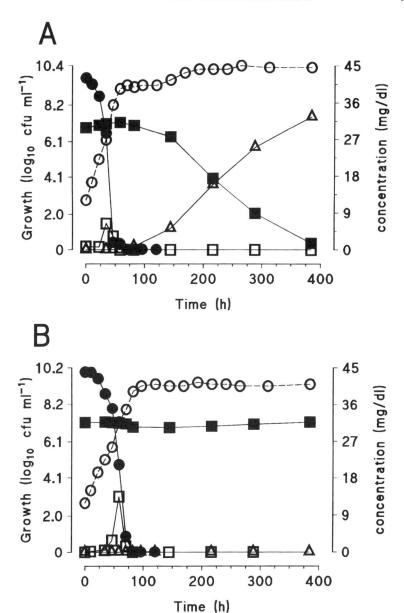

Figure 9.2 Formation of gluconate and 6-phosphogluconate by *Pseudomonas fragi*. Gluconate (□) formed during catabolism of glucose (●) either under (A) aerobic or (B) carbon dioxide-enriched atmosphere at 4 °C. Glucose 6-phosphate (■) was oxidized to 6-phosphogluconate (△) only under aerobic conditions (A). (○), Growth curve. (Reprinted from Drosinos, E.H., and Board, R.G. (1994) Metabolic activities of pseudomonads in batch cultures in extract of minced lamb. *J. Appl. Bacteriol.*, **77**, 613–20, with kind permission from Blackwell Science, Osney Mead, Oxford, OX2 0EL, UK).

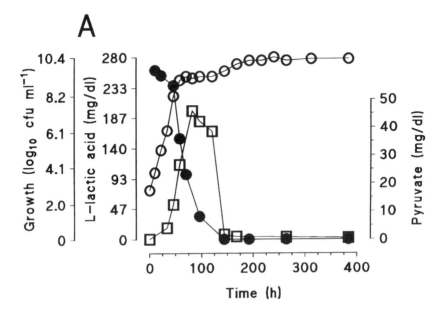

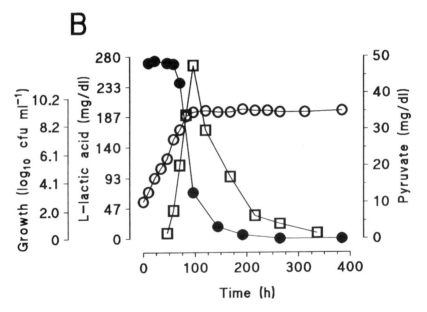

Figure 9.3 Formation of pyruvate. Pyruvate (□) accumulated during catabolism of lactic acid (●) by *Pseudomonas fragi* in a meat juice at 4 °C. The rate of lactic acid and pyruvate catabolism was less under a carbon dioxide enriched atmosphere (B) than that under an aerobic one (A). (○), Growth curve. (Reprinted from Drosinos, E.H., and Board, R.G. (1994) Metabolic activities of pseudomonads in batch cultures in extract of minced lamb. *J. Appl. Bacteriol.*, **77**, 613–20, with kind permission from Blackwell Science, Osney Mead, Oxford, OX2 0EL, UK).

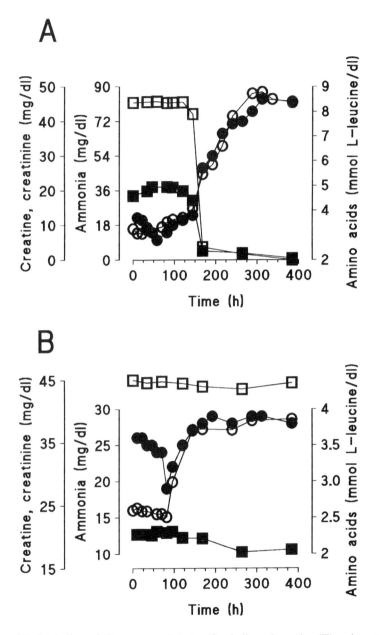

Figure 9.4 Catabolism of nitrogenous substrates. Catabolism of creatine (□) and creatinine (■) resulted in a phenomenal production of ammonia (○) and an increase in total amino acids (●) under an aerobic atmosphere (A) but the extent of these changes was restricted by the blockage of catabolism of creatine and creatinine under a carbon dioxide-enriched atmosphere (B). (Reprinted from Drosinos, E.H., and Board, R.G. (1994) Metabolic activities of pseudomonads in batch cultures in extract of minced lamb. *J. Appl. Bacteriol.*, **77**, 613–20, with kind permission from Blackwell Science, Osney Mead, Oxford, OX2 0EL, UK).

was also observed (Fig. 9.3). A slight decrease in the concentration of amino acids by the end of the exponential growth was seen. Thereafter, a marked increase in their concentration under aerobic conditions was noted. *Pseudomonas fragi* was able to catabolize creatine and creatinine under aerobic conditions. The phenomenal release of ammonia and the increase in pH was inextricably linked with the catabolism of these substrates (Fig. 9.4).

Ammonia, which is the major cause of the increase of pH, can be produced by many microbes, including pseudomonads, during amino acid metabolism. Other volatile compounds found in spoiled meat are shown in Table 9.9. The odours of such by-products (Table 9.9) are usually the first manifestation of spoilage of chilled meat or poultry stored under aerobic conditions (Dainty and Hoffman, 1983; Dainty *et al.*, 1983, 1985, 1989a,b; Edwards *et al.*, 1987). The identities of these compounds involved are consistent with amino acids as a major source. It was mentioned previously that spoilage of meat is not manifested until either the bacterial load exceeds 10^8 cells/cm^2, or the glucose/gluconate concentration at the meat surface is reduced to an undetectable amount (Gill, 1976; Nychas *et al.*, 1988). The time scale of the production of these odours is also consistent with the

Table 9.9 Major volatiles microbial metabolites detected in naturally contaminated samples of chilled meat stored in air. From McMeekin (1977); Dainty *et al.* (1985, 1989a, b); Molin and Tenström (1986); Edwards and Dainty (1987); Stutz *et al.* (1991); Jackson *et al.* (1992); Lasta *et al.* (1995).

Compound	Compound
Hydrogen sulphide	Methyl mercaptan
Ethyl acetate	Ethanol
n-propanoate	Methanethiol
iso-butanoate	Dimethylsulphide
3-methylbutanoate	Dimethyltrisulphide
n-hexanoate	Methylthioacetate
n-heptanoate	Ammonia
n-octanoate	Putrescine
crotonate	Cadaverine
3-methyl-2-butenoate	Tyramine
tiglate	Spermidine
iso-propyl acetate	Diaminopripane
iso-butyl acetate	Agmatine
n-propanoate	1,4-Heptadiene
n-hexanoate	1-Undecene
iso-pentyl acetate	1,4-Undecadiene
3-methyl butanol	Acetoin
2-methyl butanol	Diacetyl
Acetone	3-Methyl butanal
Methyl ethyl ketone	Butane
Methanol	Hexane
2-butanone	Toluene

demonstrated restriction of amino acid metabolism until after glucose/gluconate/lactate depletion at the meat surface (Gill, 1976; Nychas *et al.*, 1988; Drosinos, 1994). Pseudomonads, in particular *Pseudomonas fragi*, are the major and possibly the sole producers of ethyl esters in air-stored meat (Dainty *et al.*, 1985; Edwards *et al.*, 1987). Similar results were obtained in studies with beef and poultry inoculated with pure culture of pseudomonads (Freeman *et al.*, 1976; McMeekin, 1975, 1977; Dainty *et al.*, 1985, 1989b; Edwards *et al.*, 1987). The mechanism of production of the above mentioned sulphides by pseudomonads and *Sh. putrefaciens* is shown in Table 9.10. According to McMeekin's (1982) review, the organisms other than pseudomonads responsible for these volatiles compounds are *Sh.* (*'Alteromonas'*) *putrefaciens*, *Proteus*, *Citrobacter*, *Hafnia* and *Serratia*. In addition to the formation of malodorous compounds, the release of large amounts of ammonia contributes also to the development of spoilage odours (Dainty *et al.*, 1983; Dainty and Mackey, 1992). Schmitt and Schmidt-Lorenz (1992a,b) found that the concentration of ammonia increased in air-stored samples of broiler skin whose microflora was dominated by pseudomonads. About half of the pseudomonads and Enterobacteriaceae produced ammonia in a medium having a chemical composition similar to that of chicken skin. The concentration of four of the volatile compounds, acetone, methyl ethyl ketone, dimethyl sulfide and dimethyl disulphide, listed in Table 9.9, increased continuously during storage of minced meat stored aerobically at 5, 10 or 20 °C (Stutz *et al.*, 1991). Hydrogen sulphide, another potential indicator of spoilage (Table 9.9), is not produced by pseudomonads, while dimethylsulphide is not produced by Enterobacteriaceae (Dainty *et al.*, 1985). Hydrogen sulphide and ammonia are formed as a result of the conversion of cysteine to pyruvate by the enzyme cysteine desulphydrase (Gill, 1982). Hydrogen sulphide combines with the muscle pigment to give a green discoloration.

Putrescine, cadaverine, histamine, tyramine, spermine and spermidine (Table 9.11) were found to be present in minced pork, beef and poultry stored at chill temperatures (Nakamura *et al.*, 1979; Edwards *et al.*, 1987; Schmitt and Schmidt-Lorenz, 1992a,b). Cadaverine was the major biogenic amine in poultry stored either aerobically or in vacuum. Schmitt and Schmidt-Lorenz (1992a) reported that putrescine and cadaverine, which are detectable with colony counts of 10^5 cfu/cm^2, could indicate onset of spoilage in poultry. Pure culture experiments proved that pseudomonads were the major producers of putrescine while the Enterobacteriaceae produced most of the cadaverine.

(b)   Enterobacteriaceae. Enterobacteriaceae can be important in spoilage if the meat ecosystem favours their growth. This group utilize mainly glucose and glucose-6-phosphate as the main carbon sources (Table 9.3); the exhaustion of these substances will allow amino acid degradation (Gill and

Table 9.10 Factors and precursors affecting the production of malodorous end-products of Gram-negative bacteria (e.g. *Pseudomonas* spp., *Shewanella putrefaciens*, *Moraxella* etc.) when inoculated in broth, sterile model system and in naturally spoiled meat. Based on McMeekin (1981); Dainty *et al.* (1985, 1989a, b); Edwards and Dainty (1987); Edwards *et al.* (1987); Stutz *et al.* (1991); Schmitt and Schmidt-Lorenz (1992b); Jackson *et al.* (1992); Lasta *et al.* (1995)

End-product	Broth	Model food	Meat	Factors	Precursors
Sulphur compounds	not tested				
sulphides		+	+	temperature	cysteine, cystine, methionine
dimethylsulphide		+	+	and substrate	methanethiol, methionine
dimethyldisulphite		+	+	(glucose)	methionine
methyl mercaptan		+	+	limitation	nad
methanethiol		−/+[a]	+		methionine
Hydrogen sulphide		+	+	high pH	cystine, cysteine
dimethyltrisulphide		+	+	nad	methionine, methanothiol
Esers	not tested				
methyl esters (acetate)		+	+	glucose[c]	nad
ethyl esters (acetate)		+	+	glucose[c]	nad
Ketones	not tested				
acetone		+	+	nad	nad
2-butanone		+	+	nad	nad
acetoin/diacetyl		+/−[b]	+	nad	nad
Aromatic hydrocarbons	not tested				
diethyl benzene		+	+	nad	nad
trimethylbenzene		+	+	nad	nad
toluene		+	+	nad	nad
Aliphatic hydrocarbons	not tested				
hexane		+	+	nad	nad
2,4 dimethylhexane				nad	nad
and methyl heptone		+	+	nad	nad

Table 9.10 Continued

End-product	Broth	Model food	Meat	Factors	Precursors
Aldehydes					
2-methylbutanal	not tested	+	+	nad	*iso*-leucine
Alcohols					
methanol	not tested	+	+	nad	nad
ethanol		+	+	nad	nad
2-methylpropanol		+	+	nad	valine
2-methylbutanol		+	+	nad	iso-leucine
3-methylbutanol		nad	+	nad	leucine
Other compounds					
ammonia	not tested	+	+	glucose (1)[c]	amino acids

[a] production only by *Sh. putrefaciens*.
[b] these compounds decreased during storage.
[c] low concentration of glucose.
nad, no data available.
+, present; –, not detected.

Table 9.11 Production of biogenic amines by meat microbial flora. Based on Maijala *et al.* (1993); Maijala (1994); Dainty *et al.* (1986); Edwards *et al.* (1987); Chandler *et al.* (1989); Rawles *et al.* (1996)

Biogenic amine	Bacteria	Storage condition		Factors
		$T(^{\circ}C)$	Packaging	
Putrescine	*H. alvei, Serr. liquefaciens*	1	VP[a]	pH, ornithine (arginine) utilization
Cadaverine	*H. alvei, Serr. liquefaciens*	1	VP	pH, lysine utilization
Histamine	*Proteus morganii, Kl. pneumoniae, H. alvei, A. hydrophila*			Temperature, pH, histidine utilization
Spermine				pH, spermidine
Spermidine				pH, agmatine, arginine
Tyramine	*Lactobacillus sp., L. carnis, L. divergens, Ent. feacalis*	1	VP	
		20	Air[b]	pH
Tryptamine				pH

[a] VP, vacuum packaging.
[b] Air, aerobic storage.

Newton 1977; Gill, 1986). Some members of this family produce ammonia, volatile sulphides, including hydrogen sulphide and malodorous amines from amino acid metabolism (Hanna *et al.*, 1976; Gill and Newton, 1979). Enterobacteriaceae and *Br. thermosphacta* (Table 9.12), fail to produce ester in pure culture, although acids and alcohols are among the end-products. The production of the branch chain esters which are listed in Table 9.9 could be due to the possibility that pseudomonads catalyse the interaction of the excreted products, or that they are formed by direct chemical interaction. Inoculation experiments with Enterobacteriaceae and *Br. thermosphacta* showed an increase initially in the level of acetoin/diacetyl. These two compounds are often detected at the same time as the esters (Dainty *et al.*, 1985). As pseudomonads catabolize acetoin and diacetyl, the concentration of both diminishes with time (Molin and Ternstrom, 1986).

9.3.2 Chemical changes by Gram-positive bacteria

In general Gram-positive bacteria, especially the lactic acid bacteria, are unimportant contaminants of meat stored under aerobic conditions (Nychas and Arkoudelos, 1990; Nychas *et al.*, 1992; Dainty and Mackey 1992; see also Chapter 2). *Brochothrix thermosphacta* may have some importance on pork and lamb, particularly on fat surfaces (Barlow and Kitchell, 1966; Talon *et al.*, 1992).

Table 9.12 Factors and precursors affecting the maximum formation of end-products of *Brochothrix thermosphacta* when inoculated in broth, sterile food model system and in naturally spoiled meat. Based on Hitchener et al. (1979); Dainty and Hibbard (1980, 1983); Dainty and Hoffman (1983); Blickstad (1983); Blickstad and Molin (1984); Dainty et al. (1985, 1989a, b); Edwards et al. (1985, 1989a. b); Paron and Talon (1988); Borch and Molin (1989); Ordonez et al. (1991); Nychas and Arkoudelos (1991); Schmitt and Schmidt-Lorenz (1992b); Talon et al. (1992); Drosinos (1994)

End-product	Broth	Model food	Meat	Factors	Precursors
Aerobically:					
acetoin	+	+	+	Glucose (h), pH (h/l), T (h/l)	Glucose (mj), alanine (mn), diacetyl
acetic acid	(±)	+	+	Glucose (h), pH (h/l), T (h/l)	Glucose (mj), alanine (mn)
L-lactic acid	+	nd	+	T (h), pH (h), O_2 (l)	Glucose
formic acid	+	na	+	T (h), pH (h)	Glucose
ethanol	+	na	+	T (h), glucose	nad
CO_2	+	na	na	nad	Glucose
iso-butyric acid	+	+	nt	Glucose (l), T (l), pH (h)	Valine, leucine
iso-valeric acid	+	+	nt	Glucose (l), T (l), pH (h)	Valine, leucine
2-methylbutyric	+	nd	nd	Glucose (l), pH (h)	*iso*-leucine
3-methylbutanol	+	+	+	Glucose (h), pH (l)	nad
2-methylbutanol	na	+	+		nad
2-methylbutanol	na	na	+		*iso*-leucine
3-methylbutanol	na	na	+		Leucine
2,3-butanediol	+	+	+	Glucose (h), T (h/l)	Diacetyl
diacetyl	+	+	+		nad
2-methylpropanol	+	na	+	Glucose (h)	Valine
2-methylpropanal	na	na	+		Valine
free fatty acids	nt	nt	+	Glucose (l), pH/O_2/T (h)	Meat fat
In different gaseous atmospheres:					
L-lactic acid	+	nt	+	Glucose (h), pH (h), T (ns)	Glucose
acetic acid	+	nt	+	O_2 (h), glucose (l)	Glucose
ethanol	+	nt	+	T (h), pH (h)	nad
formic	+	nt	+	T (h), pH (h)	nad

(h), high pH, concentration of glucose or storage temperature; (l), low pH, concentration of glucose or storage temperature; (h/l), contradictory results; (ns), not significant factor; (mj), major contribution; (mn), minor contribution; (±), no production under strictly aerobic conditions: nd, not determined. nt: not tested; na, not analysed; nad, no available data; T, temperature; +, present at end of storage.

Table 9.13 Factors and precursors affecting the maximum formation of end products of isolates from meat of lactic acid bacteria (*Lactobacillus* sp. *Leuconostoc* sp. *Carnobacterium* sp.), and inoculated in broth, sterile model system and naturally spoiled meat. Based on Montville *et al.* (1987); Borch and Molin (1989); Benito de Cardinas *et al.* (1989); Tseng and Montville (1990); Borch *et al.* (1991); Ordonez *et al.* (1991); Borch and Agerhem (1992); Nychas *et al.* (1994); Arkoudelos and Nychas (1995); Drosinos (1994); Tassou *et al.* (1996); Loubiere *et al.* (1997)

End-product	Broth	Model food	Meat	Factors	Precursors
Homofermentative strains – aerobic storage:					
L-lactic acid	+	+	+	nad	Glucose
D-lactic acid	+	+	nd	nad	Glucose
Acetic acid	+	+	+	Glucose (l), O₂ (h), E	Glucose, lactate, pyruvate
Acetoin/diacetyl	+	+	nad	pH (l), glucose (h)	Pyruvate
Hydrogen peroxide	+	nad	+	nad	nad
Formic acid	+	+	+	nad	Glucose, acetic acid
Ethanol	+	nad	+	nad	Glucose
Heterofermentative strains – aerobic storage:					
L-lactic acid	+	+	+	nad	Glucose
D-lactic acid	+	+	nd	nad	Glucose
Acetic acid	+	+	+	nad	Glucose
Acetoin/diacetyl	+	nad	+	pH (l), glucose (h)	Pyruvate
Hydrogen peroxide	+	nad	nad	nad	nad
Formic acid	nd	+	+	nad	nad
Ethanol	+	nad	+	nad	nad
Homofermentative strains - various gaseous conditions:					
L-lactic acid	+	+	+	nad	Glucose
D-lactic acid	+	+	+	nad	Glucose
Acetic acid	+	+	nad	Glucose (l), O₂ (h), E	Glucose, lactate, pyruvate
Acetoin	+	nad	nad	pH (l)	Pyruvate
Formic acid	+	nd	+	nad	Glucose, acetic acid
Ethanol	+	nd	+	nad	nad

Table 9.13 Continued

End-product	Broth	Model food	Meat	Factors	Precursors
Heterofermentative strains – various gaseous conditions					
L-lactic acid	+	+	+	nad	Glucose
D-lactic acid	+	+	+	nad	Glucose
Acetic acid	+	+	+	nad	Glucose
Formic acid	nd	nd	+	nad	Glucose
Ethanol	nd	+	+	nad	nad

(h), high oxygen; (l), low concentration of glucose; E, appropriate enzymes – LDH, NADH peroxidase, lactate or pyruvate oxidase; nad, no available data; nd, not detected; +, present at end of storage.

The physiological attributes – growth rates, end products of metabolism and the ecological role of glucose and oxygen limitation, pH and incubation temperature – of the lactic acid bacteria and *Br. thermosphacta* isolated from meat and meat products, have been studied in culture media, in meat juice and on sterile meat blocks (Gill, 1976; Dainty and Hibbard, 1980, 1983; Blickstad and Molin, 1984; Dainty and Hoffman, 1983; Blickstad, 1983; Borch and Molin, 1989; Nychas and Arkoudelos 1991; Borch *et al.*, 1991; Borch and Agerhem, 1992; Nychas *et al.*, 1994; Drosinos and Board, 1994, 1995b; Arkoudelos and Nychas, 1995). The general conclusion is that the oxygen tension, glucose concentration, and the initial pH have a major influence on the physiology of these organisms, and hence on end-product formation (Tables 9.12 and 9.13).

Brochothrix thermosphacta has a much greater spoilage potential than lactobacilli and can be important in both aerobic and anaerobic spoilage of meat. This organism utilizes glucose and glutamate but no other amino acids during aerobic incubation (Gill and Newton, 1977). It produces a mixture (Table 9.12) of end-products including acetoin, acetic, *iso*-butyric and *iso*-valeric acids, 2,3-butanediol, diacetyl, 3-methylbutanal, 2-methylpropanol and 3-methylbutanol during its aerobic metabolism in media containing glucose, ribose or glycerol as the main carbon and energy source (Dainty and Hibbard, 1980). The precise proportions of these end-products is affected by the glucose concentration, pH and temperature (Table 9.12). For example the ratio of the molar concentration of acetic acid to acetoin was greatest at low as opposed to high glucose levels (Dainty and Hoffman, 1983). Similarly the production of the two other acids, *iso*-butyric and *iso*-valeric, was enhanced by a low glucose concentration. The reverse was true for the corresponding alcohols (2-methylpropanol, 3-methylbutanol and 2,3-butanediol). Glucose is the main precursor for acetoin and acetic acid (Dainty and Hibbard, 1983) while *iso*-butyric, *iso*-valeric and 2-methyl-butyric are produced from valine, leucine and *iso*-leucine respectively. Alanine could also play a minor role in the production of acetoin and acetic acid (Dainty and Hibbard, 1983).

9.4   Chemical changes in meat ecosystems stored under vacuum or modified atmosphere packaging

It was evident in another chapter (pp. 183–193) that the final composition of the microbial associations differs significantly as a consequence of the packaging treatment used for meat. For example lactic acid bacteria and *Br. thermosphacta* rather than pseudomonads are dominant on meat (beef, pork, lamb, and poultry) stored in vacuum pack, or in atmospheres enriched with carbon dioxide, nitrogen or oxygen. The shift from a very diverse initial flora to one consisting predominantly of Gram-positive facultative anaerobic microflora and dominated by *Lactobacillus* spp. and *Br. thermosphacta* occurs commonly in muscle foods during MAP storage (Davies, 1995). Chemical changes, especially the increase in the concentration of D-lactic and acetic acids, offer reliable evidence of the quality of the food. Acetic acid, a product of the oxidation of lactic acid, may be used for the construction of models (Kakouri and Nychas, 1994). Hence the homofermentative or heterofermentative type of metabolism and the ecological determinants that affect these are of great importance (Borch and Agerhem, 1992; Kakouri and Nychas, 1994).

The microbial metabolites depend not only on the storage conditions but also on other environmental factors such as aeration, glucose and lactate availability, and pH (Tables 9.12, 9.13). For example the ethyl esters of acetic acid, propanoic, *n*-butanoic, *iso*-pentanoic and hexanoic acids, were found in meat stored in air (Dainty *et al.*, 1985) while none of the esters containing the branched chain alcohol and acid components was observed in meat under vacuum packaging (Table 9.14). The dairy/cheesy odour found in samples stored in gas mixtures with carbon dioxide, was produced by *Br. thermosphacta* and lactic acid bacteria both of which can produce diacetyl/acetoin and alcohols (Dainty and Hibbard, 1983). The microbial metabolites detected in naturally contaminated samples of chilled meat stored in vacuum and modified atmosphere packs are shown in Table 9.14. Under these conditions the putrid odours associated with storage in air are replaced by relatively inoffensive sour/acid odours. Such odours have been assumed to arise from the acid products of glucose fermentation, the primary generator of energy for growth (Gill, 1976, 1983; Gill and Newton, 1978). The production of such off-odours is difficult to explain in terms of the accumulation of acetic, *iso*-butanoic, *iso*-pentanol and D-lactic because the amounts are relatively small compared with the endogenous L-lactic acid in muscle of normal pH (Dainty, 1981; de Pablo *et al.*, 1989).

9.4.1 *Chemical changes caused by Gram-negative bacteria*

The spoilage of meat stored under oxygen limitation or in carbon dioxide-enriched atmospheres is due to the undefined actions of lactic acid bacteria and/or *Br. thermosphacta* since Gram-negative bacteria, especially pseudomonads, are inhibited. The presence of sulphuryl compounds such as propyl ester, 3-methylbutanol compounds (Table 9.14) raises again the crucial question: is this inhibition due to carbon dioxide enrichment or to oxygen limitation? It is well known that pseudomonad species are very sensitive to carbon dioxide at low storage temperatures due to the fact that at low temperatures the solubility of this gas is high. As pseudomonads have very high affinity for oxygen this could be the reason for the findings of Molin (1985) and Drosinos and Board (1994) who reported that pseudomonads can grow in relatively low oxygen tension without any

Table 9.14 Volatiles present in packaged beef, pork and poultry (minced or not) stored under different packaging conditions. From McMeekin (1981), Dainty *et al.* (1985, 1989a, b); Edwards and Dainty (1987); Edwards *et al.* (1987); Stutz *et al.* (1991); Jackson *et al.* (1992); Dainty and Mackey (1992); Nychas *et al.* (1994); Lasta *et al.* (1995)

Volatile	VP[a]	20%/80%[b] CO_2/O_2	50%/50%[b] N_2/CO_2	100%[b] CO_2	100%[b] N_2	60%/40%[b] N_2/CO_2
Ethanol	+	+	+	+	+	−
Acetone	+	+	−	+	−	+
Propan-2-ol	−	+	+	+	−	−
Dimethylsulphide	+	+	−	+	−	+
Propan-1-ol	+	+	−	−	−	+
Ethyl acetate	+	+	+	+	−	−
2,3-butandione	+	+	−	+	−	−
Acetic acid	+	+	+	+	+	−
Diacetyl	−	−	−	+	+	+
Hexane	−	+	+	+	−	−
Heptane	+	+	+	+	−	−
Pentanol	−	+	+	−	−	−
2-methylpropanol	+	+	−	−	−	−
2-methylbutanol	+	−	+	−	−	−
Pentanal	+	−	−	−	−	−
Heptadiene	−	+	−	−	−	−
Acetoin	+	+	−	+	−	−
3-methylbutan-1-ol	+	+	−	−	−	−
2-methylbutan-1-ol	+	+	−	−	−	−
Dimethyldisulphide	+	+	+	−	−	−
Octane	−	−	+	+	−	+
2,3-butandiol	−	+	−	−	−	−
3-ethyl pentane	−	−	−	+	−	−
Ammonia	+	+	+	+	+	−

+, present at the end of storage; −, absent at the end of storage.
[a] VP, vacuum packaging.
[b] modified atmosphere with the indicated gaseous mixtures.

significant differences, apart from the rate of increase in their metabolic activity. Tyramine, putrescine and cadaverine were also present in meat stored under vacuum pack–modified atmosphere (Dainty *et al.*, 1986; Smith *et al.*, 1993).

9.4.2   Chemical changes caused by Gram-positive bacteria

(a)   *Lactic acid bacteria.* Changes in lactate concentration are evident in all types of studies (pure cultures/food model system and natural eco-system). Nassos *et al.* (1983, 1985, 1988) recommended the use of lactate as a spoilage index of ground beef, having found an increase of lactate during storage. They analysed their samples with HPLC and consequently they were unable to distinguish between D- and L-lactate. On the other hand Nychas (1984), Nychas and Arkoudelos (1990), Borch and Agerhem (1992) and Drosinos (1994) have reported that L-lactate decreased during storage under aerobic or modified atmosphere conditions. When the acid profile of

Table 9.15 Compounds which may be important for the determination/prediction or assessment of the remaining shelf-life of raw meat under different conditions

Compound	Test	Storage conditions	References
Glucose	Enzymatic kit	Air, VP, MAP	1, 2, 3
Acetate	Enzymatic kit, HPLC	VP-MAP	4, 5, 6, 7
Gluconate	Enzymatic kit	Air, VP-MAP	7, 8, 9, 10
Total lactate	HPLC	VP-MAP	11
D-lactate	Enzymatic kit	VP-MAP	7, 10, 12, 13, 14, 15, 16, 17
Ethanol	Enzymatic kit, GLC	VP-MAP	5, 10, 16, 17
Free amino acids	Colourimetric	Air	18, 19
Ammonia	Enzymatic, colourimetric	Air	20, 21, 22
Acetone, methyl ethyl ketone, dimethyl sulphide dimethyldisulphide	GLC	VP-MAP	23
Diacetyl, acetoin	Colorimetric	VP-MAP	4, 14, 20
Biogenic amines	HPLC, sensors, enzymic test, GLC, enzyme electrodes, test strips	Air, VP, MAP	15, 24, 25, 26, 27, 28, 29
microbial activity	Enzymic	Air	3, 30, 31

(1) Nychas *et al.* (1988, 1992); (2) Boers *et al.* (1994); (3) Seymour *et al.* (1994); (4) Ordonez *et al.* (1991); (5) Borch and Agerham (1992); (6) Kakouri and Nychas (1994); (7) Lambropoulou *et al.* (1996); (8) Nychas (1984); (9) Drosinos (1994); (10) Dainty (1996); (11) Nassos *et al.* (1983, 1985, 1988); (12) Sinell and Lucke (1978); (13) Schneider *et al.* (1983); (14) de Pablo *et al.* (1989); (15) Ordonez *et al.* (1991); (16) Nychas *et al.* (1994); (17) Drosinos and Board (1995b); (18) Adamcic and Clark (1970); (19) Schmitt and Schmidt-Lorenz (1992b); (20) Nychas and Arkoudelos (1990); (21) Lea *et al.* (1969); (22) Nychas (1984); (23) Stutz *et al.* (1991); (24) Edwards *et al.* (1983, 1987); (25) Schmitt and Schmidt-Lorenz (1992a); (26) Yano *et al.* (1995); (27) Smith *et al.* (1993); (27) Dainty *et al.* (1987); (28) Krizek *et al.* (1995); (29) Rawles *et al* (1996); (30) de Castro *et al.* (1988); (31) Alvarado *et al.* (1992).

the water soluble compounds was analysed with HPLC (Nychas *et al.*, 1994; Lambropoulou, 1995) it was confirmed that the chromatographic area of the lactic acid did not change significantly, compared with the changes found with the enzymatic method used in these studies.

This could be due to the fact that D-lactate was formed during storage. When D- and L-lactate were analysed enzymatically in these studies, it was found that L-lactate decreased while D-lactate increased during storage of meat under different treatments (Nychas *et al.*, 1994; Drosinos, 1994; Lambropoulou *et al.*, 1996). The rate of decrease differed significantly between samples stored in 100% CO_2, 100% N_2, vacuum pack or 20%:80% CO_2/O_2. Similarly Dainty (1981), Ordonez *et al.* (1991) and de Pablo *et al.* (1989) found that D-lactate increased during storage under VP/MAP. The D-lactic acid isomer increased during storage under different gaseous conditions (Table 9.4). This compound does not arise, however, from the metabolic activity of *Br. thermosphacta* or endogenous anaerobic glycolysis because in both cases only L-lactate would be produced (Hitchener *et al.*, 1979;

Blickstad and Molin, 1984; Ordonez *et al.*, 1991). Therefore the increase in the concentration of this compound is due to metabolism of lactic acid bacteria, particularly *Carnobacterium, Leuconostoc* or *Weisella* which generate D-, L- or DL-lactate (Kandler, 1983; Collins *et al.*, 1987, 1993).

Lactic acid bacteria (LAB) may produce exclusively L-lactic, D-lactic, approximately equal amounts of L- and D- or predominantly L- or D- with trace amounts of the other isomer (Garvie 1980; Kandler and Weiss, 1986; Schleifer, 1986). Precisely whether or not micro-organisms are capable of producing L- or D- depends on the presence of D-nLDH and/or L-nLDH (specific NAD^+ dependent lactate dehydrogenases). A few LAB species (e.g. *Lact. curvatus, Lact. sake*) produce a racemase which converts L-lactic acid to D-lactic acid (Garvie, 1980). L-lactic acid induces the racemase, which results in a mixture of D- and L-lactic acids. Generally, L-lactic acid is the major form produced in the early growth phase and D-lactic acid in the late to stationary phase (Garvie, 1980). The formation of the different isomeric forms of lactic acid during fermentation of glucose can be used to distinguish between leuconostocs and most heterofermentative lactobacilli. The former produces only D-lactic acid and the latter a racemic mixture (D/L-lactic acid). Ordonez *et al.* (1991) reported that no consistent patterns were obtained for L-lactic acid concentrations in pork packed in 20% CO_2/80% air and 20% CO_2/80% O_2, whereas D-lactic acid concentration was reported to increase along with the lactobacilli counts. These authors did not detect D-lactic acid at the start of storage of their samples while low levels were detected after 5 d storage in both atmospheres. After 20 d, the levels had risen to 12–18 mg/100 g meat. Similar patterns have been reported by Nychas *et al.* (1994), Drosinos and Board (1995a,b) and Lambropoulou *et al.* (1996) all of whom reported that this compound was produced in beef (minced), lamb, chicken, pork and dry ham samples stored under MAP/VP conditions. Ordonez *et al.* (1991) concluded that D-lactic acid was a product of lactic acid bacteria metabolism since anaerobic glycolysis produces only L-lactic acid. The formation of D-lactate, according to Drosinos and Board (1995b), has only been attributed to heterofermentative lactic acid bacteria. Indeed they reported that the concentration of D-lactate increased only when *Leuconostoc* spp. were the dominant organisms in lamb meat stored under different conditions. Similar results were obtained when the physico-chemical changes were monitored in meat broth. In poultry samples, higher concentration of this acid was always associated with samples inoculated with the homofermentative *Lact. plantarum* rather than with the heterofermentative *W. minor*. Similar results have been reported by Borch and Agerhem (1992) in inoculated beef slices.

Church *et al.* (1992) reported that endogenous meat enzymes may be responsible, at least in part, for D-lactic acid production, since similar concentrations were found in slices taken at progressively increasing depths from the meat surface. If D-lactic acid production had been due solely to microbial metabolism, higher levels would have been expected at the surface, and increasingly lower levels at increasing depth from the meat surface. Meyns et al. (1992) found 10 mg/100 g at a depth of 1 cm and 40 mg/g at 5 cm in beef stored in vacuum packs, thereby providing further support for the hypothesis that endogenous meat enzymes produce D-lactic acid in the early stages of storage. The concentration of D-lactic acid at a given storage time has also been found to be independent of the packaging atmosphere and differences in LAB microbial flora, endorsing the theory that D-lactic acid production is due mainly to meat metabolism. It is the present authors' opinion that the origin of D-lactic acid in the meat eco-system is due to microbial metabolism.

In studies with dry cured ham, Church et al. (1992) reported that total lactic acid content showed little change with time. However, whilst D-lactic acid was found to increase in one of the dry cured hams, no D-lactic acid was detected at the lower a_w (0.90 compared to 0.95), higher salt-on-water ham (9.91% compared to 8.73%). Spiking experiments with D-lactic acid showed no interference with the recovery, indicating inhibition of meat enzymes or microflora responsible for D-lactic production due to the lower a_w and higher sodium chloride levels. Kudryashov et al. (1989) reported an increasing inhibition of meat enzymes as sodium chloride and sodium nitrite concentrations increased.

The absence of D-lactic acid in freshly slaughtered meat of different species and the gradual increase thereof during storage has been reported (Borch and Agerhem, 1992; Nychas et al., 1994; Drosinos and Board, 1995b; Lambropoulou et al., 1996). Furthermore when poultry meat was inoculated with homofermentative or heterofermentative lactic acid bacteria in vacuum packs or in 100% carbon dioxide and stored at 10 °C, the concentration of D-lactic acid was always higher in inoculated samples than that found in uninoculated poultry meat (Borch and Agerhem, 1992; Nychas et al., 1994). Further investigations using meat the surfaces of which have been sterilized and/or inoculated with LAB should clarify this matter.

The levels of acetic acid increased at various times throughout the storage period of meat flushed with nitrogen, carbon dioxide, or oxygen or vacuum packed. In general the storage of meat in MAP not only selects a microbial flora (lactic acid bacteria) on meat different from that stored in air, but it could also influence the metabolic activity of members of this flora (Nychas 1984, 1994). It is well known that the metabolism of lactic acid bacteria is affected by environmental factors such as pO_2, pH, glucose limitation etc. (Bobillo and Marshall, 1991, 1992; Marshall 1992). The increase of acetate in meat samples (beef, pork and certain meat products) stored under differ-ent VP/MAP conditions or in vacuum pack could be attributed either to a shift from homo- to heterofermentative metabolism of the lactic acid bac-teria or to the predominance of another organism, e.g. Br. thermosphacta, in such systems (Dainty, 1981; de Pablo et al., 1989; Ordonez et al., 1991;

Borch and Agerhem, 1992; Drosinos and Board, 1994; Nychas *et al.*, 1994; Lambropoulou *et al.*, 1996).

Acetate is a product which lactic acid bacteria and *Br. thermosphacta* would be expected to produce in various quantities under both aerobic or modified atmospheres conditions (Kandler, 1983; Dainty and Hibbard, 1983; Murphy and Condon, 1984a,b; Sedewitz *et al.*, 1984; Murphy *et al.*, 1985; Borch and Molin, 1989; Cogan *et al.*, 1989; Tseng and Montville, 1990; Cselovszky *et al.*, 1992; Ramos *et al.*, 1994). Until recently (Thomas *et al.*, 1979) there was a consensus of opinion that lactic acid bacteria were either homo- or heterofermentative in the sense that the former produced about two lactic acid molecules from one glucose molecule and the latter produced one lactic acid, one acetic acid (ethanol) and one carbon dioxide molecule. It is now recognized, however, that environmental changes may cause the fermentation of the former to 'shift or switch' to that of the latter (Sedewitz *et al.*, 1984; Borch *et al.*, 1991; Marshall, 1992). This 'shift or switch' is now well recognized by dairy microbiologists (Thomas *et al.*, 1979; Ramos *et al.*, 1994). Indeed the type of energy source (glucose or galactose), glucose limitation, the degree of aeration, the concentration of lactate dehydrogenase (*i*LDH), NADH peroxidase or fructose 1,6-diphosphate, the stereo-specific of NAD-independent flavin-containing lactate dehydrogenase, lactate oxidase or pyruvate oxidase can all influence the conversion of lactate or pyruvate to acetate (Garvie, 1980; Kandler, 1983; Sedewitz *et al.*, 1984; Thomas *et al.*, 1979; Murphy and Condon, 1984a,b; Borch and Molin, 1989; Cogan *et al.*, 1989; Tseng and Montville, 1990; Axelsson, 1993; Sakamoto and Komagata, 1996). For example, both glucose or oxygen limitation may well cause the switch noted above (Thomas *et al.*, 1979; Sedewitz *et al.*, 1984; Murphy *et al.*, 1985; Condon, 1987; Borch *et al.*, 1991; Cselovszky *et al.*, 1992) in *Lact. plantarum*, *Lactobacillus* sp. and *Lact. pentosus*. Thus the metabolism of lactic acid bacteria in meat products may well be affected by environmental factors, thereby influencing their beneficial/detrimental contribution to changes in such products. For example, acetic acid has a different flavour from, and a greater antibacterial action than, that of lactic acid (Reddy *et al.*, 1975).

Nychas *et al.* (1994) have reported that alcohols (particularly ethanol and propanol) appear to be the most promising compounds as indicators of spoilage in meat samples stored under VP/MAP. As mentioned above ethanol could be a fermentation product of the heterofermentative leuconostocs and carnobacteria, or a product from the 'switch' of homofermentative lactic acid bacteria. Results from liquid culture experiments in the laboratory suggest that carnobacteria will also produce formic acid in VP/MAP meat (Holzapfel and Gerber, 1983; Tassou *et al.*, 1996).

(b) Brochothrix thermosphacta. The metabolic products of *Brochothrix thermosphacta* under different gaseous atmospheres are different from

those under strictly aerobic conditions. When the oxygen tension is low (<0.2 μM oxygen) L-lactate and ethanol are the main metabolic end-products of this bacterium (Hitchener *et al.*, 1979; Blickstad and Molin, 1984; Borch and Molin, 1989). There was no production of acetic acid, D-lactic, 2,3-butanediol, *iso*-valeric *iso*-butyric or acetoin in broth samples flushed with gases other than oxygen (Hitchener *et al.*, 1979; Blickstad and Molin, 1984). Formic acid was among the end-products regardless of the gaseous atmosphere used. It was suggested by Hitchener *et al.* (1979) that glucose metabolism by this bacterium could be through (i) the Embden–Meyerhof glycolytic pathway, in which after the conversion of glucose to 2 mol of pyruvate, this latter compound is metabolized to lactate and/or ethanol plus carbon dioxide; (ii) glucose being converted *via* 6-phosphogluconate and pentose phosphate to equimolar amounts of lactate, ethanol and carbon dioxide and (iii) the Entner–Doudoroff pathway where glucose is converted *via* 6-phosphogluconate and 2-keto-3-deoxy-6-phosphogluconate to pyruvate. *Brochothrix thermosphacta* behaves as a heterofermentative bacterium under glucose limited conditions.

9.5 Evaluation of spoilage

Time-consuming microbiological analyses may well be replaced by the analysis of chemical changes associated with microbial growth on meat. More than 40 chemical, physical and microbiological methods have been proposed for the detection and measurement of bacterial spoilage in meats (Jay, 1986b; Sheridan, 1995) there is not as yet a single one available to assess meat quality. Spoilage is a subjective evaluation and therefore a sound definition is required to develop a suitable method of detection. The lack of general agreement on the early signs of spoilage for meat and the changes in the technology of meat preservation (e.g. vacuum, modified atmosphere etc.) makes the task of identifying spoilage indicators more difficult.

As far as the spoilage indicators or microbial metabolites are concerned it is generally accepted that these should meet (among others) the following criteria (Jay, 1986a): (i) the compound should be absent or at least occur in low levels in meat, (ii) it should increase in concentration with storage, (iii) it should be produced by the dominant flora and have good correlation with organoleptic tests.

Numerous attempts have been made over the last two decades to associate given metabolites with the microbial spoilage of meat (Table 9.15). The idea for these methods is that as the bacteria grow on meat they utilize nutrients and produce byproducts. The determination of the quantity of these metabolites could provide information about the degree of spoilage. The identification of the ideal metabolite that can be used for spoilage assessment has proved a difficult task for the following reasons: (i) most metabolites are specific to certain organisms (e.g. gluconate to pseudomonads) and when these organisms are either not present or are inhibited by the natural or imposed environmental factors from man, food ecology, this provides incorrect spoilage information, (ii) the metabolites are the result of the consumption of a specific substrate but the absence of the given substrate or its presence in low quantities does not preclude spoilage, (iii) the rate of microbial metabolite production and the metabolic pathways of these bacteria are affected by the imposed environmental conditions (e.g. pH, oxygen tension, temperature etc.), (iv) the accurate detection and their measurements requires sophisticated procedures, highly educated personnel, time and equipment, and (v) many compounds give retrospective information which is unsatisfactory. The potential use of the many indicators shown in Table 9.15 is under consideration. Most of these indicators showed good correlations with microbial numbers but there is lack of information on the organoleptic characteristics of meat. This is a task for the future.

References

Adamcic, M., Clark, D.S. and Yaguchi, M. (1970) Effect of psychrotolerant bacteria on the amino acid content of chicken skin. *J. Food Sci.*, **35**, 272–5.

Alvarado, R., Rodriguez-Yunta, M.A., Hoz, L., Garcia de Fernando, G.D. and Ordonez, J.A. (1992) Rapid p-nitroaniline test for assessing the microbial quality of refrigerated meat. *J. Food Sci.*, **57**, 1330–1.

Arkoudelos, J.S. and Nychas, G.J.E. (1995) Comparative studies of the growth of *Staphylococcus carnosus* with or without glucose *Letts Appl. Microbiol.*, **20**, 19–24.

Axelsson, L.T. (1993) Lactic acid bacteria: Classification and physiology, in *Lactic Acid Bacteria* (eds S. Salminen and A. Von Wright), Marcel Dekker, New York, pp. 1–63.

Baldwin, S.A. and Henderson, P.J.F. (1989) Homologies between sugar transporters from eukaryotes and prokaryotes. *Ann. Rev. Physiol.*, **51**, 459–71.

Barlow, J. and Kitchell, A.G. (1966) A note on the spoilage of prepacked lamb chops by *Microbacterium thermosphactum*. *J.Appl. Bacteriol.*, **29**, 185–8.

Barua, M. and Shelef, L.A. (1980) Growth suppression of pseudomonads by glucose utilization. *J. Food Sci.*, **45**, 349–51.

Benito de Cardinas, I.L., Ledesma, O. and Oliver, G. (1989) Effect of pyruvate on diacetyl and acetoin production by *Lactobacillus casei* and *Lactobacillus plantarum*. *Milchwissenschaft*, **44**, 347–50.

Blickstad, E. (1983) Growth and end-product formation of two psychrotrophic *Lactobacillus* spp. and *Brochothrix thermosphacta* ATCC 11509^T at different pH values and temperatures. *Appl. Environ. Microbiol.*, **46**, 1345–50.

Blickstad, E. and Molin, G. (1984) Growth and end-product formation in fermenter cultures of *Brochothrix thermosphacta* ATCC 11509^T and two psychrotrophic *Lactobacillus* spp. in different gaseous atmospheres. *J. Appl. Bacteriol.*, **57**, 213–20.

Bobillo, M. and Marshall, V.M. (1991) Effect of salt and culture aeration on lactate and acetate production by *Lactobacillus plantarum. Food Microbiol.*, **8**, 153–60.

Bobillo, M. and Marshall, V.M. (1992) Effect of acidic pH and salt on acid end-products by *Lactobacillus plantarum* in aerated, glucose-limited continuous culture. *J. Appl. Bacteriol.*, **73**, 67–70.

Boers, R.H., Dijkmann, K.E. and Wijngaards, G. (1994) Shelf-life of vacuum–packaged wild boar meat in relation to that of vacuum-packaged pork: relevance of intrinsic factors. *Meat Sci.*, **37**, 91–102.

Boethling, R.S. (1975) Regulation of extracellular protease secretion by *Pseudomonas maltophila*. *J. Bacteriol.*, **123**, 954–61.

Borch, E. and Agerhem, H. (1992) Chemical, microbial and sensory changes during the anaerobic cold storage of beef inoculated with a homoferementative *Lactobacillus* sp. or a *Leuconostoc* sp. *Int. J. Food Mircobiol.*, **15**, 99–108.

Borch, E., Berg, H. and Holst, O. (1991) Heterolactic fermentation by a homofermentative *Lactobacillus* sp. during glucose limitation in anaerobic continuous culture with complete cell recycle. *J. Appl. Bacteriol.*, **71**, 265–9.

Borch, E. and Molin, G. (1989) The aerobic growth and product formation of *Lactobacillus*, *Leuconostoc*, *Brochothrix* and *Carnobacterium* in batch cultures. *Appl. Microbiol. Biotechnol.*, **30**, 81–8.

Borton, R.J., Bratzler, L.J. and Price, J.F. (1970a) Effects of four species of bacteria on porcine muscle. 1. Protein solubility and emulsifying capacity. *J. Food Sci.*, **35**, 779–82.

Borton, R.J., Bratzler, L.J. and Price, J.F. (1970b) Effects of four species of bacteria on porcine muscle. 2. Electrophoretic patterns of extracts of salt soluble protein. *J. Food Sci.*, **35**, 783–6.

Button, D.K. (1983) Differences between the kinetics of nutrient uptake by micro-organisms, growth and enzyme kinetics. *Trends Biochem. Sci.*, **8**, 121–4.

Chandler, H., Batish, V.K. and Singh, R.S. (1989) Factors affecting amine production by a selected strain of *Lactobacillus bulgaricus*. *J. Food Sci.*, **54**, 940–2.

Church, P.N., Davies, A.R., Slade, A., Hart, R.J. and Gibbs, P.A. (1992) Improving the safety and quality of meat products by modified atmosphere and assessment by novel methods. *FLAIR Proposal no. 89055 Interim 2nd Year Report*, EEC DGXII, 1992.

Cogan, J.F., Walsh, D. and Condon, S. (1989) Impact of aeration on the metabolic end-products formed from glucose and galactose by *Streptococcus lactis*. *J. Appl. Bacteriol.*, **66**, 77–84.

Collins, M.D., Farrow, J.A.E., Phillips, B.A., Ferusu, S. and Jones, D. (1987) Classification of *Lactobacillus divergens*, *Lactobacillus piscicola*, and some catalase-negative, asporogenous, rod-shaped bacteria from poultry in a new genus, *Carnobacterium*. *Int. J. System. Bacteriol.*, **37**, 310–16.

Collins, M.D., Samelis, J., Metaxopoulos, J. and Wallbanks, S. (1993) Taxonomic studies on some leuconostoc-like organisms from fermented sausages: description of a new genus *Weissella* for the *Leuconostoc paramesenteroides* group of species. *J. Appl. Bacteriol.*, **75**, 595–603.

Condon, S. (1987) Responses of lactic acid bacteria to oxygen. *FEMS Microbiology Reviews*, **46**, 269–80.

Cselovszky, J., Wolf, G. and Hammes, W.P. (1992) Production of formate, acetate and succinate by anaerobic fermentation of *Lactobacillus pentosus* in the presence of citrate. *Appl. Microbiol. Biotechnol.*, **37**, 94–7.

Dainty, R.H. (1981) *Volatile Fatty Acid Detected in Vacuum Packed Beef During Storage at Chill Temperatures*, in Proceedings of 27th Meeting of European Meat Workers, University of Vienna, August 1981, pp. 688–90.

Dainty, R.H. (1996) Chemical/biochemical detection of spoilage. *Int. J. Food Microbiol.*, **33**, 19–34.

Dainty, R.H., Edwards, R.A. and Hibbard, C.M. (1985) Time course of volatile compound formation during refrigerated storage of naturally contaminated beef in air. *J. Appl. Bacteriol.*, **59**, 303–9.

Dainty, R.H., Edwards, R.A. and Hibbard, C.M. (1989a) Spoilage of vacuum-packed beef by a *Clostridium* sp. *J. Sci. Food Agric.*, **49**, 473–86.

Dainty, R.H., Edwards, R.A., Hibbard, C.M. and Marnewick, J.J. (1989b) Volatile compounds associated with microbial growth on normal and high pH beef stored at chill temperatures. *J. Appl. Bacteriol.*, **66**, 281–9.

Dainty, R.H., Edwards, R.A., Hibbard, C.M. and Ramantanis, S.V. (1986) Bacterial sources of putrescine and cadaverine in chill stored vacuum-packaged beef. *J. Appl. Bacteriol.*, **61**, 117–23.

Dainty, R.H. and Hibbard, C.M. (1980) Aerobic metabolism of *Brochothrix thermosphacta* growing on meat surfaces and in laboratory media. *J. Appl. Bacteriol.*, **48**, 387–96.

Dainty, R.H. and Hibbard, C.M. (1983) Precursors of the major end products of aerobic metabolism of *Brochothrix thermosphacta*. *J. Appl. Bacteriol.*, **55**, 387–96.

Dainty, R.H. and Hoffman, F.J.K. (1983) The influence of glucose concentration and culture incubation time on end-product formation during aerobic growth of *Brochothrix thermosphacta. J. Appl. Bacteriol.*, **55**, 233–9.

Dainty, R.H. and Mackey, B.M. (1992) The relationship between the phenotypic properties of bacteria from chill-stored meat and spoilage processes. *J. Appl. Bacteriol.*, Symp. Suppl., **73**, 103S–114S.

Dainty, R.H., Shaw, B.G., de Boer, K.A. and Scheps, S.J. (1975) Protein changes caused by bacterial growth on beef. *J. Appl. Bacteriol.*, **39**, 73–81.

Dainty, R.H., Shaw, B.G. and Roberts, T.A. (1983) Microbial and chemical changes in chill-stored red meat, in *Food Microbiology: Advances and Prospects* (eds T.A. Roberts and F.A. Skinner), Academic Press, London, pp. 151–78.

Davies, A.R. (1995) Advances in modified atmosphere packaging, in *New Methods of Food Preservation* (ed. G.W. Gould), Blackie Academic and Professional, London, pp. 304–20.

de Castro, P.B., Asensio, M.A., Sanz, B. and Ordonez, J.A. (1988) A method to assess the bacterial content of refrigerated meat. *Appl. Environ. Microbiol.*, **54**, 1462–5.

de Pablo, B., Asensio, M.A., Sanz, B. and Ordonez, J.A. (1989) The D(-) lactic acid and acetoin/diacetyl as potential indicators of the microbial quality of vacuum-packed pork and meat products. *J. Appl. Bacteriol.*, **66**, 185–90.

Drosinos, E.H. (1994) Microbial associations of minced lamb and their ecophysiological attributes. Ph.D thesis, University of Bath, United Kingdom.

Drosinos, E.H. and Board, R.G. (1994) Metabolic activities of pseudomonads in batch cultures in extract of minced lamb. *J. Appl. Bacteriol.*, **77**, 613–20.

Drosinos, E.H. and Board, R.G. (1995a) A Survey of minced lamb packaged in modified atmospheres. *Fleischwirtschaft*, **75**, 281–4.

Drosinos, E.H. and Board, R.G. (1995b) Attributes of microbial associations of meat growing as xenic batch cultures in a meat juice at 4 °C. *Int. J. Food Microbiol.*, **26**, 279–93.

Drosinos, E.H. and Board, R.G. (1995c) Microbial and physico-chemical attributes of minced lamb: sources of contamination with pseudomonads. *Food Microbiol.*, **12**, 189–97.

Edwards, R.A. and Dainty, R.H. (1987) Volatile compounds associated with spoilage of normal and high pH vacuum-packed pork. *J. Sci. Food Agric.*, **38**, 57–66.

Edwards, R.A., Dainty, R.H. and Hibbard, C.M. (1983) The relationship of bacterial numbers and types to diamine concentration in fresh and aerobically stored beef, pork and lamb. *J. Food Technol.*, **18**, 777–88.

Edwards, R.A., Dainty, R.H. and Hibbard, C.M. (1987) Volatile compounds produced by meat pseudomonads and related reference strains during growth in air at chill temperatures. *J. Appl. Bacteriol.*, **62**, 403–12.

Eisenberg, R.C., Butters, S.J., Quay, S.C., Friedman, S.B. (1974) Glucose uptake and phosphorylation in *Pseudomonas fluorescens. J. Bacteriol.*, **120**, 147–53.

Fairburn, D.J. and Law, B.A. (1986) Proteinases of psychrotrophic bacteria: their production, properties, effects and control. *J. Dairy Res.*, **53**, 139–45.

Farber, J.M. and Idziak, E.S. (1982) Detection of glucose oxidation products in chilled fresh beef undergoing spoilage. *Appl. Environ. Microbiol.*, **44**, 521–4.

Fredrickson, A.G. and Stephanopoulos, G. (1981) Microbial competition. *Science*, **213**, 972–9.

Freeman, L.R., Silverman, G.J., Angelini, P., Merrit, C. and Esselen, W.B. (1976) Volatiles produced by microorganisms isolated from refrigerated chickens at spoilage. *Appl. Environ. Microbiol.*, **32**, 222–31.

Garvie, E.I. (1980) Bacterial lactate dehydrogenase. *Microbiol. Rev.*, **44**, 106–39.

Gill, C.O. (1976) Substrate limitation of bacterial growth at meat surfaces. *J. Appl. Bacteriol.*, **41**, 401–10.

Gill, C.O. (1982) Microbial interaction with meats, in *Meat Microbiology*, (ed. M.H. Brown). Applied Science, London, pp. 225–64.

Gill, C.O. (1983) Meat spoilage and evaluation of the potential storage life of fresh meat. *J. Food Protect.*, **46**, 444–52.

Gill, C.O. (1986) The control of microbial spoilage in fresh meats, in *Advances in Meat Research: Meat and Poultry Microbiology*, (eds A.M. Pearson and T.R. Dutson), MacMillan, New York, pp. 49–88.

Gill, C.O. and Molin, G. (1991) Modified atmospheres and vacuum packaging, in *Food Preservatives* (eds N.J. Russell and G.W. Gould), Blackie, Glasgow, pp. 172–99.

Gill, C.O. and Newton, K.G. (1977) The development of aerobic spoilage flora on meat stored at chill temperatures. *J. Appl. Bacteriol.*, **43**, 189–95.

Gill, C.O. and Newton, K.G. (1978) The ecology of bacterial spoilage of fresh meat at chill temperatures. *Meat Sci.*, **2**, 207–17.

Gill, C.O. and Newton, K.G. (1979) Spoilage of vacuum-packaged dark, firm, dry meat at chill temperatures. *Appl. Environ. Microbiol.*, **37**, 362–4.

Gill, C.O. and Penney, N. (1977) Penetration of bacteria into meat. *Appl. Environ. Microbiol.*, **33**, 1284–6.

Glenn, A.R. (1976) Production of extracellular proteins by bacteria. *Ann. Rev. Microbiol.*, **30**, 41–62.

Gupta, L.K. and Nagamohini, Y. (1992) Penetration of poultry meat by *Pseudomonas* and *Lactobacillus* spp. *World J. Microbiol. Biotechnol.*, **8**, 212–13.

Hanna, M.O., Zink, D.L., Carpenter, Z.L. and Vanderzant, C. (1976) *Yersinia enterocolitica*-like organisms from vacuum-packed beef and lamb. *J. Food Sci.*, **41**, 1254–6.

Harder, W. (1979) Regulation of the synthesis of extracellular enzymes in microorganisms. *Soc. Gen. Microbiol. Quart.*, **6**, 139–40.

Hasegawa, T., Pearson, A.M., Price, J.F. and Lechowich, R.V. (1970a) Action of bacterial growth on the sarcoplasmic and urea soluble proteins from muscle. I. Effects of *Clostridium perfringens*, *Salmonella enteritidis*, *Achromobacter liquefaciens*, *Streptococcus feacalis* and *Kurthia zophii*. *Appl. Microbiol.*, **20**, 117–22.

Hasegawa, T., Pearson, A.M., Rampton, J.H. and Lechowich, R.V. (1970b) Effect of microbial growth upon sarcoplasmic and urea-soluble proteins from muscle. *J. Food Sci.*, **35**, 720–4.

Hitchener, B.J., Egan, A.F. and Rogers, P.J. (1979) Energetics of *Microbacterium thermosphactum* in glucose-limited continuous culture. *Appl. Environ. Microbiol.*, **37**, 1047–52.

Holzapfel, W.H. and Gerber, E.S. (1983) *Lactobacillus divergens* sp. *nov.*, a new homofermentative *Lactobacillus* species producing L(+) lactate. *System. Appl. Microbiol.*, **4**, 522–34.

Ingram, M. (1971) Microbial changes in foods – general considerations. *J. Appl. Bacteriol.*, **34**, 1–8.

Ingram, M. and Dainty, R.H. (1971) Changes caused by microbes in spoilage of meats. *J. Appl. Bacteriol.*, **34**, 21–39.

Jackson, T.C., Acuff, G.R., Vanderzant, C., Sharp, T.R. and Savell, J.W. (1992) Identification and evaluation of volatile compounds of vacuum and modified atmosphere packaged beef strip loins. *Meat Sci.*, **31**, 175–90.

Jay, J.M. (1986a) Microbial spoilage indicators and metabolites, in *Foodborne Microorganisms and their Toxins: Developing Methodology* (eds M.D. Pierson and N.J. Sterm), Marcel Dekker, Basel, pp. 219–40.

Jay, J.M. (1986b) *Modern Food Microbiology*, 3rd edn, Van Nostrand Reinhold, New York.

Jay, J.M. and Shelef, L.A. (1976) Effect of micro-organisms on meat proteins at low temperatures. *J. Agric. Food Chem.*, **24**, 1113–16.

Juffs, H.S. (1976) Effects of temperature and nutrient on proteinase production by *Pseudomonas fluorescens* and *Pseudomonas aeruginosa* in broth and milk. *J. Appl. Bacteriol.*, **40**, 23–30.

Kakouri, A. and Nychas, G.J.E. (1994) Storage of poultry meat under modified atmospheres or vacuum packs: possible role of microbial metabolites as indicators of spoilage. *J. Appl. Bacteriol.*, **76**, 163–72.

Kandler, O. (1983) Carbohydrate metabolism in lactic acid bacteria. *Antonie van Leeuwenhoek*, **49**, 209–24.

Kandler, O. and Weiss, N. (1986) Genus *Lactobacillus*, in *Bergey's Manual of Systematic Bacteriology* (eds P.H.A. Sneath, N.S. Mair, M.E. Sharpe and J.G. Holt), Williams and Wilkins, Baltimore, pp. 1208–34.

Krizek, A.R., Smith, J.S. and Phebus, R.K. (1995) Biogenic amine formation in fresh vacuum–packaged beef stored at –2 °C and 2 °C for 100 days. *J. Food Protect.*, **58**, 284–8.

Kudryashov, L.S., Potopayeva, N.N., Neginsky, V.A. and Bolshakov, A.S. (1989) Kinetics of the inhibition of pork muscle cathepsin D with curing ingredients, in Proceedings of the 35th International Congress of Meat Science and Technology, Danish Meat Research Institute, Roskilde, Denmark, August 1989, 930 pp.

Lambropoulou, K.A. (1995) The role of glucose in meat. M.Sc. thesis, University of Humberside, UK.

Lambropoulou, K.A., Drosinos, E.H. and Nychas, G.J.E. (1996) The effect of glucose supplementation on the spoilage microflora and chemical composition of minced beef stored aerobically or under a modified atmosphere at 4 °C. *Int. J. Food Microbiol.*, **30**, 281–91.

Lasta, J.A., Pensel, N., Masana, M., Rodriguez, H.R. and Garcia, P.T. (1995) Microbial growth and biochemical changes on naturally contaminated chilled-beef subcutaneous adipose tissue stored aerobically. *Meat Sci.*, **39**, 149–58.

Law, B.A. and Kolstad, J. (1983) Proteolytic systems in lactic acid bacteria. *Antonie van Leeuwenhoek*, **49**, 225–45.

Lawrie, R.A. (1985) *Meat Science*, 3rd edn. Pergamon Press, Oxford.

Lea, C.H., Parr, L.J. and Jackson, H.F. (1969) Chemical and organoleptic changes in poultry meat resulting from growth of psychrophilic spoilage bacteria at 1°C. 3. Glutamine, glutathione, tyrosine, ammonia, lactic acid, creatine, carbohydrate, heme pigment and hydrogen sulfide. *Br. Poultry Sci.*, **10**, 229–38.

Loubiere, P., Cocaign-Bosquet, M., Matos, J., Goma, G. and Lindley, N.D. (1997) Influence of end-products inhibition and nutrient limitations on the growth of *Lactococcus lactis* subsp. *lactis. J. Appl. Microbiol.*, **82**, 95–100.

Maijala, R.L. (1994) Histamine and tyramine production by a *Lactobacillus* strain subjected to external pH decrease, *J. Food Protect.*, **57**, 259–62.

Maijala, R.L., Eerola, S.H., Aho, M.A. and Hirn, J.A. (1993) The effect of GDL-induced pH decrease on the formation of biogenic amines in meat. *J. Food Protect.*, **56**, 125–9.

Marshall, V.M. (1992) Inoculated ecosystems in a milk environment. *J. Appl. Bacteriol.*, Symp. Suppl., **73**, 127S–135S.

McMeekin, T.A. (1975) Spoilage association of chicken breast muscle. *Appl. Microbiol.*, **29**, 44–7.

McMeekin, T.A. (1977) Spoilage association of chicken leg muscle. *Appl. Microbiol.*, **33**, 1244–6.

McMeekin, T.A. (1981) Microbial spoilage of meats, in *Developments in Food Microbiology* (ed. R. Davies), Applied Science, London, pp. 1–40.

Meynes, B., Begazo, N. and Schmidt-Lorenzo, W. (1992) Concentration changes of glucose, glycogen, L(+) and D(−) lactic acid during microbial spoilage of beef. *Mitt. Gebiete Lebensmitten Hygiene*, **83**, 121–5.

Midgley, M., Dawes, E.A. (1973) The regulation of transport of glucose and methyl α-glucoside in *Pseudomonas aeruginosa. Biochem. J.*, **132**, 141–55.

Mitchell, G.C. and Dawes, E.A. (1982) The role of oxygen in the regulation of glucose metabolism, transport and the tricarboxylic acid cycle in *Pseudomonas aeruginosa. J. Gen. Microbiol.*, **128**, 49–59.

Molin, G. (1985) Mixed carbon source utilization of meat-spoiling *Pseudomonas fragi* 72 in relation to oxygen limitation and carbon dioxide inhibition. *Appl. Environ. Microbiol.*, **49**, 1442–7.

Molin, G. and Tenström, A. (1986) Phenotypically based taxonomy of psychrotrophic *Pseudomonas* isolated from spoiled meat, water and soil. *Int. J. System. Bacteriol.*, **36**, 257–74.

Montville, T.J., Hsu, A.H.M., and Meyer, M. (1987) High-efficiency conversion of pyruvate to acetoin by *Lactobacillus plantarum* during pH-controlled and fed-batch fermentations. *Appl. Environ. Microbiol.*, **53**, 1798–802.

Mossel, D.A.A. and Ingram, M. (1955) The physiology of the microbial spoilage of foods. *J. Appl. Bacteriol.*, **18**, 232–68.

Murphy, M.G. and Condon, S. (1984a) Correlation of oxygen utilization and H_2O_2 accumulation with oxygen induced enzymes in *Lactobacillus plantarum* cultures. *Arch. Microbiol.*, **138**, 44–8.

Murphy, M.G. and Condon, S. (1984b) Comparison of aerobic and anaerobic growth of *Lactobacillus plantarum* in a glucose medium. *Arch. Microbiol.*, **138**, 49–53.

Murphy, M.G., O'Connor, L., Walsh, D. and Condon, S. (1985) Oxygen dependent lactate utilization by *Lactobacillus plantarum. Arch. Microbiol.*, **141**, 75–9.

Nakamura, M., Wada, Y., Sawaya, H. and Kawabata, T. (1979) Polyamine content in fresh and processed pork. *J. Food Sci.*, **44**, 515–17, 523.

Nassos, P.S., King, Jr., A.D. and Stafford, A.E. (1983) Relationship between lactic acid concentration and bacterial spoilage in ground beef. *Appl. Environ. Microbiol.*, **46**, 894–900.

Nassos, P.S., King, Jr., A.D. and Stafford, A.E. (1985) Lactic acid concentration and microbial spoilage in anaerobically and aerobically stored ground beef. *J. Food Sci.*, **50**, 710–12, 715.

Nassos, P.S., King, Jr., A.D. and Stafford, A.E. (1988) Lactic acid concentration as an indicator of acceptability in refrigerated or freeze-thawed ground beef. *Appl. Environ. Microbiol.*, **54**, 822–3.

Newton, K.G. and Gill, C.O. (1978) Storage quality of dark, firm, dry meat. *Appl. Environ. Microbiol.*, **36**, 375–6.

Newton, K.G. and Rigg, W.J. (1979) The effect of film permeability on the storage life and microbiology of vacuum packaged meat. *J. Appl. Bacteriol.*, **47**, 433–41.

Nychas, G.J.E. (1984) Microbial growth in minced meat. Ph.D thesis, University of Bath, Bath, UK.

Nychas, G.J.E. (1994) Modified atmosphere packaging of meats, in *Minimal Processing of Foods and Process optimization; An Interface* (eds R.P. Singh and F.A.R. Oliveira), CRC Press, London, pp. 417–35.

Nychas, G.J. and Arkoudelos, J.S. (1990) Microbiological and physicochemical changes in minced meats under carbon dioxide, nitrogen or air at 3 °C. *Int. J. Food Sci. Technol.*, **25**, 389–98.

Nychas, G.J.E. and Arkoudelos, J.S. (1991) The influence of *Brochothrix thermosphacta* on the quality of minced meat. *Agric. Res.*, **15**, 103–15 (in Greek).

Nychas, G.J., Dillon, V.M. and Board, R.G. (1988) Glucose the key substrate in the microbiological changes occurring in meat and certain meat products. *Biotechnol. Appl. Biochem.*, **10**, 203–31.

Nychas, G.J.E., Gibbs, P.A., Board, R.G. and Sheridan, J.J. (1994) Improving the safety and quality of meat and meat products by modified atmosphere and assessment by novel methods. FLAIR proposal No 89055, Contract No AGRF/0024 (SCP), Final Report, EU, DGXII, Brussels, Belgium.

Nychas, G.J., Robinson, A. and Board, R.G. (1992) Microbiological and physico-chemical evaluation of ground beef from retail shops. *Fleischwirtsch. Int.*, **1**, 49–53.

Nychas, G.J.E. and Tassou, C.C. (1997) Spoilage processes and proteolysis in chicken as detected by HPLC. *J. Sci. Food Agric.*, **74**, 199–208.

Ockerman, H.W., Cahill, V.R., Weiser, H.H., Davis, C.E. and Sifker, J.R. (1969) Comparison of sterile and inoculated beef tissue. *J. Food Sci.*, **34**, 93–9.

Ordonez, J.A., de Pablo, B., de Castro, B.P., Asensio, M.A. and Sanz, B. (1991) Selected chemical and microbiological changes in refrigerated pork stored in carbon dioxide and oxygen enriched atmospheres. *J. Agric. Food Chem.*, **39**, 668–72.

Paron, M. and Talon, R. (1988) Factors affecting growth and lipase production be meat lactobacilli strains and *Brochothrix thermosphacta*. *J. Appl. Bacteriol.*, **64**, 107–15.

Passmore, R. and Eastwood, M.A. (1986) *Human Nutrition and Dietetics*. Churchill Livingstone, New York.

Peterson, A.C. and Gunderson, M.G. (1960) Some characteristics of proteolytic enzymes from *Pseudomonas fluorescens*. *Appl. Microbiol.*, **8**, 98–103.

Pollock, M.R. (1963) Exoenzymes, in *The Bacteria* (eds I.C. Gunsalus and R.Y. Stanier), Academic Press, New York, pp. 45–62.

Ramos, A., Jordan, K.N., Cogan, T.M. and Santos, H. (1994) [13]C Nuclear Magnetic Resonance Studies of citrate and glucose cometabolism by *Lactobacillus lactis*. *Appl. Environ. Microbiol.*, **60**, 1739–48.

Rawles, D.D., Flick, G.J. and Martin, R.E. (1996) Biogenic amines in fish and shellfish. *Adv. Food Nutrition Res.*, **39**, 329–65.

Reddy, S.G., Chen, M.L. and Patel, P.J. (1975) Influence of lactic cultures on the biochemical bacterial and organoleptic changes in beef. *J. Food Sci.*, **40**, 314–18.

Sakamoto, M. and Komagata, K. (1996) Aerobic growth of and activities of NADH oxidase and NADH peroxidase in lactic acid bacteria. *J. Ferment. Bioeng.*, **82**, 210–16.

Schleifer, K.H. (1986) Gram-positive cocci, in *Bergey's Manual of Systematic Bacteriology*, vol. 2 (eds P.H.A. Sneath, N.S. Mair, M.E. Sharpe and J.G. Holt), Williams and Wilkins, Baltimore, pp. 999–1103.

Schmitt, R.E. and Schmidt-Lorenz, W. (1992a) Formation of ammonia and amines during microbial spoilage of refrigerated broilers. *Lebensmittel-Wissenschaft und-Technologie*, 25, 6–10.

Schmitt, R.E. and Schmidt-Lorenz, W. (1992b) Degradation of amino acids and protein changes during microbial spoilage of chilled unpacked and packed chicken carcasses. *Lebensmittel-Wissenschaft und-Technlogie*, 25, 11–20.

Schneider, W., Hildebrandt, G. and Sinell, H.J. (1983) D(–) lactate concentration as a parameter for evaluating the freshness of pre-packed, heat treated meat products. *Fleischwirtschaft*, 63, 1198–205.

Scott, J.H. and Nealson, K.H. (1994) A biochemical study of the intermediary carbon metabolism of *Shewanella putrefaciens*. *J. Bacteriol.*, 176, 3408–11.

Sedewitz, B., Schleifer, K.H. and Gotz, F. (1984) Physiological role of pyruvate oxidase in the aerobic metabolism of *Lactobacillus plantarum*. *J. Bacteriol.*, 160, 462–5.

Seymour, I.J., Cole, M.B. and Coote, P.J. (1994) A substrate-mediated assay of bacterial proton efflux/influx to predict the degree of spoilage of beef mince stored at chill temperatures. *J. Appl. Bacteriol.*, 76, 608–15.

Shelef, L.A. (1977) Effect of glucose on the bacterial spoilage of beef. *J. Food Sci.*, 42, 1172–5.

Sheridan, J.J. (1995) The role of indicator systems in HACCP operations. *J. Food Safety*, 15, 157–80.

Sinell, H.J. and Luke, K. (1978) D(–) lactate as parameter for microbial spoilage in frankfurter type sausages, in Proceedings of the 24th European Meeting Meat Research Workers, Kulbach, Germany, pp. C11:1–C11:6.

Smith, J.S., Kenney, P.B., Kastner, C.L. and Moore, M.M. (1993) Biogenic amine formation in fresh vacuum-packaged beef during storage at 1 °C for 120 days. *J. Food Protect.*, 56, 497–500, 532.

Stiles, M.E. (1996) Biopreservation by lactic acid bacteria, in *Lactic acid bacteria: Genetics, Metabolism and Applications* (eds G. Venema, J.H.J. Huis in't Veld and J. Hugenholtz), Kluwer Academic Publishers, London, pp. 235–49.

Stutz, H.K., Silverman, G.J., Angelini, P. and Levin, R.E. (1991) Bacteria and volatile compounds associated with ground beef spoilage. *J. Food Sci.*, 56, 1147–53.

Talon, R., Paron, M., Bauchart, D., Duboisset, F. and Montel, M-C. (1992) Lipolytic activity of *Brochothrix thermosphacta* on natural triglycerides. *Letts Appl. Microbiol.*, 14, 153–7.

Tarrant, P.J.V., Jenkins, N., Pearson, A.M. and Dutson, T.R. (1973) Proteolytic enzymes preparation from *Pseudomonas fragi* and its action on pig muscle. *Appl. Microbiol.*, 25, 996–1001.

Tassou, C., Aletras, V. and Nychas, G.J.E. (1996) The use of HPLC to monitor changes in the organic acid profile extracted from poultry stored under different storage conditions. Proceedings of the 17th National Chemistry Conference, Patras Greek Association of Chemists, Patra, 1–5 December 1996, pp. 496–99.

Thomas, T.D., Ellwood, D.C. and Longyear, V.M.C. (1979) Change from homo- to heterolactic fermentation by *Streptococcus lactis* resulting from glucose limitation in anaerobic chemostat cultures. *J. Bacteriol.*, 138, 109–17.

Tseng, C.P. and Montville, T.J. (1990) Enzyme activities affecting end-product distribution by *Lactobacillus plantarum* in response to changes in pH and O_2. *Appl. Environ. Microbiol.*, 56, 2761–3.

Venugopal, V. (1990) Extracellular proteases of contaminant bacteria in fish spoilage, a review. *J. Food Protect.*, 53, 341–50.

Yano, Y., Kataho, N., Watanabe, M. and Nakamura, T. (1995) Changes in the concentration of biogenic amines and application of tyramine sensor during storage of beef. *Food Chem.*, 54, 155–9.

Index

Numbers in *italics* refer to tables, numbers in **bold** refer to figures.